TO THE EDGES OF THE EARTH

TO THE EDGES OF THE EARTH

A JOURNEY INTO WILD LAND

PETER PICKFORD

BOOK**STORM**

ISBN: 978-1-928257-84-4
e-ISBN: 978-1-928257-85-1

First edition, first impression 2020

Published by Bookstorm (Pty) Ltd
PO Box 4532
Northcliff 2115
Johannesburg
South Africa
www.bookstorm.co.za

The quote on page vii from *Amsterdam* by Ian McEwan, published by Vintage.
Reproduced by permission of The Random House Group Ltd.

Edited by Beverly Pickford, Alfred LeMaitre and Sean Fraser
Proofread by Wesley Thompson
Cover design by publicide
Front cover image, Kaokoland Desert, Namibia
Back cover image, Svalbard Archipelago
Book design and typesetting by Triple M Design
Maps by Justin J Fox
Printed in the USA

For Beverly
My North Star

THE COMPANION PHOTOGRAPHIC BOOK

Wild Land

Southern African edition

World edition

AVAILABLE IN ALL GOOD BOOKSTORES AND ONLINE

*Soon human meaning would be bleached from the rocks,
the landscape would assume its beauty and draw him in;
the unimaginable age of the mountains and the fine mesh of living
things that lay across them would remind him that he was part of this
order and insignificant within it, and he would be set free.*

Ian McEwan, 1999

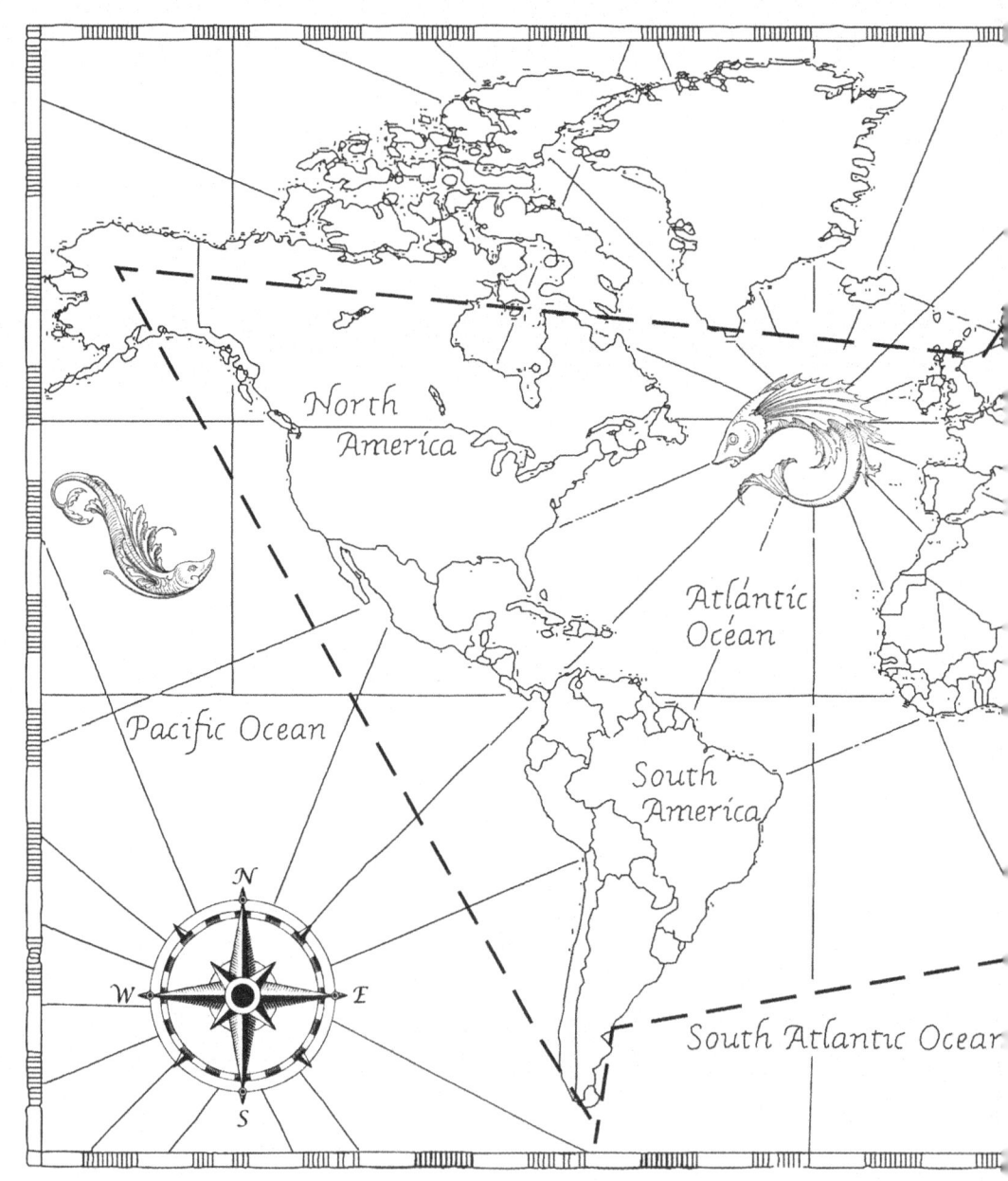

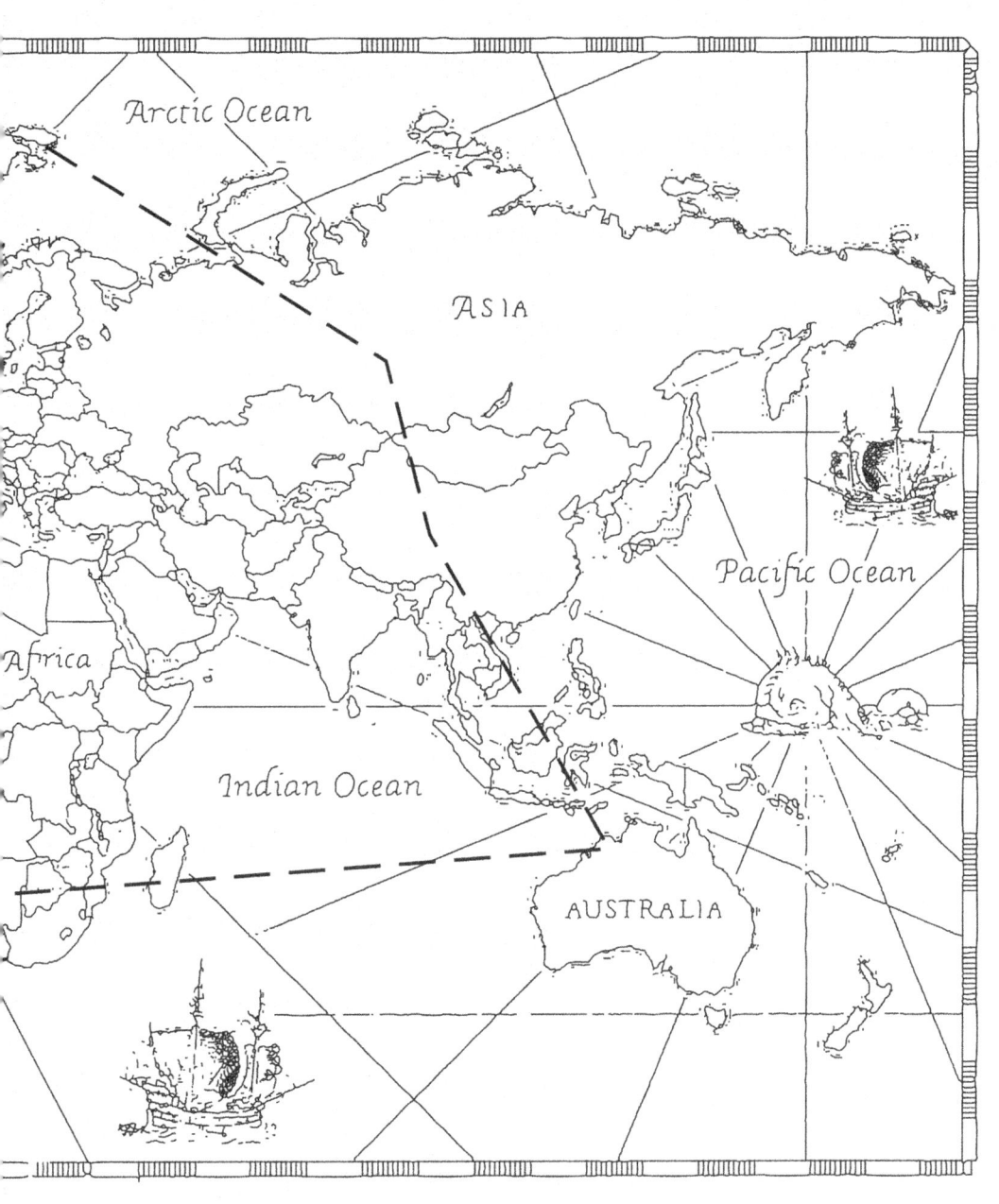

Contents

Introduction

This book does not begin at the beginning. Not because the beginning is not worth writing about – quite the contrary, it was extraordinary – but simply because I had no plans to write a book. Now, when I look back, my memory is patchy and often indistinct. I can tell you that our ship ran aground in the caldera of an old volcano in Antarctica. I remember that well.

Passengers were not permitted on the bridge of the ship during technical or difficult operating conditions. The captain was directing the officers towards a waypoint where they had anchored previously in the Deception Island caldera. I was standing outside on the flybridge. The ship was edging forward at dead slow. The depth-sounder alarm started beeping.

'Six metres of water, Captain,' said Cheli Larsen, our expedition leader.

The ship continued, slow ahead. The depth-sounder alarm does not get louder as the water gets shallower; it just continues to beep.

'Four metres of water, Captain.' Cheli's voice, unlike the depth sounder, had risen a notch. The ship continued, slow ahead.

'Captain, we are drawing three metres.' Cheli's voice was urgent. But a ship is a slow thing to stop. By the time the captain saw that the waypoint he considered a safe anchorage was now subject to a very different tide, the depth sounder showed just under two metres of water and Cheli had thrown her hands in the air and stalked from the bridge. Less than half a minute later, we ran aground. Fortunately, the bottom was muddy, but the tide was ebbing and so for the next six hours our ship lay in the shallows off an ice-strewn beach, rising higher and higher out of the water, like a beached whale.

That memory is clear, as are others: the shocking power of the wind when it blows at more than 100 kilometres per hour, or the babble of half a million penguins gathered in one place, or pouring a nip of Scotch onto the graves of Frank Wild and Ernest Shackleton and toasting their courage and tenacity. But memories are like photographs tossed in a basket; they soon become muddled. Some become bleached by the sun, so that one must squint at them to see what they are about. On others the colour has run; or they have been torn in some accident. At the bottom are a few that are clumped together; when you try to prise them apart they peel and then tear, leaving faces stuck to the backs of other photos.

I have a memory just like that. The face is missing but I recall the conversation with clarity. One of the staff on the ship was bleary-eyed. I told him he looked the worse for wear; my statement was as much a question as a relaying of fact.

'What happens on the aft deck of the ship stays on the aft deck,' he told me with a sideways look. For the rest of the trip I was tempted to steal aft in the dark of night and learn the secrets there, but I never did. If I had, and had begun this book at the beginning, then those secrets might not have stayed on the aft deck, so it is probably just as well that this account begins a little later.

It begins on 15 December 2011. It was in no respect an auspicious day except that the sun was out and the day was still, which, in the southernmost reaches of the island of Tierra del Fuego, at the extreme southern tip of South America, is notable. My wife, Beverly, and I were driving a quiet, rural dirt road sandwiched between the Canal Beagle (Beagle Channel) and the close, sheer rise of the island's fringing mountains. A tiny settlement of a few scattered houses occupied a swathe of flat ground. The forest crowded into their backyards and the sea rose to the verge of the road in front. Hanging askew on a single nail at one of the front gates, a hand-painted sign advertised 'Ate Centolla' (king crab). We stopped, and crossed the small yard, littered with broken crab traps and faded orange buoys, to the front door. A man answered our knocking, holding both hands in the air to stop his fingers dripping, and welcomed us inside, kicking the door closed with his heel. He was chopping garlic. More than a kilogram lay finely diced on a wide board on the small kitchen table. While he washed his hands, I looked around.

The house was tiny, made of lapped, waney-edged wooden planks, with sash windows that looked out onto the bleak, grey waters of the Canal Beagle. The smell of garlic permeated the house, and although the stove was not lit, the aroma of resin-scented woodsmoke hung close in the confined environment. It felt balmy and cosy, a haven. Beneath one of the windows, on the crocheted blanket of a bed seat, a young girl was drawing.

'Hola,' I said, walking over to her. She smiled at me shyly and, as I drew close, returned to her drawing. It was of a house and a huge yellow sun whose beams rained down on the land and seagulls floated adrift in the sky. It felt as the room felt: certain of its tenure, secure and, somewhere behind all its parts, infused with joy.

We lingered over the buying of the crab, as much out of our rudimentary Spanish as our peace at being there. As we made to leave, I reached for the low wooden gate, paused and looked back at the house. I should write about this, I thought to myself.

It began as notes. Notes to myself to fix the fading detail that time gradually leaches from our memory. Notes that would perhaps help in the captioning of a photograph, or just simply to read back to myself in years to come.

The notes grew longer. They changed from a record of details into description, until finally I was writing as much as we were photographing. And then, one day, as the rain pattered softly onto the tarpaulin strung over my head between two trees, I realised that what I had begun was a book. And so, at the end, I have had to come back to the beginning and find a beginning where none really existed.

But a journey does not begin on a single day; it evolves into being. Such evolutions may seem superficially quick, a single decision, but they are not. They are rooted in our personal histories, an evolution of our experiences that the random chaos of chance lays down in the path of our lives. They are refined by what we choose, what we embrace and what we shun. I would distil my own experience into further polarisation by those choices whose result stayed with me, lived on. I consider myself fortunate to have defined my life by the pursuit of that which is memorable, something that makes the day remarkable. The deeper I moved into this way of life, the more I chose to step off the paths of convention and into those I found curious and compelling, the richer my life became. Until one day, I became aware that I stood on the outside of the convention of our society, looking in. Somewhat like the proverbial goldfish that has leapt out of the bowl, but instead of landing on the floor gasping for air and eyeing the cat's sudden crouch in the corner, I had landed in the sea, stupefied and overwhelmed by the sudden magnitude of the ocean.

I have looked back over my history to try to find what it was that shook the blinkers of perception from my eyes. That made them fall, then shatter as brittle as china into the myriad vain castles of humankind, leaving in the crisp air of their wake so sharp a vision of the magnificence and scope of the opportunity to live that it pierced and stung every sense in me.

Perhaps it was the day my father taught me to fly-fish. The rivers of the Drakensberg in South Africa are the liquid distillation of mountains, clear and cold, with a translucent purity that yet holds something back; a jealous secret that compels one to pause, look down and in. The rivers cascade and glide between giant boulders and ancient, gnarl-stemmed protea trees. Dragonflies hawk from bent-over grass stems and shy small birds flit and fuss among the shadows of the river's margins, their movement quick and nervous. The pools are deep, hollowed out by the water's rush from the heights, their yellow-pebble bottoms morphing to a grey granite-blue.

My father showed me how to use the rod's whip for power to drive the cast. Back stroke, pause, forward stroke, pause – the line describing a perfect tight arc as the cast unfolded to straight and settled with a light touch on the water. My effort was not the same. I got snarled in the grass behind, hooked fast on a tall protea tree, or landed the line in a crumple at my feet. I spent more time undoing knots than casting. My father retreated to the shade tree where our family picnic was laid out on a blanket. I kept trying. Eventually, I managed to get a few yards of line onto the water where the current arced it downstream. I sighed with relief and then started to pull it in to try again. The line came wet to my fingers, a smooth glide that went suddenly taut. My fingers closed instinctively on the line and the rod quivered and shook under the sudden tension. The fish broke free but it was not the fish that was hooked, it was me. I had touched a wild life, made contact with some ephemeral unseen thing, whose presence was a sense as ancient as life itself.

I passed through a portal then to a view of the world from which I could not return. That touch of something wild was seared in my psyche like a branded tattoo.

Years later I was on a school trip to the Okavango Delta in Botswana, two masters and six boys crammed into a wide Ford F250 4x4. It was 1973. The road from Nata to Maun through the Makgadikgadi Pans was a 300-kilometre trial of heavy sand, axle-shattering corrugations and lunar-like potholes. We were running late, but any attempt at speed shook the vehicle so violently that it seemed sure to disintegrate. The sun set, the sky faded from orange to mauve, then purple and finally black. The stars came out and we pressed on with our headlights describing a stuttering dance in the featureless dark. A mist formed low on the ground and grew thicker as we advanced. But the mist was dry and it smelt like earth-dust. And then the first zebra appeared like a wraith on the periphery of our vehicle's beam of light. The zebra bolted to the side and vanished in the opaque mist. Another appeared and then another, all moving across the road from north to south. A few became more and then many, and our snail's pace slowed even further for fear of hitting one of the striped djinns, which morphed from apparition to solid form before our eyes as they stepped out of the curtain of dust. We could hear the herd now, a dull thunder, like the echo of some distant avalanche, a thousand hooves drumming the ground. Their press grew so dense that we had to stop, and as the motor died and the lights went out, we floated suddenly adrift into an eerie other-realm where even gravity seemed to become skewed and inverted. The dust was chokingly thick and produced a disconcertedly immediate sense of the edge of the world. Yet it remained malleable and porous as the heart of the zebra migration flowed around us. The stars were blotted out, there was no up, no down, just the rumble of the hooves and the taste of the dust and the smell of something wild, which, although I had never encountered it before, my young senses recognised in an instant, growing both animated and wary.

We slept right there on the ground on the leeward side of the truck, in the middle of the road, the zebra flowing past like a tide come suddenly across the land. I fluttered on the edge of sleep, fearful of letting myself fall into the void of this strange place. I could feel with every sense, but the dust and the dull drum of the hooves and the half-seen striped bodies in the dark seemed to float, weightless and unbound by restraint. I reached outside my sleeping bag and touched the ground with my steepled fingertips to hold anchor in the world where it seemed as if creation was gushing forth from the very earth itself.

I found myself compelled, bound in the very core of both my thinking and my heart, to what I perceived as the essential versus the imagined value to be found in life. In nature, each time I looked, I encountered something solid, the strength of whose root or foundation grew exponentially more real the deeper I delved. It is for this reason that, still to this day, stepping outside I become plagued by curiosity and vitalised by the immediate association with the elements. The facets of the elements, so often determinedly stamped out in human-created environments, converse to their supposed discomfort, draw me in. I become engaged, aware and seeing. And it is how we see that defines us.

The stimuli are so multifaceted that it becomes impossible to assign my joy to just one aspect. It's a cumulative effect and different every moment. It might be the flight of a bird that closes it wings and for a moment describes an arc through the air, before flaring to land. It lies in the twinkle of sunlight reflected off wind-ruffled water; the myriad sparkling gems whose only recipient is me. It lies at my feet in the tiny horizontal slit in the earth with its small evidence of excavation, which a Bushman of the Kgalagadi recently revealed, by patient and slow digging, to be the zigzag descending tunnel of a scorpion. It lies in the echo of the song that the Bushman sang as he dug.

I am not alone in knowing this place; it has been given many names: earth-song, mother nature, the call of the wild. But this naming implies an association of recognition, something reciprocal, but like all things sublime it neither seeks nor endorses recognition, it just is. The only awareness is within us and I stand convinced that it is our greatest gift.

But not all of us are aware. Superficially, we are; the wind whips our hair into our eyes and eddies a cold shaft of air down the gap in our collar. Our discomfort makes us aware, but many fail to see beyond their discomfort, their desire only to be away. They fail to look up and watch the trees weather the same tempest; fail to see the last seed shake free and fall to the ground, or the sparrow that follows it down, its wings burnished briefly silver by the light.

I remember as a young game ranger, fresh to the world of seeing fully and deep in the grip of the writings of Kant, Nietzsche and Hess, making an experiment of what is seen and what unseen. I asked people to sit down, handed them a leaf plucked fresh from a nearby tree and asked them what they saw. Many, no doubt made uncomfortable by the intensity of my scrutiny, fidgeted, held the leaf up, turned it over. 'A leaf?' An answer rhetorical with doubt. My mentor, Magqubu Ntombela, saw more. He saw the leaf boiled and applied as a salve to spider bites or infected scratches. He spoke of which birds were drawn to the tree when it fruited. Of the dense shade the leaf thickets provided and how suni and duiker sheltered in the often cavern-like bole of the tree. He put the leaf in his mouth and chewed it, grimaced and spat it out; it was bitter.

There were, however, some who saw what I did. And that was the whole remarkable process of evolution wrapped up in this single example, plucked randomly from the immediate. For me it released a dam burst of thought. The remarkable process of photosynthesis – of producing matter from light. The wonder of life without a brain not only propagating itself but refining its niche in the world, so that it created shelter for animals at the base of its trunk. Animals who then fertilised its roots and carried off its ingested seeds to other suitable sites to germinate. The form of the leaf itself: variegated with thin veins, others dry and papery to the touch, or smooth and waxy; all different, all perfect. How remarkable a history one held in one's hand, how much time, how many deviations occurred on the path to species specialisation, how random chance had performed change akin to miracles. And this was just one of a myriad examples that surround us all the time. I would shake my head, overwhelmed by the magnitude.

There was a young woman, not yet twenty, who saw the leaf as I did. She held it up to the sunlight, its veins etched yellow against the translucent green.

'Beautiful,' she paused, 'such symmetry.'

I nodded my encouragement. She put the leaf between her lips and slowly pulled it out.

'Fresh.'

But I was so distracted by the sensuality of the gesture that I had lost concentration.

'Fresh,' she repeated. 'In the way that the smell of a mown lawn is fresh, as if sunshine and summer are caught up in the scent. It's like all of nature is here in this one small thing.'

It was the same young woman who had, a month or two earlier, asked me one of the most important questions of my life: 'What do you want to do?' I had never asked myself that. What I wanted to do was for holidays and weekends, what I had to do was choose a career. I had never considered that the two could be the same.

Her question changed my life, radically altering its course. It released me

7

to my passion: the world outside. It was as if a star rose into my heaven, and no matter what cloud or foul weather might obscure it, it holds my compass pointed north.

It has held true for forty-two years and we both still see leaves in the same way and she continues to ask me questions that knock on the doors I am subconsciously closing. And that constant probing has kept our path close to the truth of our seeing. Wild is the crucible of our passion: the earth outside the management of humankind. But wild is changing. Each day it is less.

I know that it can be argued that wild is actually beside us all the time: in the sapling that pushes up through the pavement cracks, in the owl that raises its chicks in an abandoned shed, in the wolf packs that patrol the deserted streets of towns around Chernobyl – it is there, waiting to reclaim its sovereignty. But everywhere men are more and when men occupy land what is wild is the first to retreat.

Wild is to me something vast, sufficient to overwhelm us in the arrogance of our possession and induce a cognisance of something whose sum is greater than our own. Something that induces a humility, a wonder and a fear, for there is little as healthy to the perspective of our egos than to realise that a landscape would be indifferent to our suffering or death.

But wild has something further, a character. It is the character of anticipation, of tension, and its hinge is life. When life enters a landscape, it brings to it a charge that crackles like electricity. It changes the measure of time from the aeons of geological evolution to something sudden and immediate. Evidence of life has this same quality: find the fallen feather of a hawk, or a fresh hoof print in soft loamy mud and one looks up with a quickened heart.

It remains beyond my ability to comprehend why we would threaten this, but we do. I would ask you to elaborate to yourself on your own idea of heaven – not just your space or place in it, but the whole idea of heaven. How would it be? Then I would ask you to compare it to our planet – humankind aside – to the natural, wild earth, and then to measure how far what we have exceeds even the wildest exercise of our imagination of nirvana. Yet, we deem it acceptable to destroy it. And what is even more remarkable still is that our right to destroy the gift of heaven has one single justification: profit.

Throughout our lives, Beverly and I have watched the retreat of wild. In some places even though the vestiges of wilderness have been preserved, what was wild

8

became so trammelled with people that it mutated into something else entirely. The loss of wilderness has been proportionally exponential to the increase of the world's human population, and as we moved into the 21st Century, I had a sense of its accelerating disappearance. Ancient forests were cut down, watery worlds were drained or dammed or poisoned by pollution, the earth was dug up and sifted through, vast herds slaughtered for meat and precious commodities like ivory used to fund and support mercenary wars. Plastic became ubiquitous in the oceans. We poisoned our air.

By 2010, as we completed our ninth book, Beverly and I had become increasingly distressed by two ideas simultaneously. The first was a sense of panic as to how rapidly wild space and the life that thrived there was diminishing. The second was that we felt compelled to act, to do something about it. I was haunted by the words of Gandhi, 'Be the change you want to see in the world.'

We decided then that there was only one meaningful way we could do this, and that would be to embark on a journey to see for ourselves and photograph the last wild land left on earth. It was apparent to us that we were amongst the last generations who might know the privilege of wild land, and that in the face of rampant change our idea was possessed of an urgency.

For the purpose of our project we chose to define wilderness as vast, contiguous stretches of land in as natural a state as possible. It could include protected areas such as national parks, but we wanted something larger, beyond human-defined boundaries, where the land in its natural state was of a magnitude too great to hold as a single entity in one's mind.

In regard to the presence of people in wilderness, we applied a simple principle: their presence should be the same as any other creature found in a wild place. That people's relationship with the land should be one of association not domination.

To make the magnitude of the task we were setting ourselves manageable and to give the work both definition and credibility, we further refined our search to one destination per continent. With our criteria in place, we set about trying to find landscapes that might fit. We spent hours making notes from books on wildlife and wild places, we read articles, scrutinised maps, trawled through hundreds of Google searches, and wrote and spoke to friends and colleagues around the world. But wilderness is by its nature remote, removed from the broad experience of humankind, distant to the common and the known. Conclusive material, sufficient to encompass a vast area of a little-known place, was more difficult than we had imagined. As we moved away from the spheres of human activity, so the available information became progressively thinner, less substantive, and finally only anecdotal. In places where few had ventured, even fewer have given definition to a map of such places. Even on satellite imagery it became impossible to follow a track through a forest, to find a settlement of a few houses between a myriad of lakes, unless it was personally known to someone. Our quest to find suitable land was becoming as challenging as trying to pin down a fluttering moth with the tip of a pencil.

One night after dinner, we sat on the floor in front of the fire, as was our custom, a candle burning on the dining table behind us. I looked out upon the wide vista afforded us by the high mountain our home rested on. More than 80 kilometres distant, I could see the town of Robertson, its lights twinkling with the warm promise of sanctuary, home and hearth. I sat up. A shiver ran down my spine.

'I've got it!' I cried.

'What?'

'Lights!'

'You've got what lights?' Beverly asked, frowning.

'Lights at night will show us more clearly than anything else where people are, how many there are, whether they use electricity or fires. Wilderness will be the darkest places on the map!'

It worked.

But one hurdle remained, how would we fund it? Seven continents. The fieldwork alone for this would take at least four years – four years of massive expenses and no income.

One night, as we pored over the impossible numbers, Beverly took my hand and waited for me to look up into her eyes. 'We could sell the farm,' she said.

I pulled my hand away. 'No. Never.' And, after a pause, 'Your parents are buried here.'

'My parents will always be buried here.'

'But we bought it to conserve it! Have you forgotten what that took? Every last cent!'

Beverly's eyes were soft. 'We can choose to sell it to people who will honour the nature reserve. It's conserved, it's done. But imagine if we could use the funds to help conserve all the things we're talking about, to realise all these dreams that have us tossing in our sleep. Imagine!'

It took me nearly a year to agree.

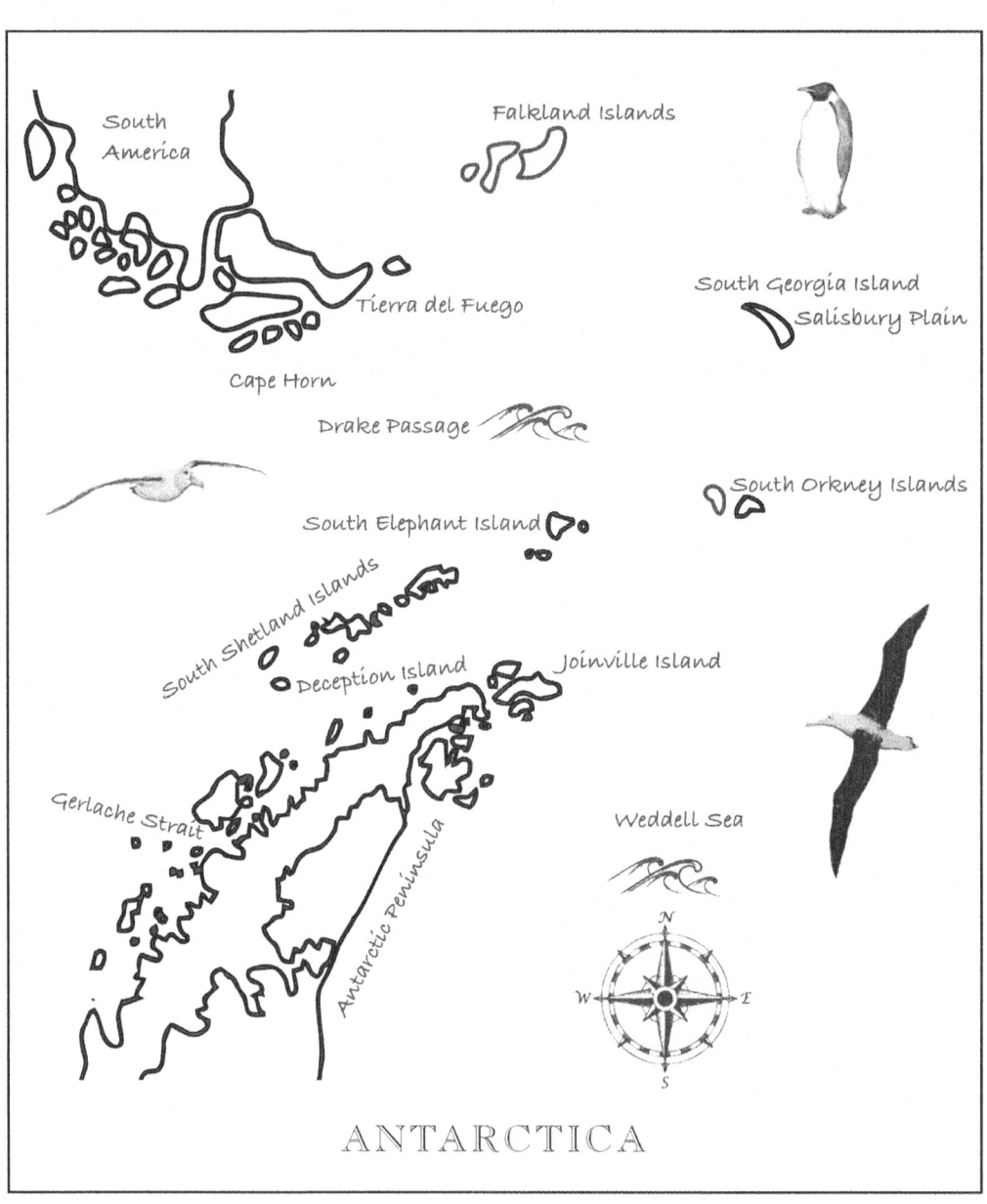

South
America

Falkland Islands

Tierra del Fuego

Cape Horn

South Georgia Island
Salisbury Plain

Drake Passage

South Orkney Islands

South Elephant Island

South Shetland Islands

Deception Island

Joinville Island

Gerlache Strait

Weddell Sea

Antarctic Peninsula

N

W   E

S

ANTARCTICA

I

ANTARCTICA
And the Subantarctic Islands

SUMMER 2015

Seldom have I known such a madness of weather. It is as angry, driven and unrelenting as a swarm of killer bees. There is no respite, no place to hide.

NOTE TO THE READER: *Although we made two journeys to Antarctica, by the first trip in 2011 I had not yet considered writing a book, so I start this story with the second, which took place in 2015. Apart from the fact that the power of the Antarctic landscape could draw one back year after year, if it were possible, our first journey fell closer to the spring thaw rather than the lingering drama of winter's last storms. For this reason, we needed to return earlier in the season; with more snow on the ground and sea, with territorial disputes between seals at the height of the action and mating displays, and nest building of the penguins in urgent frenzy.*

18 NOVEMBER 2015

Beneath a thick grey sky, snow blows in flurries between the unkempt yard and the half-finished hotel that separates our room from the harbour. A huge Argentinean flag flaps soggily in the wind. Traffic throws up arcs of muddy water that pool again in the poorly drained street. Through a tangle of wires and cables I can see the flame of the monument of the Islas Malvinas, a tribute to the Argentinean soldiers who fell in the Falklands War of 1982. The wind buffets the flame, but it does not go out.

Beverly and I are in Ushuaia, the southernmost town in South America, on the southern shore of Tierra del Fuego, 'The Land of Fire'. But the world is all snow, sleet and cold, and the only fire is the monument's flame and the passion of the Argentinean people.

On the official map, given out by the information office in Ushuaia, the eastern half of Tierra del Fuego, the whole of the Antarctic Peninsula, the Islas Malvinas, South Georgia Island, the South Sandwich Islands, the South Shetland Islands and others are shown as Argentinean territory. It is a bold declaration, one that is not recognised by any other country. It does not matter that the Falkland Islands are a British possession; to Argentineans they are the Islas Malvinas, and to call them otherwise is to arouse scowling condescension.

Despite Argentina's claims to the contrary, Antarctica is the only continent on earth that is not owned by any country. It was the United States that, in an effort to erase mounting tensions three years after the Second World War over who owned what in Antarctica, first proposed that the continent should fall under the trusteeship of the United Nations. It was, however, only eleven years later that the Antarctic Treaty was first signed by twelve nations, declaring that Antarctica was to remain a sovereignty- and nuclear-free continent, dedicated to science and exploration. Today, there are fifty-four signatories to the Treaty.

By mid-morning the clouds have lifted to reveal some of the snow-covered mountains that hold Ushuaia pressed close to the discoloured grey sea of the Canal Beagle. The wind, however, has not abated and it drives the snow, stinging, into our eyes, as we head down the pier towards our ship. We pass the *Ocean Endeavour* tied to the wharf. An almost eight-metre-long gash right through the steel of her hull, caused by a collision with an iceberg, is a sobering reminder of the power of the elements. South of Tierra del Fuego lies the most savage ocean on earth.

At the end of the pier we board the Russian ice vessel, *Akademik Sergey Vavilov*. The gangway door closes solidly behind us, shutting out the bitter wind. There are old friends and warm embraces, as this is not our first voyage on the *Vavilov*. In 2013 we had travelled on the same vessel to the Arctic.

In the late afternoon we cast off, the wind pushing us clear of the pier. The Canal Beagle is calm, even though the wind whips whitecaps that run like flocks of frightened sheep, overtaking the ship. Snow-shrouded mountains stand implacable sentinels as Ushuaia fades to a speck and the sun burnishes the sea to a blinding silver. The channel pilot climbs down the ladder and boards the pilot boat as we turn northeast towards the open sea. The remnant swells of yesterday's storm lift the *Vavilov*'s bow and she rolls wide and slow. The troughs grow deeper and I tighten the knotted tourniquet on my wrist that is my only defence against debilitating seasickness.

19 NOVEMBER

To the south, Isla de los Estados appears from beneath layered cloud, and I stand alone at the stern for an hour photographing its emergence as albatross and petrels drift on stiff wings through my view. The birds are fearless of my presence, some coming so close that I can touch them with my outstretched hand. The smallest: the Wilson's storm petrel, little bigger than a large swallow. The largest: a pure-white wandering albatross with a wingspan wider than I am tall. All drift and course, dip and rise, on wings held still against the buffeting wind. It is a symphony without sound, each bird an individual note. It is only the biting cold and my dangerously numb fingertips that drive me indoors. I sit by the window clasping a mug of hot coffee, warming my fingers while another warmth runs to that indefinable place in my core. It is fed by the sense of my privilege to have stood alone in such a place as this, a king for an hour of our extraordinary earth.

By mid-morning we have lost sight of land. Our heading is east-northeast, towards the Falkland Islands. The swell has abated to a calm sea.

Ferdinand Magellan, the remarkable Portuguese sailor who was the first European to cross the Pacific Ocean, first sailed through these unknown seas in the early 1500s. The sea remains as he described it: an alternation of calm and fury, attended by bitter cold and treacherous currents.

In the afternoon, we find a rare fin whale, which survives in the southern oceans by a frangible thread. The second-largest creature on earth is shy to reveal itself. A tall, upright blow gives away its position, and through the binoculars I can see the long, dark-black back terminating in the briefly revealed re-curved dorsal fin. In the early history of the whaling industry, the fin whale was simply too fast a swimmer to be harpooned by whalers from rowing boats. Capable of speeds of up to 25 knots, these 22-metre-long, 50-tonne giants were finally caught up with by motorised vessels equipped with explosive harpoons. In the 20th Century a staggering 725 000 were killed, leaving the population in the Antarctic region in a tenuous balance that continues to today. For an hour, we sail between their distant, isolated blows, which the wind carries quickly away. As the last blow fades behind the stern of the ship, I find myself wondering why humankind, capable of rational thought, is so historically quick to learn and adapt from experience in every single aspect of life, except when it comes to selfish greed. Hundreds of thousands of dead whales brought oil for lamps, baleen for Victorian fashion and profit for a small company of men for a brief flicker in the history of the world. But it left the earth impoverished on a scale that is measured by the endlessness of time.

20 NOVEMBER

A flurry of birds crowds the stern of the ship as the sun breaks free of the clouds that rim the horizon. The Cape petrels dip into the frigid waters, seeking whatever the propellers churn to the surface. The giant petrels are more patient as they fly to windward, without a single flap of their wings, to drift past the ship, turn, and come back again from far astern. Through the lens they are so close that I can see the droplet of water that forms at the tip of their robust bill. They have mistaken us for a long-line fishing vessel, expecting their patience to be rewarded with the offal tossed overboard. We have passed three long-liners in our two days at sea. They are here to catch Patagonian toothfish, also known as Chilean sea bass, a pretty name for a rather plain fish, hauled up from depths of 50 to 3 000 metres, to appear as sought-after fillets all over the globe. With the Patagonian toothfish having a lifespan of up to forty years, and only reaching sexual maturity at about eight years of age, the sustainability of the fishery is questionable.

In the afternoon, land looms low on the horizon: the Falkland Islands, Islas Malvinas if you are Argentinean. They are yet distant, but via the ship's communication network we become caught up in a drama unfolding there. The cruise vessel *Le Boreal* has caught fire. The tense communications we have intercepted seem to indicate that the fire started in the engine room and could not be extinguished. The more than three hundred and forty souls on board were evacuated into lifeboats and the less fortunate into life rafts. Helicopters based on the Falklands lifted the last few stalwarts from the decks. *Le Boreal*'s sister ship, *Le Austral*, turned about to rescue those adrift at sea. Carrying twice her normal complement, she docked today in Port Stanley.

21 NOVEMBER

Port Stanley looks like a Scottish village in the Hebrides. Mud-spattered Land Rovers drive beneath low grey skies, and single-storey stone houses crowd the calm bay of the port, where locals sit with their elbows on the counter of a pub.

The *Vavilov* is held tight against the dock by a brisk northerly. The dock is actually a temporary structure of anchored barges and pontoons, built as a makeshift facility for large vessels after the Falklands War. The shore is a litter of debris: discarded wood, greyed by the weather, is jumbled between tangled rope, torn shreds of netting, plastic, old hard hats, threadbare canvas and rusting steel. Black rocks, layered with bright yellow lichen, rim the tidal zone, where flightless Falkland steamer ducks rest with their bills tucked snug beneath their stubby wings. Turkey vultures patrol the detritus of human endeavour, cruising low along the shore on wide, black wings. Between the shore and the rim of grey gravel roads, upland geese wander the grassy banks with gaggles of fluffy brown chicks, between rotting boats and abandoned machinery that is sinking slowly into the peaty ground. In the small craft harbour, 30% of the boats have sunk or are sinking. Masts, bows and old cabin roofs act as roosts for rock shags and perches for kelp gulls.

Crossing a steel expansion bridge, we come to a brief copse of less disturbed country, where a sponge of stunted shrub grows dense and close to the ground with low ferns crowding in between. The ground is black, loamy and soft underfoot. Beverly and I lie prone on the cold stone, beside the slowly rising flood of the tide, and capture images of the wreck of the *Mary Elizabeth* with a family of steamer ducks in the foreground. The *Mary Elizabeth* was a steel-hulled sailing vessel, built for bulk cargo, and her tall sides and high masts make a graphic counterpoint to the tiny balls of fluff of the newly hatched steamer duck chicks. The chicks are so fearless that they probe the rising sea for food right beneath our cameras, and we must lift our lenses for fear of harming them.

In the late afternoon, we slip our lines quietly and sail into a close, fog-bound world, where the only sign of our passage is the widening ripple of our wake, drawing dark lines in the glass-like, smoky-green surface of the sea.

22 NOVEMBER

The morning is a grey cast of light in a world that terminates less than a hundred metres from the ship. We move through the low veil of cloud, rolling like a bow-legged man, as the ocean swells travel under and past the ship. An occasional bird finds us in the fog, but there is little wind and they fall behind us to vanish into the featureless world. It leaves me wondering how they find their way. Perhaps for some it does not matter, but many of the Antarctic Ocean sea-birds are nesting at this time of year and need to return to their brood. Without a compass, it would be impossible to keep a direction. Between the Falklands and South Georgia, with the exception of Shag Rocks, lie 800 nautical miles of featureless ocean (a nautical mile is 1 852 metres). On a day like today one would be dangerously lost. But the birds are not, and we have yet to understand how they manage. I like things like that. It puts a lid on our smug cleverness: a few million seabirds have a secret that we have yet to crack.

In the evening, Chris Packham, the BBC presenter accompanying us as a specialist guide, gives a talk on the persecution of predators that inhabit grouse moors of England. In an effort to make the numbers of grouse sufficient to be driven by beaters into a line of guns, all natural predators of the moors are exterminated. Badgers, weasels, stoats and raptors – including the hen harrier, perilously close to local extinction – are killed. But the practice does not stop there. Wetlands are drained, and even the indigenous mountain hares are eradi-cated because they carry a parasite that can possibly adversely affect the grouse. Grouse hunting is the province of the wealthy and Chris is questioning its valid-ity, not only in the face of the hen harrier's slide towards extinction in England, but in the broadest terms of sensible environmental practice.

Chris's talk is eloquent, considered and passionate, and at its conclusion there is a protracted silence from the audience. It reminds me of the first time I heard the wilderness mentor of my youth, Dr Ian Player, speak about the need to protect, conserve and look to the future of rhinoceros in Africa. When words convey an obvious truth, there is nothing to say. Dr Player achieved such success with the rhinoceros that for nearly half a century they were a common sight in all southern African wilderness. Chris Packham's quest is rooted in the same authenticity, and although the money of wealthy grouse hunters can raise a pow-erful lobby, it cannot, finally, disguise the truth.

23 NOVEMBER

The wind is blowing snow at 14 metres per second. The air temperature is 0°C. The wind is gusting to 35 knots, which on the Beaufort Scale is a near severe gale. The swell increases with the wind, and we pitch into a cross-sea, where swells from the northwest run confusedly into smaller swells from the south. From time to time the bow shudders as it stalls against the combined crest or trough of meeting swells.

During the morning all the gear, clothing and bags of every passenger are thoroughly wiped down, vacuumed in every recess and crevice and scrubbed with stiff-bristle brushes to remove any foreign matter, mud, seeds or insects that might be lurking there. It is a requirement of every single visitor to the Antarctic continent and its associated islands, to avoid contamination of the pristine environments by alien species. We are also required to watch a short film outlining the behavioural standards expected of each visitor to South Georgia, with special emphasis on the approach and treatment of wildlife and historical artefacts. The film dwells for a time on the current eradication programme of South Georgia's rats and mice. Introduced, accidentally, by the multitude of whaling ships that used the island as a rendering station for their catches during the 20th Century, rats and mice almost devastated the island's nesting birds.

24 NOVEMBER

Land ho! The shout brings the ship's complement out on deck despite the icy wind. South Georgia's snow-capped peaks sit low on the horizon ahead, beneath dense cloud. By mid-morning the mountains have grown in stature, until their peaks loom almost a thousand metres above us as we sail the comparatively calm waters in the eastern lee of the land. The mountains rend the cloud that has hidden the sky for the past few days, and in the bright sunshine the ocean is a crystal-clear royal blue. The wind, however, has not let up, and four species of albatross soar and dip against the backdrop of cliffs and snow. The swell is huge, and as it crashes against the reefs and cliffs, manes of white sweep out behind the crests, like giant horses rising from the sea. The ship turns hard to starboard, into the ancient glacial valley that forms Right Whale Bay. From the relative calm of the fringing sea the bay becomes increasingly hostile. Katabatic winds, sweeping down from the peaks, accelerate over the snowy slopes and valley bottoms to blast across the water. Snow, driven like shotgun pellets, stings what little of my face is exposed. The wind lifts sheets of white water off the ocean surface, drawing it in vertical curtains out to sea.

There are penguins and Antarctic fur seals on the shore, but they remain a distant fuzzy blur as one 50-knot gust after another slams into the ship. We drop anchor and wait. The wind becomes more furious still, until it is impossible to look windward without squinting. The ship heels over, held by the anchor chain. It becomes clear that we will not be able to reach the beach by Zodiac inflatable craft here, and the anchor chain clatters back into the bow. The ship heels to the wind as we turn away.

A little further out to sea, the localised effects of being in the narrow valley fall away, and we are once again in sunshine, the wind having dropped back to a settled 15 to 20 knots. The coastline is rugged and steep. Surf pounds the black rock along the shore. The bases of the peaks are a band of green between the sea and the snow-bound peaks. Once again, we turn to starboard and navigate between small islands to drop anchor off Salisbury Plain. Here, for some peculiar but wonderful reason, the sea is mirror calm and windless. A steep shore-break delivers the Zodiac in a wet rush onto a beach of coarse dark gravel.

I am jittery with excitement, and walk away from the landing area to collect myself before I begin working. To the south, the church-steeple summit of a mountain towers high above the surround of the tall peaks that encircle the plain. The flat land is covered in short grass, punctuated by small ponds, and a stream runs through it down to the sea. The entire place is strewn with sleeping Antarctic fur seal bulls, each defending a small territory of open ground around them. Along the shore, groups of king penguins stand preening, or march in single file in their peculiar straight-armed gait towards the tussocked hillside, where a colony of thousands of their species are nesting.

The penguins utter a wheezing, nasal call, and I smile to hear it again. The Antarctic fur seal bulls have a thin, high-pitched whine, at odds with their bulk and aggression. They are here to establish mating territories, and they assert themselves against every intruder, charging other seals and humans alike. The expedition team has cleared an area between them for us to land, and shoo-ed them to create an open passage leading away from the beach to the penguin nesting colony. The bulls have reluctantly conceded their ground and lie along the fringe of our presence, watching us with moist, blinking eyes.

I walk away from the group. There is a lone elephant seal on the beach, and every few minutes it raises itself to lumber a few metres over the smooth stones of the upper beach towards the west. I walk beyond it, picking my way between the fur seals so as to least intrude upon their space, until I can circle back and lie down in the elephant seal's path. Twice it moves in its caterpillar-like, hunching crawl, which brings it surprisingly swiftly across the stones. It is coming straight towards me. The elephant seal is a small one, but it is nonetheless perhaps four metres long and weighs in excess of a tonne. I edge to the side, staying low on my stomach. It advances again and this time stops only two metres and slightly to the side of where I am lying. When I lift the camera to make an image of it, it regards me with a deeply bloodshot eye. It utters a deep, guttural drawn-out grunt that shakes the ground like a passing train. It closes its eyes and the soft, trunk-like proboscis flutters as it breathes. Its neck is scarred with a crisscross of healed cuts. Each time I move to adjust my composition of its giant form against the backdrop of the peaks, its eyes flutter open. After five minutes, it raises itself again and moves past, a few fur seals grudgingly giving way to its superior bulk.

I work my way back to join Beverly, and we start out towards the colony of nesting king penguins. We are photographing a group of penguins crossing a small stream when I notice one of the expedition staff hurry past carrying the emergency plastic stretcher. I look up. A hundred metres away there is a tight huddle of staff and passengers. My stomach tightens. Something has happened. I see the stretcher positioned on the ground. When it is lifted I cannot see who is on it, or what the trouble might be, for the press of people trying to help carry it. I stand to join them but change my mind, for there are already many wanting to help. The news spreads like flame along the beach. A fellow passenger, Andrew Pringle, has been bitten by a fur seal and is bleeding profusely.

In circumstances such as these, the remoteness of our situation becomes critical. Although the ship provides a haven of comfort and security, it belies our absolute isolation. In emergencies, we are bound by two critical factors: the facilities available on board and the speed of the ship, which is affected by the Antarctic weather, if evacuation is required.

What follows is a remarkable instance of providence in the wake of adversity, as related to me afterwards. Andy had been lying on the ground, photographing a group of king penguins, when he heard the Antarctic fur seal approaching him from behind. He turned and for a moment lay face to face with a testosterone-laden seal growling its aggression. Andy scrambled. The seal did not hesitate and, charging, bit him on the shoulder and thigh, then caught hold of his arm as he flailed against its attack. As is typical of the seal, it shook its head vigorously as it bit into the arm, severing an artery. Blood sprayed into the air. Incredibly, the person closest to the attack was the ship's physician, Dr Matthew Crank. Together with another staff member, he charged the seal, smacking his hiking poles together, and drove it off. As the seal released Andy's arm, he crumpled to the ground. When an artery is severed one has only a few minutes, at most, to live if the bleeding is not stemmed. Matt Crank seized the injured arm and pinned it beneath his hands with all his weight. The blood stopped gushing, the pressure he exerted having squeezed the artery shut, but he could not then take his hands away. He had already made a radio call for assistance as he rushed at the seal. Blood was oozing between his fingers. Thirty seconds had passed since the seal's teeth tore through the artery. Three of the ship's crew, trained in first response, dashed the few hundred metres from the landing site to the scene, bringing the emergency medical kit with them. They placed their hands where Matt's had been. Even in the split second it took to change hands, blood squirted into their faces and onto the clothes of the rescuers. Matt was then able to apply wads of gauze over the two places were the artery had been severed, and he bound them as tightly as he could, forming a tourniquet that temporarily stemmed the dangerous bleeding. While he waited for the spinal-support pallet to be brought to the scene, he bandaged the wound at the top of the shoulder, which was also bleeding badly. In the meantime, the expedition leader had put out a call for any medical expertise that might be found among the passengers. By the time Andy was carried to the beach landing, where a Zodiac waited to transport him to the ship, a remarkable complement had already assembled. Out of fewer than a hundred possible candidates, an intensive-care nurse, a theatre nurse, a general practitioner, an anaesthetist and, most remarkably of all, a retired Scottish vascular surgeon, were all on the beach for the urgent Zodiac voyage to the ship. Andy was set down on the floor of the Zodiac, the others on the pontoons on either side. At the *Vavilov*, the surgeon, anaesthetist and nurses were offloaded up the gangway, while Matt Crank and one of the medically trained expedition staff remained on the Zodiac with the injured man as the pallet was lifted aboard by the ship's cranes and set down gently on the open aft deck of Deck Five, where the ship's clinic is located.

When the surgeon arrived at the clinic he was met by the ship's on-board Russian doctor, who knew the clinic intimately. Instruments were arranged on one theatre bed, while Andy was brought in and laid carefully on the other. His injured arm was set out on the side on a special support mechanism fitted to the bed. He was in shock. It was now seventy minutes since the seal had bitten through his arm, and as a result of the tourniquet he could no longer feel or move his fingers. The surgeon placed a second tourniquet higher up the arm and removed the field tourniquet. Vascular surgery is a bloody business in the best-equipped theatres; with inadequate tools, it is more so. The floor of the clinic became wet and slippery. It was vital that the ship did not rock. The surgeon injected the wound area with local anaesthetic, and sedated Andy intravenously with morphine and midazolam. The seal's bite had torn two horizontal gashes deep through his arm, severing the artery. The artery in the arm runs down from the shoulder, just below the bicep on the inside of the arm. Just above the elbow it curves inward towards the centre of the arm, and it was here that it had been ripped in half. A cut was made through the skin between the two tears, following the path of the artery. The severed lower section was located quickly and tied off. The upper section, however, because of the elasticity of the artery, had retreated upward and an incision was necessary to find the upper end, which was then tied off. The tourniquet was released slowly. A few small severed veins continued to bleed, but the closure of the artery held. The bleeding veins were tied closed. A critical few minutes ensued as the impromptu theatre staff waited to see whether the smaller subsidiary arteries were healthy and sufficient enough to carry blood to the lower arm and hand. Within ten minutes, the patient's blue fingers had regained their colour and he was able to feel and move his fingers. The wound was cleaned, treated with antiseptic, and bandaged closed. The bites to his shoulder and buttocks were then also cleaned and treated.

While the operation is being performed, the guides and passengers are returned aboard from the shore. As the last Zodiac is lifted from the water, the anchor is weighed, and I notice that both of the ship's engines are running. From past experience, I know that this means that the ship is being kept on full standby for an immediate and urgent departure. The captain holds the ship in the calm of the bay, using the bow thruster to keep it steady against the wind. The calm outside contrasts with the unspoken tension on the ship. We wait. The engines turn briefly in reverse, churning the water, but it is just the captain keeping the ship's position. People stare in silence at the shore. Conversations are brief and voices are kept low. Time becomes elastic, each moment a fraction longer than its predecessor. The bow thruster comes on, its distinctive whine growing stronger. It pushes the ship's bow towards the passage out to the open sea. We are moving. Clear of the mosaic of tiny shore-side islands, we turn north, the engines coming up fast to full speed. We are headed back to Port Stanley.

A debriefing is held in the ship's observation lounge. The mood is sombre and subdued. Andy is alive and stable. By the remarkable good fortune of there being a vascular surgeon on board, not only has his life been saved but there is also a good chance that he will not lose his arm. A seal bite, however, is dangerously infectious and there is a long road before Andy can be considered out of danger. What has been performed on board is only a temporary stay; further surgery will be necessary.

We are steaming at full power back to Port Stanley, the closest place, for more thorough medical attention. In good conditions, it is two and a half days away. We are running directly into a high sea, the wind gusting over 30 knots almost straight into our face. I look at the clock ticking on the wall and wonder what it is timing. I have often considered what price would be put on human life in an eventuality. For Beverly and me, we have put more than our last penny into the opportunity to visit South Georgia again, and now we are sailing away. Added to the two and a half days to get Andy to the Falkland Islands, it will take another two and a half to return. We do not have the luxury of that much time. We will not see South Georgia again. It has cost us more than we can afford for no more than two hours ashore. But we are just two of almost a hundred people, all of whom now know the same, and there is not one murmur of dissent when the briefing calls for questions. The expedition company, One Ocean, and the travel company, Exodus, have without hesitation committed themselves to five days of extra steaming. The event is catastrophic in terms of cost and yet it has been committed to without a moment's hesitation. The decision does not require consideration. There is no price on human life; you give everything you have.

25 NOVEMBER

The ship plunges like a lunging horse into the troughs of the swell. White water bursts aside from the bows, to be drawn away in trailing ribbons by the shriek of the wind. The grey skylines merge into the angry sea. The ship slows down against the tempest. The captain closes the bridge.

At breakfast, we are told that Andy's condition remains stable. We are twelve hours closer to rescue. A tiny flame of hope flickers amid the sombre mood. The dining room is filled with the murmur of muted conversation. Lectures and films distract us from the dark day outside. But even when I venture out onto the stern deck to escape the cabin fever that threatens to overtake me, there is nothing but the assault of the elements to distract me. In the dense fog, not even the birds can find the ship, to swoop back and forth across its wake in the hope of reward. There is little distinction between morning, noon and evening, except that the weather's assault has gradually lessened. Andy is out of immediate danger, and the few permitted to visit him report that he has the spirit to make a joke or two about his circumstances.

26 NOVEMBER

Sunshine and a calm sea. It is the first morning since casting off our lines in Ushuaia that we have woken to a cloudless morning. It comes like water to a thirsty plant; I can feel my spirit expand again in my chest. With a steaming mug of coffee in my hand, I stand outside on the high deck and look down into the sea. It is absolutely clear, and I watch the silver beams of sunlight pierce the deep blue, running like the shafts of a hundred arrows, until their tips can no longer be seen. The euphoric lift of the sun is tempered by an announcement requesting the entire ship's complement to attend breakfast, so that we can be brought up to speed on our situation. If the other passengers are like me, they eat by mechanical rote, heeding little the taste of the breakfast buffet, our minds toying with a litany of grave realities.

We are told that Andrew Pringle had, in his youth, been a Royal Marine, and is a veteran of the Falklands War. Although he is long retired from military service, it seems he has not been forgotten by his comrades. Today just happens to be the day that the Royal Navy has planned exercises in the seas surrounding the Falklands, and they have decided to give their exercise a real purpose: they are coming to fetch an old shipmate in distress. The announcement is met with a spontaneous burst of applause and animated exclamation.

The patrol vessel HMS *Clyde* is already on course to intercept us. It is sailing at full steam, and when within range will dispatch a rescue helicopter to lift the injured ex-marine from the *Vavilov* and transfer him to Port Stanley, to the very people he helped liberate. Everyone is smiling. The dining room is a buzz of conversation. The sunshine outside has come indoors.

While I am preparing a presentation, Beverly visits Andy on our behalf to wish him well with the medical challenges ahead and his long journey home to England, and to promise photographs of the Antarctic Peninsula.

Just after 11 am the squat, fat form of a Hercules transport aircraft appears low on the horizon and quickly closes on the ship. It circles widely, round and round, on the edge of our vision, keeping pace as we sail northwards. At 11:40 am, the distinctive thump-thump-thump of helicopter rotors brings every soul on the ship out on deck. An enormous yellow helicopter is slowly circling the *Vavilov*. The sea is thankfully the calmest we have experienced and the ship lifts and falls slightly in the light swell. There is no wind. To allow the helicopter to approach into the wind, the ship continues ahead at 10 knots. The helicopter draws close and takes up a position behind and above the ship's funnel, on the port side. A door slides open and a crewman, in orange overalls, first sits in the open doorway and then eases over the edge. He is attached by a cable to the helicopter and is lowered quickly to the stern deck. He is almost there when the stern heaves high on a swell and the airman disappears behind the ship. As the ship sinks into the trough he reappears, and this time is lowered quickly onto the stern. The helicopter retrieves its cable and flies away, returning a few minutes later to take up its seemingly stationary position above the moving ship. This time a stretcher is lowered to the deck. Again, the helicopter retrieves its cable and leaves. The Hercules continues to circle towards the limit of our horizon. Contrails in the sky, drawing circular designs above the ship, betray the presence of yet another aircraft in attendance: a jet, also keeping pace with our progress. The helicopter returns and, holding steady behind the funnel to the ship's pitch and roll, lifts the airman and the stretcher with the injured ex-marine, Andy Pringle, off the deck. They are held steady by a rope from below, until they reach the helicopter's open door. The stretcher is pulled inside, the airman clambering in beside it. The helicopter lifts away, the door slides closed, and the aircraft accelerates toward the northern horizon.

I am left standing with the sunshine warm on my back. The air seems dense with a sense of the heroic. At the evening dinner, the expedition leader names the passengers who came forward with their expertise and helped save a man's arm and probably his life. They receive a three-minute standing ovation.

27 NOVEMBER

At 6 am we are in the Scotia Sea, at 55° 53' S and 52° 15' W, heading south at just over 10 knots. The day is grey once again. Since collecting the full complement of passengers in Port Stanley, we have sailed 1 376 nautical miles and most of it has been in fog. I am African; I like the sun. In its protracted absence, my love of the day dwindles until I sense in myself some vital ingredient to be missing. It makes me restless, like a wild cat in a cage.

After Andy was airlifted off the ship we did a complete about-turn and headed almost due south towards Elephant Island. We are crossing the eastern fringe of the Drake Passage. Although the swell and wind increase as the day grows long, the sea remains almost placid by comparison to what this stretch of ocean can throw at mariners. The air becomes progressively colder, and in the afternoon we pass a massive tabular iceberg. It towers high above the ship's seven decks and surf crashes against its base, which is more than a kilometre long. Caves of turquoise-blue ice are carved into the base of its vertical cliffs, and surf explodes out of their confines. It undermines the weight above, and, as we watch, a column of ice, twice as high as Nelson's Column in London and more than 10 metres across, breaks free. It sinks in an almost silent rush into the sea, plunging straight down rather than falling outwards, and the only splash is the ocean surging over it again. The massive tableau, as big as five New York City blocks, passes behind and fades slowly into the obscuring fog like some fantasy world that for a moment came to life. I look forward for more, but there is just the fog and the constant sea. There is no measure here, neither land nor sun by which to gauge the passage of the day. Standing for a time alone on the bow, I am taken by a curious sensation that I am standing still and that time is flowing past me, as if I have died and am being arrested by a single moment in the passage of eternity, past which all of the future must come.

28 NOVEMBER

Elephant Island can barely be seen through the gloom of the fog. A dark shore is the only definition between the veiled sky and the grey sea. The narrow saddle, where, a hundred years ago, Frank Wild and the crew of the *Endurance* spent one hundred and thirty-six days while they waited against hope for 'the Boss', Ernest Shackleton, to return, is an insignificant isthmus that the sea is slowly eroding. We turn away. Yet again, the weather has foiled us, and claustrophobia is brewing on the ship. Conversations swerve toward the ridiculous.

29 NOVEMBER

Sunrise is at 3:15 am but it is only after 4 am that light filters through a break in the clouds. Brown Bluff lies under layers of cotton-wool clouds that run from white through grey to a gunmetal blue. Overhead, snow drifts down in a swept-back grey curtain.

We step onto the Antarctic continent at 7:30 am. A flat-topped mountain of bare, brown earth falls to a close shore. Thousands of Adélie and gentoo penguins nest on the bare scree. They build their nests with the only material available: pebbles. The nests are little more than a mound, just high enough to keep the eggs above the damp and freezing ground. There are thousands of nests, spaced an Adélie pecking distance apart. Nest-sized pebbles within the colony that are not already used are scarce. Those birds that go to the lengths of carrying a pebble up from the beach are invariably pirated of their prize. Pebble theft is rife. I watch skuas and kelp gulls drifting over the nesting colony, looking for unattended eggs to steal.

A cold wind-blown sleet has me hunched deep in my waterproofs. Water runs in a constant trickle through the stony ground; its odour is wet guano. My gloves are damp and soggy. My camera is wet and smeared with a green algae and streaks of pink penguin poop. Each time I wipe the droplets of sleet from the lens, I must dry it inside my coat to get rid of the lingering moisture. The ground is a soft, squelchy morass, but the penguins do not stand very high and I must lie down in it to create images that have relevant proportion. Each time, I hesitate: I am cold and wet, and in my multiple layers of clothing to keep out the cold, bending or kneeling is demanding and uncomfortable.

When the sleet finally lessens after several hours, and then stops, the penguins seem as relieved as we are. They stand on top of their nests, stretch and wave their flipper-wings rapidly and shake their coats like dogs emerging from the water.

In the afternoon, we anchor off the Argentinean research station, Esperanza, in Hope Bay. It was here that Emilio Marcos Palma, the first child to be born on the Antarctic continent, came into the world on 7 January 1978. A political move rather than one caused by isolation, the birth is used by Argentina to endorse its claims to the continent.

Hope Bay is fronted at its far end by several glaciers; the sea is a placid, wind-riffled lake as we motor in the Zodiacs past their tall faces. The water is clear to the bottom and bitingly cold. We find a leopard seal that has cornered a single penguin on a tiny iceberg grounded in the shallows. The penguin watches the giant seal with a fixated apprehension. With a short lunge the seal could reach it, but the penguin is using the block of ice as a shield, always retreating to the far side of its tiny two-by-one-metre sanctuary. The shore is just a short, desperate dash away. The seal gives up and swims leisurely away down the shore, cruising the shallows of the penguin colony that occupies the hills behind Esperanza Research Station. When I look back, the lone penguin still lingers uncertainly on its perch of ice. The wind is picking up. The penguins that arrive and leave from the black-rock shore do so at speed to avoid the lurking predator. Their small bodies rocket out of the water like a salvo of tiny torpedoes. Out to sea, the splashes of large groups are caught by the rising wind until their passage looks like lines of machine-gun fire over the water. The rising seas drive us back to the ship. Katabatic winds can rise quickly, and as the gusts hit 35 knots we shelter in the lee of the ship's stern to avoid the sudden violence while we await our turn for the gangway.

We sail out through the Antarctic Sound as the wind reaches fever pitch. I walk out to stand on the bow. The wind moans without respite in the foremast rigging. The ropes slap against the steel, like castanets. The wind continues to strengthen. Off the land the gale raises a dense sheet of snow that blows over the ice, like the contrails of a thousand jets. It hits the water so hard and fast that the entire surface of the sea is whipped to white. The ship's seven-storey structure acts like a sail and the whole ship heels to twelve degrees and more in the vicious gusts. Clouds percolate over the sea, driven overhead of the maddened earth at more than 18 metres per second. On the leeward side of the ship, the spray raised off the ocean wets the very top decks as it falls like rain in the sudden eddy.

Giant icebergs lie across the suddenly berserk ocean, like massive blocks of discarded Lego. In the sun, they shine a brilliant white beneath the clouds that hold in their bellies the dark purple and blue-grey promise of storms. In the shadow, the icebergs are saturated with the subtler hues of the light filtered through the cloud. A washed-out mustard yellow, or strange powder-blues, or plain grey, while the sea at their base is a froth of white over a brief circle of the bright turquoise water. Small flocks of Cape petrels wheel and dip over the seething cauldron, where the driven sea smashes into the ice and sets the long-trapped food particles finally free. The icebergs are monstrous. The smallest dwarfs the ship. The largest is an unbroken mass of ice six nautical miles long. The ice has the aspect of land: immovable, even by so fierce a wind; the sheer faces loom like highrises around us in the gloaming of the day made premature by the storm. I stand on the foredeck, transfixed. I feel an extraordinary energy, as if I had been plugged directly into an electric power source. John Muir clung to the top of a redwood in a California storm. I brace against the bollard of the *Vavilov* as a singular gust lifts the 20-kilogram lid from the anchor-chain hatch. The wind pummels my body from side to side, like a circle of bullies. I am elated. The wildness of wilderness has lost none of its power.

30 NOVEMBER

It is snowing when we land on tiny D'Hainaut Island, which sits off the southern tip of the far larger Trinity Island. It is little more than a small dome of land on which gentoo penguins have established a colony. The small birds haul their barrel-shaped, 5.5-kilogram bodies up the hill to the highest point in the distinctive waddle of penguins. Their bright orange feet have well-developed black nails that help them claw their way up the slope through the layer of fresh snow. The birds sit so firmly on their nests in the inclement weather that many are buried entirely by the fresh white drifts, only their heads and beaks visible through the crust. This is summer in Antarctica.

There is a *refugio* on the far side of the island, a small, red wooden building with pale-blue shutters, built to serve as an emergency shelter against the elements. The gentoos have made good use of it for exactly that, and their nests are crowded together in the lee of its walls. The shelter and density of nests have turned the ground into a quagmire of mud and penguin excrement. It smells like fish, old fish. The penguins that are relieved of their duty at the nest waddle to the sea, with chests and bellies stained and streaked with mud. Those returning from the sea are pure white.

I am told that penguins cannot taste. Chris Packham tells us that it has been found that penguins cannot taste sweet, sour or umami (savoury). It seems remarkable, therefore, that they cannot taste the fishiness of the fish or krill that they eat. Eating, by deduction, then, must be driven solely by hunger, and satisfaction achieved only by a full belly. What makes this situation more fascinating still is that fish can taste and that the small fish on which the penguins feed actually have small taste receptors in their skin or scales, so they are acutely aware of when they are surrounded by their own food source. But this progression through the food chain takes a peculiar twist when one of these small fish is caught by a penguin. The penguin cannot taste the fish that it is eating, but the fish can taste the penguin that is eating it!

In the afternoon, the ship continues further south along the western edge of the Antarctic Peninsula. We find calm shelter from the chop of the Orléans Strait in Cierva Cove and board the Zodiacs in the lee of the ship, where all-but-invisible mountains stand beneath glacier-thick blankets of snow and ice. Almost no rock is visible, and yet the shores of the narrow fjord are sheer-sided to over 500 metres. On the short, rocky, wave-washed shore of the tiny fortress-like Isla Pengüino, gentoo penguins pause at the water's edge, trying to peer into the frigid sea for signs of predators. The islet edge drops straight into deep water and, once committed, the penguins streak away from the ambush depths towards the relative sanctuary of the open sea. In the midst of a maze of icebergs we find a single humpback whale feeding. Able to dive beneath the huge obstruction of the floating ice, it soon outpaces us and we are left, as suddenly, alone again in the silent world of cold.

1 DECEMBER

Andvord is more a fjord than a bay. Mountains are hidden beneath ice and snow, their outlines shortened to curves. We go ashore at Neko Harbour, a pretty name that belies an ugly history. The *Neko* was a factory ship that hunted whales along the Antarctic Peninsula between 1911 and 1924. Neko Harbour was one of its favoured places of shelter, turning this scenic bay, with its high surround of peaks, into a fetid and bloody arena.

The landing site lies opposite a giant glacier, whose seaward face looks like a mass of castle-sized cogs of discarded machinery, frozen to blue ice by the cold. The powder-blues found in glacial ice are a result of ice subjected to great pressure. Snow falling at the head of the glacier does not ever melt. It moves extremely slowly down the side of the mountain, drawn by gravity toward the sea but held in check by the friction with the ground. Each year another layer of snow forms on top of the previous year's snow, and the glacier grows thicker. The extreme weight bearing down from above transforms the loose snow into solid ice, and at the compression required to do this the ice forms a solid through which all colours of the spectrum are absorbed, except blue, which has the shortest wavelength of light. At the base of the glacier, the towers of blue ice are more than 80 metres high.

As we come ashore, a block of ice the size of an articulated truck breaks free, not into the sea, but over one of the 200-metre-high cliffs to the side. It explodes against the stone and shatters into a white cloud that roars with a sound that makes it seem as if the earth is waking. Its power holds me still and I feel the awe of humility, the kind of fragile insignificance you feel when standing alone at the altar of a great cathedral.

The expedition crew has cut steps into the high ice bank that fringes the tidal reach, and after we have all come ashore, the penguins quickly start to make use of the easy path out of the sea. At each low step they lean forward, weigh the height, and hop up. Penguins are knee-less and the hop is produced by their ankles. For the very limited mobility this allows them on land, they are determined climbers. Once up the step, they follow paths to their rookery. Their preference is to make their nests on areas of bare rock, and some of the choicest are on a tall hill above the shore. Pausing every now and then, they climb in their short step-waddle up to the heights. They have the look of lifelong, overweight indolents out for the first time in years to walk the path up the hill on the common. Their attitude, however, lacks any sense of struggle. They have panache and, combined with their reticent gentleness, one is left, when watching them, with a tender happiness. Like seeing the radiant joy of a child who has finished a challenging race and cares not that they came last but only that they did it and someone important was watching and clapped them in.

Leaving Andvord Bay, we turn southwest, but at 64° 42' S, 63° 02' W the Gerlache Strait is choked with ice. The captain noses the *Vavilov* gently to its edge but there is no way through and we must turn around. In the water, as we head northeast, whales are feeding. Nate Small, one of the expedition team whose sharp eyes are constantly scanning the ocean, spots a pod of killer whales far ahead. As we draw closer, and definition improves, it seems that they are hunting. The pod is moving fast, all the whales together as a team. The ship closes from behind: 1.5 kilometres, one kilometre, 900 metres. The whales are at speed, surfacing only briefly.

'Blood in the water!'

Beverly and I hurry to the foredeck with our cameras. Through my longest lens I can see an oil slick forming on the surface. The orcas are milling in a tight group. As the *Vavilov* closes, so do the birds, and out of a seemingly empty sea, giant petrels and Wilson's storm petrels gather over the foaming melee, in numbers that are soon in their hundreds. There are more orcas than I can reliably count, the tall dorsal fins of the males emphasising their size. There are calves too. One lifts its head clear of the surface, a long streak of bloody meat hanging from its jaws. At times, they are all below the surface, the birds hovering in anticipation above. It is clear that they are diving after their kill and then dragging it back towards the surface, for when they do break into view again there are so many moving in a confusion of courses that the sea is whipped into a short chop. Giant petrels hold their huge wings out, facing remarkable standoffs over scraps with the whales.

For all its 117-metre length, the *Vavilov* is a quiet ship and we have edged to within 200 metres of this once-in-a-lifetime scene. I can hear clearly every breath the orcas take. Their prey, however, remains hidden below the surface. The only clue, vague as it is, is that from high on the bridge the guides have occasionally glimpsed something white. Given the numbers of humpback whales we have spotted in the vicinity, it seems likely that the kill is a humpback calf. These orcas are a distinct group called pack ice orcas, and are known as marine-mammal hunters. To kill a whale calf, the orcas will harass the mother and calf until they manage to separate the two. They then chase the young whale until it is exhausted, and in instinctive cooperative behaviour they band together and prevent the whale from surfacing. Whales must breathe. The orcas do not let it, and the young whale held down by their driving attacks finally drowns.

The oil slick that we see on the surface is likely the fat released from the blubber as the whale is torn apart by the powerful jaws of the orcas. We are close enough to see strips of meat in some of the orcas' mouths when they surface briefly. A group of four orcas breaks away and swims right below the bow of the ship. We can see them clearly beneath the surface. The white of their sides is stained the discoloured yellow of old paper. It is not their natural skin colour, which is pure black and white. The colouration is caused by diatoms, a plankton-type organism that occurs in these frigid waters and attaches itself to the skin of the whales. In a photosynthesis process, the plankton discolours the body of its host. Even cruising with full bellies, the whales have the aspect of predatory speed. They rise slowly to the surface to breathe. The narrow, upright dorsal fin of the male, sinister in its simplicity, cuts almost no wake as it scythes out of the water.

Behind them stand tall mountains, quilted in snow, and a choppy black sea scattered with chunks of white ice. A humpback whale calf weighs between six and nine tonnes, and I am surprised that, even given the number of orcas feeding, within an hour and a half the feasting has lost its frenzy and only one or two of the younger killer whales remain tussling over scraps. I have watched a pack of twenty-three wild dogs in Africa catch and consume an impala in twenty-seven seconds. It was too quick to be savage. It was not the gory panting and feeding of lions. It was death and done. Today, beneath the sea, it must have been the same. We are left finally with only the palm-sized Wilson's storm petrel, dipping daintily into the oil-slick surface for tiny titbits and globules of oil and fat. They fly with their long, thin black legs dangling like those of an African jacana, patting their feet here and there onto the surface in a dance that seems too dainty and petite to be an appropriate requiem to such bestial death.

2 DECEMBER

Cuverville Island's rocky buttress climbs sheer from the sea for 252 metres, and this morning is several metres higher with its cap of white snow. Some 6 500 pairs of gentoo penguins breed here, making it the largest gentoo colony in the world. The landing beach is a protected bay of gradually shelving boulders, the ocean made glass-calm by the shifting maze of icebergs littering the shore, which we must thread to reach the land. I wade into the shallows to make underwater images of the penguins coming ashore from the open ocean. Within seconds it becomes obvious that my borrowed waders leak. The water temperature is o°c. My feet feel like they have been packed with ice. I concentrate on the penguins. They come fast, fluid and streaking through the water. It is difficult, almost impossible, to see them until they are virtually upon me, and I find that if I turn the camera too fast as I try to follow them, they spook and accelerate away. I am shooting blind, watching from above as I hold the underwater housing with my camera beneath the surface and try to follow their darting forms. While there is action my mind is absorbed, but in the lulls my attention returns to my acute discomfort. The trickle into the waders has now filled them to my knees. My feet are at the point of acute pain, just before numbness. The cold feels like fire lapping at my toes. Another raft of penguins approaches, and I stumble unfeeling on the boulders beneath my feet as I adjust my stance to gain the best view of them, before an iceberg grounded in the shallows 2.5 metres from me. As I crouch to force the camera deeper in the water, a hole in the crotch of the waders that had not revealed itself before becomes instantly apparent. Bitter cold runs against my skin. I squeak with shock, forgetting the photography instantly and standing up. A shiver runs from my neck down to my toes.

More penguins come porpoising towards us and, like a blinkered horse, all else dissolves onto the periphery of my consciousness as I focus on their approach. For an hour and fifteen minutes I work on making underwater images. My feet, lower legs and knees have passed from cold to fire to ache, and finally to numb. When I start to shiver I know it is time to get out. I shuffle one unfeeling foot at a time onto the cobbled shore. I cannot feel the footing and check my balance each time before moving the other foot.

I find myself wondering how the penguins tolerate water cold enough to freeze and then, emerging wet onto land into sub-zero temperatures, with wind chill often in the -20°c zone, do not die of the extreme exposure. Like all cold-dwelling mammals that live in frigid water, a layer of blubber thinly laced with tiny blood vessels is the principal insulation layer of the penguin. It was this layer of blubber that Shackleton and his men were able to use to help ensure their survival in their years stuck in the ice. They used the blubber of slaughtered penguins for stove fuel, lighting, cooking fat and occasionally to embellish otherwise-impoverished fare. But blubber alone is not enough. The penguin's feathers have evolved over time to provide an external blanket to protect the blubber, which in turn protects the muscle and organs from cold. Of all birds, Antarctic penguins have among the densest concentration of feathers, at up to 45 per centimetre. Take a comparative look at the hairs on your wrist and that gives some idea of how dense this is. These feathers are further adapted into two layers. The longer, outer layer is hardly a feather at all; it is more like a spine with multiple hooks, which interlaces with all its neighbours to form a shield when wet, and protects the inner layer, which is a short, dense, downy feather. Air trapped in the downy layer adds to the insulation. Recent studies have revealed that the penguin can, at will, squeeze down these outer feathers to force some of the trapped air out. This causes a stream of tiny bubbles to erupt from the penguin's outer layer. Air has less friction than water and so, as it does this, the penguin moves markedly faster through the water. The trapped air therefore acts not just as a heat trap, like the fluff of a duvet, but also as an anti-friction grease when the penguin needs to flee, rather like water on a slide.

But the penguin has body parts that remain exposed to the water: the flippers, the feet, eyes and bill. These exposed areas receive blood directly from the heart to keep them warm. But what of the cold blood coming back? To minimise the shock of blood chilled in the exposed extremities returning to the penguin's body, the veins with the cold blood run close alongside the arteries of warm blood coming from the heart, and this performs a heat exchange and pre-warms the blood coming into the body. These adaptations work so well that the penguin's problem is not getting too cold, but rather getting too hot. Penguins do not sweat. To get rid of too much heat, caused by either hard swimming or being exposed to the sun, they are able to regulate the amount of blood going into their flippers and feet. A warm penguin allows blood to flow into both, which turns the feet and underside of the flippers pink. A cold penguin restricts the blood flow and the feet turn pale while the underside of the flipper goes white.

There still remains, however, the problem of water freezing on their feathers in the extreme snow, rain and wind chill experienced in Antarctica. Penguins are fastidious preeners. At the base of their tail is a gland that secretes an oil, which they smear over all of their feathers with their bills. This oil breaks the surface tension of water. Water on a penguin, therefore, has nothing to adhere to and falls or runs off before it can freeze.

Penguins porpoise and preen between blue-based icebergs as I button up for the return trip to the ship. The weather is deteriorating. Snow floats down out of a dark sky and the wind drives a freezing spray onto my back and over my head from a short choppy sea. I lick cold, salty drops from my lips.

At Fournier Bay the snow is falling so heavily that it blankets the sea in a slushy layer, too thick to see the ocean through. We edge forward into a world of white. White sea, white mountains, white sky. The 2 760-metre-high Mount Français, which dominates the southwestern end of the bay, is visible only for the first 50 metres of its base. Icebergs rise through snow, and it is only the ship's wake that reveals briefly that we are afloat before the white world draws it closed again.

On the bridge, voices are kept to a whisper. The eyes of the watch peer into the dangerously low visibility of the shrouded world. The sweep of the radar reveals ice all around. We are at slow-ahead. I watch the computer screens showing our position. There is land close by, all around. Shallow and foul ground is marked by a red skull-and-crossbones. It makes me marvel at the brand of madness and courage of the men who came here in wooden sailing ships more than two hundred years ago.

3 DECEMBER

In Wilhelmina Bay the dawn is an opaqueness that seeps from a night that never quite grew dark. The snow is still with us. In the calmer coves, the snow that has settled on the surface has grown thick enough to form brash ice. It is a frozen blanket that lies on the sea, with myriad small icebergs snared in its still grasp. The ocean moves beneath it and the ice blanket stirs, as if some flat-bodied reptile were slithering beneath its cover to the shore. A humpback whale surfaces and fist-sized blocks of ice rise free on its black back. It lies sleeping in the midst of a scene almost too incredible for invention. The temperature is a balmy 0°c. The sea is as cold, too cold for the snow to melt. The whale wakes and dives beneath the ice. It blows as it surfaces a minute later between two small icebergs. It lifts its snout clear of the monotone surface of white. A young kelp gull perched on an ice fragment watches it from two metres away. Flurries of snow waft down over the scene. A zephyr of wind lifts them into a sudden dance.

We cannot leave the ship in the Zodiacs because of the real possibility of losing it in the opaque grey world where definition blurs, reappears, and then vanishes completely. The wind increases exponentially as the unseen day waxes into fullness. In the Gerlache Strait, our heading south becomes a head-on battle with the growing wind. We have just turned around when a distress call is received on the ship's radio. The wind is blowing a steady 47 knots, classified as a Force 9, a strong gale on the Beaufort Scale and only one knot short of the storm classification. A sailing yacht, *Tarka*, has run into difficulty in the high winds that have now whipped the entire surface of the strait to white. It has run aground. Even in the deep water the wind is causing the larger swells to break. They have ten passengers aboard and their situation is urgent.

We are headed in their direction and offer assistance, but another ship is closer. I had seen the *Tarka* on the pier in Ushuaia. It is a robust, sleek-looking yacht that I guess to be in the region of 60 feet, built of aluminium. Its superstructure is low and clear of obstacles, making it well suited for the rigours of Antarctic weather, but the rescue vessel reports that where it has run aground the wind is blowing a steady 30 to 40 knots, with gusts reaching up to 70 and 80 knots. It is too dangerous for the ship to approach closely, and the Zodiacs that they dispatch to attempt to tow the yacht free, with their 60-horsepower motors, are insufficient for the strength of the wind, and the yacht is becoming increasingly fast with the falling tide. At this point the yacht has not yet been holed.

I stand on the lower deck as we run before the wind. An angry sea rears alongside the ship. The crests of the waves are as high as my head. The whole surface of the water is flecked with foam, like the jaws of a rabid dog. We pass an iceberg as high as the bridge of the *Vavilov*. A wave slams into its face – a wall of blue that reaches two-thirds up its height – and then explodes into a shatter of white, its mane drifting over the top of the iceberg. It is as compelling as fire and I wonder how long the yacht will last. All day the wind continues to howl through the superstructure of the ship. Seldom have I known such a madness of weather. It is as angry, driven and unrelenting as a swarm of killer bees. There is no respite, no place to hide. The waves toss the ship like an autumn leaf in a rapid.

4 DECEMBER

At 11:45 pm last night, at the height of the high tide, a small flotilla of Zodiacs braved the wild sea and succeeded in towing the yacht loose. Miraculously, the hull had not been damaged and *Tarka* floated free without taking any water. It was lucky. Each year, this place of the fiercest weather on earth claims a victim or two among those who for a moment drop their guard. For centuries, the Drake Passage has reinforced its reputation as a savage sea. Today, it keeps to its notoriety. The open seas resemble the tumult of a storm-battered shore, with the sea seething beneath a wind-torn surface.

We find a brief relief in the leeward harbour of Half Moon Island. The wind here is blowing only 28 knots. As we push against the chop toward the shore in the Zodiacs, we are drenched by showers of icy water. Onshore, it is snowing hard. Only the penguins appear comfortable. The freezing conditions keep the ground free of water. It is water that is the penguin's enemy here. It turns the ground to a quagmire and wets the down of the chicks, spoiling their insulation and causing many to freeze to death. Snow is dry. The diminutive chinstrap penguins, in their tuxedo plumage of black and white, waddle with an almost jaunty step through the stinging blasts of snow driven across the island's surface.

The wind fills the hood of my parka, driving snow inside my layered clothing. It has been snowing for days. Every few steps I take my boots break through the crust and I drop through up to my knees. Walking becomes the effort of lifting oneself free from the snow. You push down with your knuckles and those too sink in. You bend to your knee for the extra surface it gives you, lean on your elbow, and pull the sunken boot free. The camera pack makes your struggle top-heavy. You stand, straighten your beanie out of your eyes, take a few steps and fall through again. The chinstraps waddle past with their flippers out wide, like the arms of a braggart. By the time I crest the island, where their nests lie dense between the sharp exposed rock, I am warm with effort. The penguins sit on eggs, while their partners compulsively gather more small stones to add to the nest mound. The egg is cradled between their feet and kept warm by a special parting in the penguin's down that exposes the skin. There are a few gentoo penguin nests between the more prevalent chinstrap nests.

In an extraordinary adaptation to the short summer of Antarctica, the fledging period of the gentoo penguins here is an unbelievable thirty days shorter than that of exactly the same species nesting on the Falkland Islands, several hundred miles to the north, where it is much warmer for considerably longer.

On the far shore, where the driving snow is less furious in the lee of the ridge, a few Weddell seals are hauled out onto the snow banks of the shore. They have a blotchy coat that turns from grey to tan as it dries, and long whiskers that surround a short muzzle, in a face that seems frozen in an expression of sad reticence. I lie on the crust of white and look into their faces through the lens. They look back, blink, and look away. The wind hurls a shower of snow across my view. It patters onto the sea like gravel shot from a cannon.

Penguins come ashore in an airborne spurt, hurling themselves onto the beach from the backs of the small waves. The beach is of smooth, round, black, cobble-sized stones to which the snow has stuck, making them white at the centre with a rim of pure black.

Another gust rips at my outer gear, the pellets of snow pattering a tattoo against the hood beside my ear. The wind is rising. We are called back to the landing spot. The gale is beginning to gust to 36 knots, which is outside our safety zone, and we must return to the ship. The sea spray is lifted like a waterfall over the rubber bows. We duck our faces away as it drenches us. I peep aft towards the shore. A stranded wooden whaling boat pokes its prow through the snow. A lone penguin stands above it, its flippers askance, a single dot in the field of white. Men come and go to these shores. None stay. Each of us has taken something from it. Mine is awe.

Puerto Montt

Pacific Ocean

Sierra de los Andes

Atlantic Ocean

Parque Patagonia

Caleta Tortel

Reserva Nacional
Katalalixar

Parque Nacional
Bernardo O' Higgins

Gran Campo
de Hielo
Patagonico

Torres del Paine

N

W   E

S

Tierra del Fuego

Ushuaia

Canal Beagle

PATAGONIA

II

SOUTH AMERICA

Patagonia, the High Andes and the Chilean Archipelago

SPRING 2011 — SUMMER 2012

I wonder, if wild is a silent thing, what it is that makes this landscape seem so alive? Perhaps it is that it seems poised, on the edge of something, while remaining still.

The customs broker, in the plush boardroom of the clearance agency, reminded me of a character in a novel: an English dandy on holiday abroad. He wore a pale-peach shirt, beneath a brass-buttoned, double-breasted blazer, complete with a red handkerchief in the top pocket. There was a cravat at his throat and his pleated trousers fell over ornate black brogues. But his hand had a slight tremor, and I noticed that his fingernails were long and the tips of his index and middle fingers stained yellow by tobacco. He shrugged under my scrutiny, causing the sleeves of his blazer to rise, revealing worn shirt cuffs. When he stood to put his espresso cup down on the dark wood sideboard, I noticed a hand-stitched repair in the knee of his trousers. With his back turned, I could see that the heels of his shoes were worn down to the sole. The back of the left shoe had split and been repaired with white string. I tried to look into him as he returned to the table, past the immaculately combed silver hair to the centre of his deception. He smiled, and I steeled myself for more of what had already been a hard road to this place.

Five weeks earlier, on Cape Town's docks, Beverly and I had packed our Land Rover into a container. In it was all our camera equipment, specialist photographic gear, clothing, tools, camping gear, spare parts and supplies to last for the coming four years – everything that thirty-five years as professional wildlife photographers had taught us we would need. We had watched with trepidation as a crane hoisted the container aboard a ship bound for Buenos Aires. Three weeks later, when we arrived in Buenos Aires to collect it, the container was not there. A dispassionate clerk in the shipping office told me it was still sitting in Coega in South Africa.

'It should be here in two weeks,' he told us.

Three weeks later we finally had confirmation of its arrival and were meeting our faux aristocrat in a glass-panelled boardroom. He was third in a line of genial, charming men in suits who had materialised to help us. Each had their fee. In the end, it cost us five days and US$3 500 to clear customs in Buenos Aires, making it the most expensive port to enter in our entire four-year journey. Almost half of the cost went to *fees* – for which there were no invoices.

We raced south, driving nearly 4 000 kilometres in three days to make the scheduled departure of the ship for our first journey to Antarctica. Three weeks later we returned to Ushuaia on Tierra del Fuego to continue the first leg of our four-year exploration of every continent on earth in search of the last wild land.

*

Tierra del Fuego means Land of Fire, but no greater misnomer exists. At 54° 48' South we drive out of town into a land of snow-draped mountains, whose peaks lie hidden beneath an omnipresent pall of grey cloud that lashes the land with bouts of wind, cold rain and sleet.

It was Ferdinand Magellan who, in response to the multiple fires made by the local inhabitants on the shores, named Tierra del Fuego when he landed here in 1520. His journals recorded copper-coloured giants, the first of which appeared on the beach of the bay in which they were anchored and began dancing madly. Magellan picked a single sailor to venture onto the beach. The sailor beached his boat and when he climbed onto the sand, the true stature of the native became apparent: the sailor came only to his waist. The seafarer showed a daring initiative by dancing as wildly as the giant and managed by this introduction to eventually persuade the giant to join him in the boat. Back on the ship, Magellan noted that he wore clothes of neatly sewn skins. The giant's face was painted red, with a yellow surround to his eyes. Magellan, however, was most struck by the enormous guanaco fur boots he wore and this led him to name the giant 'Patagones', which loosely translated means 'big feet'. The name remains today: Patagonia.

History tells us that the fires were those of the Yámana and Selk'nam, who like so many other indigenous peoples, are gone to guns and germs. They only made contact with European colonialism in the late 1800s, but tragically, due to conflict with gold miners and sheep farmers, there were only 150 Selk'nam left by 1947. Believed to have migrated to Tierra del Fuego by canoe across the Strait of Magellan from Chile, the only tribute to their remarkable culture is the striking images of their elaborate initiation ceremonies. In male initiation ceremonies they completely transformed their appearance: going naked in this bitter-cold place, donning tall bark headdresses, and smearing their bodies with ochre and broad lines of white ash. In this adornment, they became 'Hain', spirits able to communicate with the nether world.

15 DECEMBER 2011

Further to the east of the Canal Beagle, where the mountains give way to a country of tall hills, we meet Thomas Goodhall. He has the reticent, slow-moving manner of someone who has spent his life at physical labour. He sips tea as we talk, his fingers around the cup thick and coarse, with cracks of black in the weather-hardened, broad, strong nails. He speaks English with a slight slur and tells us that he is the fourth-generation grandson of Thomas Bridges.

As a child, Thomas Bridges, like Louis Armstrong, was found abandoned beneath a bridge. He was later adopted by Anglican missionaries and at age thirteen moved with them from England to South America. In 1886, after thirty years of mission work, Bridges received Argentinean citizenship and, in recognition of his work with the Selk'nam and Yámana and the shipwrecked sailors of Cape Horn, was granted land. He established Estancia Harberton, where we sit today gazing through the window onto a calm sea, where a few gulls drift across the sky and a cormorant sits drying its wings. It is a picturesque setting, with jetties running up from the placid lagoon to the low cluster of well-kept farm buildings.

Bridges's pioneering spirit and entrepreneurial vigour marked a turning point in the history of Tierra del Fuego, for he proved that sheep and cattle farming in the area were profitable enterprises. Soon the entire island followed suit. The land was divided into *estancia*s (estates) that even today, after the divisions of inheritance, remain vast. Along the eastern shore of the island we drive for almost a hundred kilometres along the boundary of the famed Estancia José Menéndez, which covers 1 600 square kilometres and, at the height of the wool industry, carried 140 000 head of sheep.

Thomas Goodhall, however, speaks more of rivers than of land. We talk of fly-fishing and he talks of brookies and brownies and rainbows as if they are friends with whom he has shared the river. When I rise to leave, he wipes a drop of tea on the leg of his faded dungarees.

He watches from the doorway as Beverly and I climb into the Land Rover and back away to drive to a campsite he has suggested, where we might feel the scope of the land with a view to the sea. I look back once in the rear-view mirror. He is still there, and I envy him the sure tenure of his calm and certain place, the history that ties him to this farm on the edge of the sea, on the fringe of time, far from the whirlpool suck of the busy world.

THE TWELVE DAYS OF CHRISTMAS 2011 AND A FEW MORE

The road ends at Río Möat. For nearly a hundred kilometres to the north and east the land is trackless, pathless, wild. We walk in. But a day's effort brings us small reward. The hills are steep and densely forested and our going is slow. The forest is pristine but with little obvious life between the trees. Along the shore, the going is faster and our craving for animation is satisfied by night herons that skulk in the shadows, crested ducks and flightless steamer ducks that waddle into the wavelets and drift a few metres offshore. A flock of dolphin gulls bathe in a small stream that rushes over round stones into the sea. It is late by the time we return to the Land Rover, and as I sit and let that wonderful release of stopping after a day's hard excursion flow through my body, pairs of ashy-headed geese pass overhead, leaving the taint of their call on the dusk's mute colours.

Poring over our map, we see that estancias lie charted like provinces across the landscape. I find Spanish names have a luxurious feel on my tongue. Sierra Alvear: a barrier of mountains, implacable sentinels of rock that guard the divide of the Canal Beagle and the lands to the north. Lago Fagnano: on their far side, where bands of locals are camping, oblivious of the dramatic half-hourly morphing of the weather. They are vocal and cheerful, dragging logs and ourselves to the warmth of their campfires.

At Estancia Indiana, they are shearing a thousand sheep. A gaucho in traditional garb, whose dogs and horse are as animated as he is calm, rides twice around the Land Rover, bending low in the saddle to peer into the cab. He is tickled by the fact that our steering wheel is on the wrong side and shakes his head at our explanation that we have brought it from South Africa. In an old shed, one side open to the weather, the floor is saturated with the lanolin of the sheep that were sheared there. Bending down to examine the smooth, worn planks, I catch the familiar smell of it and can vividly recall the mustard-yellow tube of lanolin my mother used on us as a balm for chapped skin.

Estancia Buenos Aires, which in a country of beautiful traditional homesteads is more beautiful still, causes me to pause with my hand on the gate as I look up the gentle hill toward it. The buildings are painted the traditional iron-oxide red, with the windows and doors picked out in white. The main house has a spire-turret in the crux of its L-shaped layout. It is nestled against a hill, facing northeast, with a thin lenga forest coming down to the buildings from behind, and the snow-capped peaks of the mountains to the southwest, pale blue and black in the distance. It commands a view over the valley of the Río del Fuego, which meanders in wide, slow turns across the expansive valley floor. A mesa rises at its edges before giving way to the hills. We are given permission to fish, specifically fly-fish, by a handsome woman in workingman's clothes. Behind our conversation the dogs, kennelled in the runs above the house, keen their excitement.

At Estancia María Luisa, we stop and stare in stupefied amazement at the devastation ahead. They are logging old-growth lenga forest, leaving not a single tree standing. At the estancia boundary, the forest of lenga trees, a slow-growing birch whose gnarled and ancient reachings are an indelible signature of Tierra del Fuego, ends abruptly. The suddenly bare land is littered with the grey trunks of old fallen trees whose wood is too rotten to harvest, like the bleached bones at the site of a massacre. As we drive further in, it becomes obvious that nothing is being spared. Logs lie in piles beside the road. María Luisa, so pretty a name for a place of such a rash and selfish disregard.

Driving out of the hills on Christmas Day to a point where we might find cellphone reception to call our families, we discover that the people of Tierra del Fuego have abandoned the towns for the country. For 40 kilometres, the rural roadside has the atmosphere of a fiesta. Everywhere families have climbed the estancia fences to set up circles of tents and build fires. The smell of roasting meat and the shouts of informal soccer games fill the air.

We camp on a lonely reach of land on the periphery of a stand of lenga forest, looking down a valley where a stream provides a dinner of trout that I cook over the coals of a low fire. As the stars turn out, the wind that had remained at bay all day announces its return with a ripple, which moves like a wave through the grass heads. By the time we crawl into bed, the wind is rocking the Land Rover like a yacht at sea.

Wind is a constant companion on Tierra del Fuego. Mostly it screams, but even when it is calm one is aware of its absence, for it will come again. And, in the stillness, one is taut with anxiety, waiting for the first zephyrs that herald its return. We learn to set out into flat country walking to windward so that our energy can be well used, and when we are weary at the end of the day, the wind will be at our backs.

Camping one evening in the crescent of a lenga forest where the tall mature thicket of the trees diverts the shrieking gale over our heads, I feel a few rain-drops. Looking up, the sky is clear. Another patter of rain falls. I get up and walk clear of the shelter of the trees. Only a single drift of cloud clings to the twin peaks of the distant mountains, but sporadic patters of rain continue to fall. I turn right around in bafflement and then realise that it is water being carried on the tempest from the lake, behind the shelter of the trees. Incredulous, I pace it out. It is two hundred and eighty-six paces to the shore. The wind is blowing so hard that it is lifting water off the wavelets it has churned up and carrying the drops high over the top of a mature forest to fall as rain, almost 300 metres away.

The wind has abated on the morning we cross into Chile, on the western portion of Tierra del Fuego. I cross, unintentionally, on foot first. We find a Fuegian fox and for two hours I work my way closer to the small predator, gain-ing its confidence, until it lies sleeping not two metres away from where I lie in the grass. It is only when I stand up and look back to the distant Land Rover that I realise how far I have come, and that lugging a long camera lens through the open ground of the border country may make me look suspicious. But at the small *puesto* (post) that is the Argentinean border-control office, a kilometre down the road, we have to knock and call out several times to rouse the sleepy-eyed official.

La Frontera is a broad, shallow river, the crossing marked by a few sticks and old broken fence posts hammered into the sand. The water is clear, and small fish dart away over the coarse-gravel bottom as the bow wave of our Land Rover breaks white across the calm.

What took five days in Buenos Aires takes just fifteen minutes and costs only a cup of tea as we discuss wild areas with the enthusiastic young Chilean border official. We head for the crosses he makes on our map, negotiating two spectacular hairpin passes where guanaco graze the mountain slopes and stop to watch us with cocked-ear curiosity as we photograph them. Guanaco are the long-legged and long-bodied, robust wild cousins of the llama. Their camel-like head and muzzle are made attentive by narrow, upright-held ears, and their shaggy, curly-haired coat runs from a russet-orange to pure-white on the belly, inner legs and neck.

It is a simple matter to define wild land on paper. It is more difficult in the field, particularly in places outside officially designated wilderness or parks. Ultimately, it is a sense, a knowing, a feeling in one's gut. It comes to us as we drive into the mountains of southwestern Tierra del Fuego. No fences and no signage designate the boundary of the Karukinka wilderness, but the pastures, small puestos and estancias fall away. One can look back from the heights and see the presence of man in the low country behind, but in the mountains there comes to us, for the first time in our exploration here, the knowing that we are in untrammelled country. Forest clads the slopes. Raptors hold station on the updrafts and guanaco move in small herds on the open mountain crests. We drive slowly, compelled by each sharp turn that reveals yet more pristine mountain country ahead. We stop by a lake where not a single track or path follows the shore. But our euphoria is soon dashed by the sight of a locked boom across the road and an army barracks hunkering behind a high fence.

The officer I come before stares at me dispassionately across his desk. In my halting Spanish I deliver the best, most impassioned speech I can, requesting permission to drive into the tantalising wilderness beyond the boom. His expression never changes. He just shakes his head.

'*Caminando?*' I ask. He shakes his head again.

It is exasperation that makes me defy him. I duck under the boom and walk

200 metres down the road and raise my arms to the mountains to the south. The rare temple that we have come so far to unearth lies within our sight but is denied us. I want to shout. Not words, just an animal scream of frustration.

When I turn around, an armed soldier is watching me. His eyes never leave me – all the way to the Land Rover and until we drive away.

NEW YEAR'S DAY 2012

We cross from Tierra del Fuego to Punta Arenas, on mainland South America, on the last night of the year, aboard a ferry whose schedule is dictated by the wind. Drunks patrol the ever-lengthening queue in search of handouts, and on the ferry there is a small, battered blue truck in which sheep are penned in a homemade double-storey rig of wooden pallets. On the top, a single sheep gnaws frenetically at the coarse wood of its confinement. Between us and them a man, still wearing sunglasses despite the darkness, sits in the cocooned comfort of his canary-yellow Hummer, staring at his mobile phone.

After stocking up with supplies we press on, to arrive at dusk in Puerto Natales, where we are devastated to learn that Parque Nacional Torres del Paine is on fire. Our only option is to take the five-day ferry trip to Puerto Montt: 944 nautical miles to the north, through the otherwise inaccessible reaches of the vast Chilean Archipelago. There is no road in Chile connecting where we are to Puerto Montt, the country between covered entirely by the great Patagonian ice fields.

2–3 JANUARY

In Puerto Natales we finalise ferry reservations and pay the exorbitant fee for a AAA cabin, two beds with private bathroom, the only places available at such short notice. In the morning we wake early to find the entire cargo deck has filled up with trucks, four of which are loaded with fishing nets that pervade the rear of the ferry with the stench of rotten fish.

The day is overcast and grey, which the sailors tell me is absolutely typical. The ferry moves slowly through a tall landscape of steep hills and mountains, the tallest lost in the omnipresent clouds. The hills are either forest-clad or bare rock, and waterfalls rush down from the high cliffs. We pass the Cordillera Riesco and the Cordillera Sarmiento and are soon sailing deep within the mystical landscape of the Chilean Archipelago, revealed from time to time through the clouds. I can only imagine what lies beyond the mountains and the endless waterways that lead away from the main channel. I see deserted beaches with penguins and seals on offshore domes of rock. The forest is unbroken, and each hill and peak behind is a wilderness without any mark of humankind, not even flotsam on the shore. I yearn to step off and feel the thrill of setting foot in this wild country.

4 JANUARY

I get up twice in the night to check whether there is anything to see as we sail along Canal Messier, leaving Reserva Nacional Alacalufes behind, and pass through the heart of Parque Nacional Bernardo O'Higgins, where the channel is dangerously narrow. But it is dark, misty and cold and the only view is courtesy of the thin lights of the ship. Towards morning we pass Reserva Nacional Katalalixar to the starboard side. Parque Nacional Bernardo O'Higgins continues, unbelievably, on our port side, an endless unfolding of deep green indigenous forests covering steep, rocky mountains that fall to the sea.

Mid-afternoon we emerge into the Golfo de Penas in the open Pacific Ocean. Minke whales are everywhere, feasting on what is clearly an upwelling of food where the sea meets the channels and fjords.

5 JANUARY

For the last dinner, the AAA passengers, who have become our friends, request and are allowed to take over the officers' dining room, which has a fine view over the bow. The party goes on till late, with passengers donating bottles of champagne, Panamanian rum and single-malt whiskey. David, an Englishman, sings a Welsh ballad, looking close into the eyes of an Argentinean's wife. The song is good and well sung, but the intimacy is uncomfortable.

6 JANUARY

Disembarking at Puerto Montt, we head for the 251 000-hectare expanse of the Parque Nacional Vicente Pérez Rosales. Established in 1926, it is Chile's oldest national park, a fairy-tale landscape of soaring volcanoes, towering mountains and myriad lakes. The lakes lead to a historical pass through the Andes used by Mapuche people, which they ingeniously managed to keep secret from the Spanish for more than a century.

Here, we make our first acquaintance of the ubiquitous *tábano*, a bumble-bee-sized, black horsefly with emerald eyes and bright orange epaulettes on its behind and shoulders. It bites: a rasping sting that later swells to an angry red bump. They seem drawn to movement, so I follow the example of a local fisherman who walks without haste and uses a leafy twig to discourage their buzzing attentions.

7–8 JANUARY

We travel down the western side of the Estuario de Reloncaví, past a string of small cottages and plots, despite the map's showing it as a national park. Only on closer inspection do we notice that the coastal strip is excluded from the park, and this proves to be the case for many of the other national parks. Salmon farms abound along the sea edge. Our retreat to Lago Chapo, sandwiched between two volcano national parks, is equally disappointing. The lake is clearly excluded, and salmon hatcheries, squat and ugly, dot its expanse.

9 JANUARY

It is raining in earnest as we depart Hornopirén by ferry, headed south. We emerge into a brief wash of sunshine as we sail beside the dense and jungled forests of Parque Pumalín to the east, but then the clouds close in again. Steep, almost sheer-sided granite rock emerges from the sea in a vertical rush to bare snow-flecked tops occasionally glimpsed through the blanketing cloud. The slopes are pristine forest, but in the sea, salmon farms are plentiful. For the entire passage, four Roman Catholic priests on vacation spend their time ensconced in their small car reading their Bibles. I am perplexed at the priests' absolute absorption in the word of God, while the work of God passes unacknowledged.

The ferry journey is in two parts, with an overland transit of about 10 kilometres through Parque Pumalín in between. The park was established by the American conservationists Douglas and Kristine Tompkins, and its nearly 2 900 square kilometres consist of pristine rainforest and untouched wilderness. The second ferry we board emerges from the shelter of the landing into a stormy sea. The going is slow as we slam into the swell. Arriving at the landing at Caleta Gonzalo, there are hundreds of dead fish in the water, some bloated, some still feebly swimming. No one can explain the cause. The jetty ramp is completely exposed, and we make a turbulent landing as waves pound the ferry against the shore.

10 JANUARY

Parque Pumalín intrigues and draws us in. The forest is a riotous tangle of wet trunks, bamboo and a profusion of moss. The alerce, the longest-living tree in the world, which may attain an age of 4 000 years, grows in quiet seclusion here. Wilderness stretches away in every direction, on a vast and overwhelming scale. A scale whose endless possibility ignites my wild imagining. We would need at least a month to begin to know it.

The rain falls as a torrent and the trail becomes a stream flowing through the polished surface of the tangled roots, which are decorated in the abstract colour of fallen leaves. We are saturated, but are rewarded by the sight of waterfalls in spate, with dense spray driven from their edges by the wind. We camp beside Lago Blanco and find our bedding sodden. Everything in the vehicle is musty and damp.

12 JANUARY

The Río Yelcho and Lago Yelcho take our breath away. The huge river is tinged turquoise by glacial meltwater. It pushes in fast runs through forest-bound banks, with mountain peaks high about. The lake is mirror-calm, reflecting the sheer-sided mountains that contain it. The edges are reed-fringed, and a few small hand-made boats drift along the edge, probing for fish.

The road is stony and potholed, and late in the afternoon the passenger window of our vehicle shatters. Then, despite our binding it with tape, it collapses. We painstakingly replace it with taped-together zip-lock bags. By late evening we are worn out and still searching for a campsite. We finally choose a large grassy field beside the Río Futaleufú, where a sleek black mare with a beautiful gait frolics with her foal. Just as we are set up, a gaucho arrives driving a small group of cattle. I walk out to meet him. He is a small man with pinched features and speaks through his nose with a strange nasal lisp, and I understand almost nothing of what he says. A half-hour negotiation ensues before we agree on 3 000 pesos to camp in his field.

14 JANUARY

Hoping to photograph the Valle Futaleufú in the dawn light, we rise early, but our efforts are in vain as the sun touches only the tops of the steep-sided surrounds. For a while we drive beside the aquamarine-blue swathe of the Río Futaleufú. It is the embodiment of a powerful river, with deep rapids carving through a picturesque country of steeply forested mountains and granite cliffs.

By breakfast we turn southeast towards Palena. At the tiny settlement of El Malíto we head down to yet another Patagonian giant, the Río Palena. It is as impressive as the Río Futaleufú, but set in a wider valley. A puttering ferry-barge, on a cable, offers access to the road on the far side to the settlement of El Tranquilo. Outside the El Diablo Social Club, pick-ups are parked in a cluster, blocking access to the river, and saddled horses are tethered in the shade of the trees. Music issues loudly from the octagonal, wooden central building, and men wearing Stetson-like leather hats stand at the door looking in.

We turn up a side road that leads us deeper and deeper into the country. It leaves the river course and takes us past two small lakes. On the map, this route is marked only as a hiking trail. We stop to swim and fish at the second lake, making our way to the deep water beyond the reeds on the wide trunks of fallen trees.

Half an hour further the road ends at a lovely estancia. Cradled between the slopes and hilltops, and run through with small streams, it lies in a basin of the Cordón Blanco and is surrounded by sharp-edged, snow-capped peaks. The smiling, white-haired farmer leaves his sheep pen and joins his nervous but friendly wife to greet us. Ciro and Luisa tell us that we are welcome to stay as long as we like and show us to a charming campsite between old crab-apple trees on the knoll of a hill where every view is contained by a snowy peak. Over maté and bread with *miel de montante* (mountain honey) we learn that they have lived here for twenty-three years. They live without electricity or refrigerators. Water is channelled from a spring and falls constantly from a spout at the corner of their small veranda. Its sound, a pleasant gurgle, seems to breathe happiness into the air. The road here was completed fifty days ago, he tells us. At the beginning of December, she confirms. No, they do not have a car. He offers to take us on horseback to the Río Palena in the morning.

15 JANUARY

We ride with Ciro to the Palena and while Beverly explores the banks for birds to photograph I begin a day of fishing. The river is sweeping with a strong flow and the fish are fit. I kill one for Ciro and Luisa for dinner. Without a refrigerator, slaughtering livestock in summer is a problem, so meat is a summer rarity at their table. Ciro leaves us at midday, and we spend the warm afternoon swimming and fishing in the Palena and, towards evening, return with the horses through the tall, ancient forests. The trail is muddy and a tangle of roots, and with two deep river crossings before home, I am glad to be on horseback. The gaucho saddle, layers of sheepskin tied with a girth strap over a simple saddle frame, is surprisingly comfortable.

Back at the estancia over tea, in the untreated wood-board kitchen, we discover that Ciro Barrientos Rademacher is eighty-one. I had guessed him to be sixty. And Luisa Espinoza Gallardo is seventy-seven, although her skin and demeanour suggest someone in their late fifties. The farm, Antijull-Loma del Sol, is 700 hectares and set in the Cordón Blanco. They show us pictures of the family of six grown children and a multitude of grandchildren. After tea, I help gather the hay that was mown the day before. I bundle it by hand and tie it with rope before lugging it on my back to the barn, where, climbing a hand-made ladder, I deliver it to Ciro, who hauls it into the loft and spreads it to dry over the open-slatted floor. Just before dark, Ciro arrives at our campsite bearing a tub of fresh yoghurt and a bowl of raspberries.

16 JANUARY

In the early morning, I write beneath the shade of the crab-apple trees, with their small, green fruit turning pink-red. Chilean flickers land in the tree above me and give their plaintive whistling call. A flock of sheep approaches, grazing slowly, and the horseflies multiply with their approach. The mountains are bright green and white with snow in the sunlight. We are sad to be leaving Ciro and Luisa, and our parting is heartfelt, for despite the barrier of language, we have felt at home in their company.

I am captivated by their choice, for they are clearly intelligent people and yet have opted for complete simplicity. They tell us that they need very little money, for they buy only a few essential items such as sugar and salt. Ciro says that Villa Palena is only a day's ride away. Their farm, despite its remoteness, is a model of efficiency, with a variety of crops and two hothouse tunnels with a jungle of produce flourishing inside. The buildings, corrals and fences are neat and orderly, unlike most of the ramshackle rural enterprises we encounter along the road. I sense in their good health and humour some deeper recourse. It lies beneath layers somewhere in the back of my mind, linked to the final survival of all humankind. Our shift from greed to another essential element, or ingredient, that is not destructive but constructive. It gives me hope.

17–19 JANUARY

Before La Junta we turn east to Lago Claro Solar, simply because of its alluring name. We are greeted by a valley dominated by a serration of jagged, bare peaks whose names we do not know and are not marked on the map yet their majesty rivals any we have seen. The valley is narrow, and after two steep but short passes through tight, granite cliffs, we emerge beside the Río Figueroa. It is another spectacular body of water that is deep as well as swift, and our first view of it, from the high bend of a pass with vertical drops, draws exclamations from both of us. Its sheer size and energy are magnetic and we stop often to gaze into the aqua turmoil of its passage, cut steeply through the valley. At one point, where the road rises through another small pass, the land grows even wilder and our anticipation rises. But this is short-lived, as, just after we pass a sign that declares this a zone for the protection of fauna – pudu, puma, zorro and more – we emerge again into estancias, where logs lie piled beside the road and the forest has been destroyed for pasture and firewood.

Looking at the mountains, I wonder just how many puma (South American mountain lion) do survive here, for, although the sides and the tops of this greatly folded country remain for the most part inaccessible, I have seen little here that would constitute adequate prey. Beverly suggests rabbits, for in the introduced pastures of the valley around the estancias they are not uncommon. But this would bring the puma into contact with humans, and the puma legend and the animal's unquestioned persecution remain strong here. Every gaucho we have seen rides with at least two dogs in attendance.

21 JANUARY

Just outside La Junta, we find people of all ages walking or on horseback, making their way along the short distance from the town to the bridge over the Río Rosselot. The air is thick with the dust of the slow procession of pick-ups and cars. There is a fiesta today and there will be a rowing race on the river, and a barbecue. Thin-limbed children, in bright bathing outfits, dig in the dark wet sand of the shore. Men standing in boats bellow instructions to large women on the bank. Horses tethered between pick-ups stand on three legs, their heads drooping in the soporific warmth. A band in shiny purple shirts, with an accordion and a violin, shouts encouragement over blown speakers.

22 JANUARY

Turning off the Carretera Austral, we take a gravel road following the Río Palena to the coast. After a time, we notice that the estancias have ended. For the next 50 kilometres, we travel through undisturbed forest, the road leaving the river to climb through the hills, then returning to the sheer banks, where it is cut from vertical granite cliffs. We stop several times to savour the sense of a pristine landscape. It does not seem to be a declared park, just wild country, lying south of the Parque Nacional Corcovado and northwest of Reserva Nacional Lago Rosselot. It is what we have come to find, and that it should be here where we least expect it, open land in the valley of a massive river that would usually be heavily exploited for agriculture, makes it doubly enticing.

The road, with its somewhat potholed, somewhat corrugated surface, covered with a thick layer of uncrushed river pebbles, and crowded to a single track by the riotous growth of the forest, is typical of the region. Fortunately, the going is slow, for on a hairpin bend we come face to face with virtually the only other car we are to encounter all day. Both cars swerve and slew to a halt on the loose stone, and after a moment to recover, edge past each other, brushing through the roadside vegetation.

Just past the ferry crossing of the Palena, 10 kilometres outside Puerto Raúl Marin Balmaceda, a sign proclaims Reserva Añihue and the wilderness remains intact all the way to Raúl Marin Balmaceda, where due to my poor Spanish, I can only establish that Añihue is a privately funded reserve.

Raúl Marin Balmaceda is a tiny smattering of wooden cabins, separated by a few sand tracks that have not been gravelled. Yards are unkempt, trees abound, and Beverly observes that it seems that the town is being driven out by the wilderness and not the wilderness by the town. The people are shy and curious and the small craft harbour with its red-railed jetty is pretty, and we are drawn to stay.

We find a campsite on a spit of sand where the Brazo Pallin enters the sea and spend the evening wandering the bank and photographing the shags, gulls, dolphins and seals that pass by against the smooth, bottle-green reflection of the hills of Parque Nacional Corcovado that rise on the far side.

23 JANUARY

Despite a cloudy, overcast morning, we are so enchanted with our campsite on the fjord estuary, with its passage of wildlife and wooden boats, that we opt to stay longer. In the afternoon, we walk to the mouth where the push of the out-going tide raises standing waves in the narrowest neck of the fjord mouth, which is nonetheless 250 metres across. On the way back, we watch a pod of Austral dolphins herd fish into the shallows and then charge them down. Close by our camp, women from the village are gathering wild strawberries in the dunes. As the tide drops, the Austral dolphins are replaced by a pair of Chilean dolphins that slowly patrol the shallows up and down from our camp, their rounded dorsal fins soundlessly breaking the surface. We have missed the presence of wildlife, for, although the forests are filled with birdsong, the birds themselves are hard to spot, and we have yet to see a wild land animal other than the hares in the pastureland. We sit in the evening wrapped against a thin, cool breeze and enjoy the spectacle of wild mountains, an unfettered river and its parade of life.

24 JANUARY

At the ferry crossing I reset the odometer to measure the exact extent of the wild country. I am deep in thought as we drive slowly through the forest and I seek adjectives for what I feel. After some searching, 'lonely' seems the most apt. Not lonely in a desperate sense, but in terms of some undefined longing. Perhaps it is the singular absence of obvious life, for even though birdsong emerges like flow-ers from the trees, this life, in a visual sense, is but a glimpsed flutter, or rapid darting, among the otherwise-still vegetation.

Earlier, when walking away from the road down a path to the Río Palena, entering the dark, vaulted interior of the Valdivian rainforest, it felt as though we were making our way through a catacomb. The trees formed grottos and caves, and wandering between them I moved with a softer tread, as one does in anticipation. But here the anticipation remains unsatisfied. In our slow drive through this wild country I have come to realise that I am accustomed to wilder-ness in the possession of animals. And that the evolution of the day, the time it takes, becomes measured by the urgency of their needs, movement and tension making all quick and immediate. Here, the rugged earth is possessed by the forest, and time is measured by a far slower gauge: growth. I cannot reduce my perception to grasp it, but by looking at this wilderness in terms of the evolu-tion of a huge, rotted trunk, lying moss-coated on the forest floor, I begin to understand it more closely: that its evolution, its movement, its event, is beyond the scope of my short tenure. That what I see of it is just a thin slice, which, like the glaciers that abound here, moves so slowly that it seems static. Beverly suggests that the country feels like it could be shattered at any time by a cataclysmic event. That the rugged, sheer mountains, the giant rivers, the complete envelopment of the land by forest seem to hide the forces responsible for so dramatically shaping the land. It is how the land feels, and it enforces my idea of its evolution being too slow to be observed in terms of my own urgent time. And yet, when understood, it is as exciting in its way as the more immediately observable life of animals, for the land here feels young, growing and dangerous.

In the midst of these thoughts, an animal appears in the road. A scramble for the binoculars reveals a pudu: the world's smallest deer, with a fluffy ginger-red coat and the tentative delicate tread of a duiker. It pauses briefly to glance back before vanishing in a quick step into the roadside thicket. It is the first wild animal we have seen on the mainland of Chile and the first since the guanaco and foxes of Tierra del Fuego.

26 JANUARY

Back on the Carretera Austral, we drive south toward Coyhaique and then turn east, following the Río Cisnes, on the road to Villa La Tapera and Alto Río Cisnes. Just before Villa La Tapera, as the valley widens, and the road descends to the river once again, we find ourselves to the east of the Cordillera de los Andes, where the saturating rain of the Pacific is cut off by the peaks. The change in the country is immediate and dramatic, the forest replaced by short trees and open grassland that is tussocked and stony. Moving away from the mountains, the land becomes more arid still, with trees present only in the gullies and the bed of the river. The vegetation is scrub, and large, bare domes of basalt puncture the landscape of stony hills. The mountains recede behind us and the land becomes flatter, and it feels strange that the Río Cisnes, now not much more than a wide spring creek, flows west towards the barrier of the mountains.

The land here is farmed in the Argentinean estancia sense: vast tracts upon which the livestock is loosed and patrolled by gauchos, a complete contrast to the tiny subsistence fundos that are the norm in the mountains to the west. The road becomes less and less used, until it is little more than a farm track, and cara-cara fly up at our approach from the road-kill hares they have been feeding on.

We camp beside the Río Ñirehuao in a country with a mood entirely different to which we have been accustomed. In the last light of day, a night heron lands in the tree beside our camp, then flies down to the river. My last glimpse of it is of the bird walking hurriedly across the stones toward the water, like a stiff old English gentleman trying to catch the end of a rowing race.

28 JANUARY

Having returned to the west of the Cordillera de los Andes, and now being back in the rain belt, we wake to find that the Land Rover has leaked during the night and our sleeping bags, jackets, mattress and camping gear are sodden. The ground is dew wet and nothing is drying, but I rig a washing line in the hope that we will get at least some sun between the scudding clouds. Faced with the prospect of the wait for the down sleeping bags and duvet to dry, I change the rear tyres, as the rough roads have eaten the tyres to a coarse roughness. While I am working, three young gauchos arrive on horseback and they pass close by in silence, their dogs stopping to lift their legs against a newly changed tyre.

By mid-afternoon rain threatens again, but the linen and sleeping bags are dry so we pack up in preparation to leave. The young gauchos return, leading a string of horses, and this time stop to chat. They are teenagers and excited by the prospect of the Fiestas Costumbristas tonight, celebrated by all the small towns of Patagonia at this time of year. They leave at a canter, the led horses swerving to avoid stumps and logs in the cleared field. Later, one of their dogs shows up limping badly and comes close, to fall asleep under the vehicle. We feed him before we leave, appeasing the strong unspoken yearning his coming has brought upon us for our own dog.

We leave through Ñirehuao and stop on the roadside beyond the town to pick cherries. Beyond us rain is falling in iron-grey sheets.

30 JANUARY–2 FEBRUARY

In Coyhaique we spend three days repairing the vehicle's smashed window, restocking and editing images. We rent a small cabana set in attractive gardens, with a wonderful view over the mountains. Aptly named Mirador (viewpoint), it is close to the heart of town and the owner is enthusiastic, willing and helpful. When I tell him that his is the prettiest garden I have seen in Chile, which has a singular lack of any gardens at all, he is bashfully delighted. He sings opera while he works.

Our time in Coyhaique is unremarkable except, perhaps, for a street drunk who sleeps in doorways with a collection of devoted stray dogs. He loudly abuses the gentry of the town and shop-owners annoyed by his entourage. At midday, the church plays a recording of bells chiming 'Jingle Bells', a little too loudly for the rooftop speakers' capacity.

4 FEBRUARY

Where the road turns away from Lago Elizalde, the Reserva Nacional Cerro Castillo comes down on both sides of the road, but here too, farms, fences and locked gates line the road through the reserve valley. We cannot get used to the Chilean notion of a national park that excludes any areas of possible human recreation: essentially the flatter land of the valleys, rivers and lakes. It leads us to question whether the country's national parks are indeed a sincere effort at conservation or a concession to an ideal that has international appeal. Chile has set aside 19% of its land as national parks or natural areas, but if this is to include only the inaccessible or unusable land and specifically exclude rivers and valleys, what is the point? Surely, the rivers and lakes and their associated ecosystems are critical to conservation efforts?

Chilean brochures and information booklets acknowledge that the parks are mostly the domain of foreigners. Except for a father and son in the company of a Canadian friend, we have not seen many Chileans camping. Perhaps the Chileans do not have a camping culture, or perhaps because of the weather they have been unable to develop one. Certainly, in Argentina, where accessible public areas abound, camping is very popular.

The road crosses a narrow bridge over a deep gorge, and below it the Río de la Paloma, which issues from Lago La Paloma and Lago Desíerto above, spreads out as it runs its fast course to Lago Claro, cutting a wide, boulder-strewn, snaking path through the valley. Again, the valley is heavily agricultural, and where the natural forest has been removed to points high up their steep sides, the mountains are scarred by bare earth and landslides.

We drive to a puesto to get permission to camp. It is set deep in the shadow of a copse of trees and the yard is littered with scrap. Seven dogs rush out barking at our arrival, one of them with the heavy teats of a bitch suckling pups. On the packed-earth entranceway a plastic Chinese-made tricycle lies on its side. No one is home. In the yard a lump of meat hangs on a rope, turning in the wind beyond the dogs' reach.

7 FEBRUARY

In a clear dawn, we attempt to catch the soft light on the Cerro Castillo massif, with pink clouds piled up behind the spires and a myriad of islands peppering the lakes in the foreground. For an hour we battle a wind gusting in excess of 80 knots, until our eyes water and our fingers grow stiff with the cold. The clouds, however, are not long in closing in and our view is highlighted by a rainbow before the day falls to grey and the all-too-familiar Patagonian rain sets in. We stick to the back roads, finding a remarkable rough, rumpled country of granite outcroppings in which countless small lakes lie trapped by the irregular valleys. The dirt road is ungravelled and, with the rain, is in places a slippery quagmire. On the edge of precipitous drops, we keep our eyes more on the narrow track than on the spectacular views of the braided Valle Río Ibáñez, which from the heights is so wide and braided that it looks like a delta. It takes several hours to cover just over 20 kilometres.

At Puerto Río Tranquilo, we turn west to Bahia Exploradores. What starts as a pretty drive soon becomes spectacular, and as the clouds dissipate and the sun comes through we find ourselves in a Yosemite-like valley, where waterfalls lace domed, forest-clad granite mountains. Behind these rise yet higher peaks, bare of vegetation and covered in snow and ice. The sun is deliciously hot, and I feel its long absence keenly. As we drink our tea a pair of rare Andean condors sail along the edge of the mountains directly above us, and then thermal across to the far side to vanish in the shadowed upper valley.

Photographing a waterfall, we catch a brief glimpse of a long-legged, cat-like creature of close, dark fur that leaps from a streamside boulder onto a log to cross the river's torrent. The closest match I can find in our mammal guide is an Austral spotted cat, but it is impossible to be certain.

9 FEBRUARY

Our dawn rising is greeted by the spectacular sight of the full moon over Monte San Valentín and Cerro Leones with the peaks gathering trailings of clouds tinged with the warm light of the rising sun. We drive upriver and are fortunate to join a party of hikers and their guides just setting off for Lago Leones. It is an easy hike up the floodplain of the river. The arrival at the saddle of the source of the river, issuing from the lake, is one of the most singular views we have seen in all of Chile. In the 180-degree view before us, mountains with bare spires cut short the sky, their high slopes deep with snow and from whose centre the scalloped blue and dirt-grey surface of a glacier winds a sinuous path to a lake pale-grey with silt. To the right a second valley issues a darker river that meets the first among a tumble of weathered granite boulders and stunted lenga trees.

On the way back I fall into conversation with Gonzalo Marin, who studies avian vision at the University of Santiago. Having spent six years studying in New York, he speaks fluent English. In answer to my questions about national parks and their river-excluding boundaries, he explains that the problem lies with CONAF, or Corporación Nacional Forestal, the body that governs the national parks. CONAF, he points out, was established to manage forests, with national parks being added to its portfolio only recently, though with no change to its management structure, policy or planning. CONAF is simply not up to the task and lacks the singular vision and drive required to align the Chilean national parks system not only with the Chilean people as a recreational resource but also with its protection and wise and appropriate management, in short, the guarantee of its protected land's long-term survival. He agrees with my observations that land that can be utilised has been excluded from parks, and that this seems an insane policy. He blames the exclusion of valleys with rivers and lakes on two factors. First, in many places there were homesteaders in place before a national park was established, and the government was unwilling to expropriate them. Second, the government actively seeks and encourages rural settlement and development.

Faced with its own inadequate policies, vision and support, CONAF plays virtually no role in what would be a hotly contested debate anywhere else in the world. Gonzalo also points to the monopolistic control of Chile's wealth by only a few families. It is advantageous to have the areas of national parks fragmented because that destroys the concept of an irrevocable whole, and makes it easier for the wealthy to exploit them should the need arise. This alone would be reason enough, he argues, to keep CONAF ineffectual, and to divide land that may be utilised for human need from that included in a national park. In my observation of Chilean wilderness, it would seem that if this were a plan, or even a spinoff of some more basic idea, then it has succeeded, because the national parks of Chile, despite their proportionate size, are only minimally enjoyed by the Chilean people because they are, at the most basic level, inaccessible. This means that Chileans are unlikely to place an especial value on one particular park or another, because it has not become entrenched in their hearts. Only a park that is possessed by the people of a country in the depths of their passion for their land is truly safe as a national park.

11 FEBRUARY

In Cochrane the single supermercado, which faces onto the paved and leafy plaza, reminds me of the trading stores of old in Lesotho, and I browse happily while Beverly procures our short list of needs. There are chainsaws and blankets, a butcher section whose impressive display fridges look bleakly empty with the half-diminished content of a single beef carcass, and beside which stands a beautiful upright cabinet of glass and wood that displays ukuleles and harmonicas. There is shampoo, imitation Oreos, imported plastic snakeskin cowboy boots and beautiful enamelled cast-iron stoves. On sloping, green-painted wooden shelves, new produce is piled atop that which is now too rotten to sell. From the ceiling hang ladles and a winged Jesus, and rawhide halters and bridles made with the most delicate inlay, beside cheap Chinese toys in thin cellophane wrappers. The shop is busy and the assistants almost outnumber the shoppers.

On the way out of town we stop at the showgrounds, where the Fiestas Costumbristas is just beginning. The arena is filled with gauchos on horseback, and speeches are being made from the podium. I photograph children falling asleep on their horse's necks, and an old lady with ill-fitting false teeth dressed in finery with a yellow ribbon pinned to her breast. She sits on a sawn-off stump, chomping her jaw habitually, and smiles brightly at a handsome young gaucho dressed in a gallant, stiff, broad black hat and wide silvered belt who greets her as he passes. He is carrying an enormous Chilean flag that drags in the dust over his horse's flanks. The horse shies each time the wind lifts it wide. A crowd mills along the edges, where lamb roasts on open-pit fires and men lay money on the ground between two points where they toss weights with a blunt blade into a demarcated slurry of stiff mud. The loser is consistent and frowns with each bad throw. I suspect the shouts of the crowd and the loud haggle of a bookmaker put him off. We photograph the gauchos riding wild-caught broncos, a practice that has too cruel and crude an edge for us to enjoy properly.

12–13 FEBRUARY

Caleta Tortel is a largely roadless port accessed by a network of wooden walkways built on stilts. We take a long walk to the far side of the village, and on the way back inquire about hiring one of the many boats we see moored. It is the only way to get out into the fjord areas of the Parque Nacional Laguna San Rafael, the Parque Nacional Bernardo O'Higgins and the Reserva Nacional Katalalixar we had passed through on our ferry trip from Puerto Natales to Puerto Montt. The prices, however, are outrageously high, US$800 per day for a four-metre fibreglass skiff with an outboard, and well beyond our budget.

It is Beverly's suggestion that we approach the local CONAF officer with our project and see whether we cannot join him on his patrols. At 2 pm I meet Orlando Beltrán and, in my rudimentary Spanish, explain our quest and present him with the credentials of our book and letters of introduction. By 7 pm I have had a long conversation with his superior in Cochrane, met Captain Luis Paredes Samoloval of the CONAF launch, and agreed to pay for the 450 litres of fuel needed to take us to the remote outer regions of the Bernardo O'Higgins and the Katalalixar. We hand Orlando the US$900 for fuel in cash that evening, without a receipt, and depart for our campsite outside of town, animated and thrilled by our success. We are to return in two days' time to overnight in anticipation of an early-morning departure.

14 FEBRUARY

The Río Bravo roils beneath a mirror-smooth skin and is absolutely silent, although its power and speed remain apparent. To look at it feels like watching a giant cobra move across the land, for there is something sinister in its absolute quiet and the strength it keeps in abeyance. We have two days and decide to make the drive to Villa O'Higgins, but our plans are thwarted when we miss the 12-o'clock ferry from Puerto Yungay to Río Bravo by ten minutes. It is still raining when we make the next crossing at 6 pm and the rain does not let up as we drive the road to Río Pascua.

For the first time in a long while we are in pure wild country, with dense vegetation that reaches dripping into the roadway. Tight shrubs are laden with pendulous red flowers, and hummingbirds dart into the dark canopy at our approach. We pass the temporary shelter of woodcutters, where hand-hewn shingles of orange wood are stacked in circular bundles, but even here the forest is omnipresent and pristine. It is dramatic country and even a single home-steader, with a small pasture wrested from the forest, and dogs that chase our tyres, cannot dampen our enthusiasm for the place. The land in its natural state reaches to the tops of the mountains, glimpsed through the mist and persistent rain. As far as we can see through it into far valleys, with hills tied by ropes of white mist, the forest is unbroken. There are no roads or tracks off the main road, which ends abruptly, without warning, in a locked gate across the road. We have been told that the area beyond is a military zone and so turn around and camp a few metres off the road. Not a single car passes in either direction for our entire time on this road.

15 FEBRUARY

It rains heavily all night and does not abate in the morning. We drive down the road towards Villa O'Higgins and celebrate the landscape of a wild country. A riotous growth that stands so tall, hemming the margins of the road, that our view of the sky is a narrow grey ribbon. Water runs everywhere and flat ground is a bog of moss mounds. Slender, dead cypress trees have the aspect of a fallen volley of arrows, pale green moss fluttering from their flanks like broken, tattered flight feathers. We pass only two homesteaders, their crude buildings forlorn in the tiny patch of cutaway forest. But they are the forerunners and others will come.

On the ferry back to Puerto Yungay, I note that the advantage of being on a boat is that one has an unobscured view of the entire land: the mountains and valleys and the change of vegetation from the deep shadowed ravines to the bald outcroppings of granite, too steep or thinly soiled to hold vegetation, and I am exhilarated by the prospect of our planned trip with Orlando into the fjords and islands of the national parks.

Orlando is full of bubbling enthusiasm when we meet in his office. He has procured the fuel in our absence and we discuss our needs for the voyage. At our request, he points out in detail the planned route on the relief map on his wall.

In anticipation of an early start, we decide to camp in the parking area at the access point to the town. Parked in the far corner, away from the entrance walkways, I am drinking a glass of beer and watching a man with a very young child holding his hand walk slowly up the road in the thin drizzle. A piebald kitten scampers along with them like a dog. Each time it comes within reach, the child tries to kick it. They return a few minutes later carrying a laden plastic bag, the kitten in tow filled with the playful bounce and whirl of the very young. Just before the car, the child attempts another kick, falling over his own feet. The man joins the sport and connects with the tiny kitten, lifting it with a thump into the air, to tumble against the roughly unfinished road edge.

16 FEBRUARY

The constant rain is eating at my sanity. Droplets of condensation run down the windows of the Land Rover to form small puddles. Everything is damp. I must make a conscious effort to remain cheerful. Orlando is strangely reticent when I knock at his door at 7:30 am. He says it will rain all day, but this does not deter my will to depart. He tells me that they must still load the fuel, which is sitting in the yard of the shipbuilders at the airport because it is too bulky to lug down the steep stairways to the boat. They will take the boat around to load it there, and I propose that we do the same with our gear, for which I get the thumbs-up. Orlando suggests breakfast before we start and I reluctantly agree, aware of the prospect of our damp, close space in a public parking lot. Orlando is working on his computer in his office when I return, and I feel immediately that our plans for the day are in jeopardy. He greets me with a long explanation, and only at the third telling do I fully grasp that the port captain of Caleta Tortel is away for the day in Cochrane and that our voyage requires his clearance before we can set off. We will have to wait another day in the rain. I struggle to maintain my composure. I ask when we will load the fuel and again Orlando calls on the radio without response. He rises and we set off to find the captain, while Beverly returns to the vehicle to ensure that we are completely prepared.

We walk a circuitous route through the town, eventually arriving at the captain's house. 'Captain' is the term Orlando uses for the driver of the CONAF launch – a five-metre, partially enclosed, moulded plastic, V-hulled boat with a tall, black Mercury 115-horsepower outboard bolted onto the stern. The captain's house sits at the entrance to the jetty and is divided into upper and lower parts. We knock at the lower part, which, like most rural Chilean houses, is squat, small and ramshackle, with lace curtains at the few windows. A young girl emerges and disappears upstairs to the upper house, which stands on stilts on the steep hillside, with two bay windows facing onto the harbour and fjord. Washing is strung beneath the support posts, and rivulets of water run over the bare exposed earth. The girl reappears and ushers us upstairs, where I step into a large single room with several doors leading off toward the rear of the house. The captain, Luis Paredes Samoloval, emerges from one of the doors, tucking his shirt in. It is 11:30 am and the import of Orlando's circuitous delays dawns on me. I am fighting a resistance that is beyond my status as guest to topple.

We collect Beverly and our stores and camping gear and drive around to meet the boat at the landing near the airport, parking in the sawdust of the boat-maker's yard. We are invited into the house and drink maté around the white enamelled stove. The captain is boisterously cheerful and drinking beer, a crumpled can already on the floor beside him. The persistent rain and our host's apparent indifference to our urgency to be under way leave me sighing with frustration. We load the heavy fuel drums, sliding them down an angled jetty of thick but worn cypress beams, where a dog stands gingerly eating the discarded intestines of a slaughtered animal, the green sludge of the stomach contents spilling into the water. When the tanks are full we load the last drum into the stern and our food and camping gear into the cabin. Orlando says we will leave at 7 am the next morning, and I repeat my understanding of it several times. We return to Caleta Tortel, where the clouds have thinned and the sun drifts in patches over the dripping town.

17 FEBRUARY

We wake Orlando at 6:45 am, and when we get down to the dock with our camera equipment, Luis is there and the motor is running. We divide the reflection of an almost cloudless sky as we depart, trailings of mist clinging in layers to the mountainsides. As soon as we lose sight of the town, we are in the embrace of a wild, unblemished land. We pass through several narrow and sheer passages where the tide runs like a river, and I am overcome by the magnitude and scope of the wilderness about me.

To the north lies Parque Nacional Laguna San Rafael, 1 742 000 hectares in extent, to the south Parque Nacional Bernardo O'Higgins (921 000 hectares), and to the west Reserva Nacional Katalalixar (674 500 hectares). In response to my query about salmon farming in the area, Orlando tells me that the sea here is a marine reserve where only recreational fishing is permitted.

In the great stillness of this place, the moments of beauty that each gaze takes in, the huge scale of the country, where each buttress that rises from the sea is gigantic by any measure, I feel the sense that has so long evaded me take hold in the core of my being. Perhaps wild is only something you know with your soul. It is definitely not the product of an equation. I feel it as a heat in my bones, and at my relief to finally come upon my destination, I must clench my jaw to work back the tears it brings to my eyes. Sitting outside, hunched against the cabin housing, with my hands pushed deep into the pockets of my parka, I let my gaze rest on the passage of the landscape.

Gnarled, contorted trunks punctuate the sharp fringe of the tidal reach, beneath which fissured rock gives way to the cold, clear sea. Forest, tall, dark, green and dense in the deep valleys, growing thinner and paler until it finally peters out on the domes of granite, or on the faces too sheer to hold soil. The rock is a white-grey, as if bleached by its exposure, the darker faces holding something closer to purple, more than black. The peaks are domed, occasionally jagged, and with the light wind that is now blowing, snow is everywhere. Apart from a few scattered sea birds, we see no life.

I am disappointed when Luis and Orlando open a bottle of beer. They mix it half and half with Fanta, and I decline their offer to make one for me. It is 10:30 am. Two of the four quarts of beer we have brought along, at their request, disappear over the next few hours, but by the time we stop to photograph a small colony of South American sea lions, in the early afternoon, the drinking has slowed. At the head of the Fjordo Ofhidro, we are joined by a pod of Peale's dolphins. When we slow, they surf the wake of the launch until we come to shallow water, where we must idle. We resort to oars to bring us close enough to shore to photograph two condors, which stand on the mudflats at the edge of the tidal zone.

At the head of the fjord we must disembark, amid clouds of mosquitoes, to drag the boat over the shallows at the mouth of a small river. We walk upstream through the river course, which varies from a coarse mat of pebbles, covered in brittle dry moss, to peaty bogs where water pools around our boots. After half an hour of pathless advance, Orlando calls to me and points out our first huemul (South Andean deer). It is disappointingly far away, a small speck across a fast river where great chunks of ice, half the size of our boat, grind loudly against the rocks as they are swept downstream. But the scene is spectacular. We take a few pictures of the deer between tumbled, moss-covered boulders with a forest of dead trees behind. The glacial lake is spectacular, its silt-grey waters littered with giant icebergs, with the wide, blue-white tongue of ice, a kilometre across, retreating into the clouded heights to the west. The sound of the ice moving comes like peals of thunder through the valley. Where we stand at the mouth, the bergs twist and groan as if alive, broken by the meltwater into progressively smaller chunks.

It is cold near the ice but Luis wears only a thin vest, his arms red with weals from the sharp stinging scratch of the calafate bushes and the persistent attentions of the mosquitoes. Back at the boat, the tide has come in. We motor over the sandbar, and forty minutes later, tie up to the ladder of the refugio at the foot of the Bernardo O'Higgins Glacier, which is twice as big as the first glacier. There is a smell of dead fish in the air. The refugio is a double-storey wooden building, impressively new and securely locked. Orlando does not have the key. To my relief, he has a unique solution to access the sealed building that does not involve smashing our way in. The interior is surprisingly bright and warm and clearly little used. Although fittings are in place, they are not connected, and the remnants of construction lie piled about the floor. Beverly and I make our bed upstairs where it is warm and the floor clear. We eat dinner on the boat, where Orlando and Luis will sleep. Our offer of wine to accompany the dinner is turned down, and while we sip a fine Chilean red blend, they are now drinking water.

18 FEBRUARY

I wake to the sound of Orlando and Luis in conversation. The sky is still lightening, and when I head down to the boat to make coffee it is only 6:15 am, very early by Chilean standards. The boat, however, has been left high and dry by the outgoing tide and is canted at an uncomfortable angle. I set the water to boil and return to pack up the sleeping bags, but am interrupted halfway by Orlando who says that there is a huemul near the boat. Coming down the stairs, I see Beverly on her knees on the rear deck with a camera, a huemul peering at her transfixed from not three metres away. Luis is delighted by my surprise and laughs heartily, making loud comment to Orlando. The huemul shows not the least disturbance at the intrusion.

For the next two hours, I work without pause around the most curiously tame wild animal I have ever encountered. It tolerates me without concern to within two metres, and when I push closer gives way without alarm. It is as curious about me, and when I lie still on the damp ground, it approaches closely, look-ing down on my prone form with cocked-head interest. Eventually, Orlando has to be quite firm about his intention to move on or drag me away. The tide has returned and the boat is floating again. There are dead fish everywhere, account-ing for the smell. Orlando cannot offer any definite cause for their demise, citing a glut of fresh water, or silt, or perhaps a lack of oxygen.

As we motor out into the bay, three condors drift silently in to land on the top edge of a cliff face high above. We do not motor far. Entering the mouth of a glacial river, wide but with only small chunks of drifting ice, we tie up at a steep bank. Setting off with our packs, we almost immediately encounter a group of huemul that are as tame and approachable as the one near the boat this morning. The sun is warm on our backs and we shed our parkas and leave them draped over a bush to collect on our return. We encounter more huemul on our way to the glacier, and each time Orlando smilingly encourages us away after a few minutes' work, assuring us that there will be more.

The foot of the Bernardo O'Higgins Glacier and its associated mirror-calm lake, adrift with sculpted icebergs, is breathtaking; a scene beyond conjuring in the imagination. And, as if to keep Orlando's promise, a small herd of seven huemul browse on the very hem of the shore. It is an utterly singular setting, and several times I force myself to put down my camera and take in my sur-roundings. The ice is silent here, the land and the huemul too. I can hear my own breathing and the faraway call of kelp gulls in the estuary. I wonder that if wild is a silent thing, what it is that makes this landscape seem so alive. Perhaps it is that it seems poised, on the edge of something, while remaining still. A contentment to wait that for all the world I cannot emulate within myself. It seems at peace, and yet wherever I look there is vital aliveness, even in the inanimate: the mountains and the gigantic valley of ice are in a dynamic state of tension. Perhaps wild is the land in possession of itself, with our-selves reduced to proper proportion, without the sense of proprietorship that we bestow upon ourselves, with what has become an almost instinctive right when we exert our domination. I feel it here, myself reduced, as if I have found the place where God lived. I cannot ignore the reverence and sense of humility that rises unbidden within me in that presence. I am in awe, overwhelmed to the point of joy.

It is early afternoon when we have finally exhausted ourselves with the photographic possibilities of this unspoiled heartland, and on the walk back to the launch, Orlando persuades me to eat calafate and chaura berries. The sun is on my back, warm in my hair, and with the pack heavy on my shoulders and my limbs already weary with exertion, the day tastes as sweet as the red and purple berries I pop into my mouth.

We depart much later than planned, and it is 10:30 pm and dark when we tie up at the jetty in Caleta Tortel, after hours of uncomfortable passage through the short chop of a windswept sea. We are greeted by Luis's severe-faced wife, standing in the thin pool of light of the street lamp on the jetty. She frowns at the sight of their glasses of whiskey and Coke, in which large chunks of ice that we had fished from the river are floating, but kisses him on the cheek and sits down to eat one of the *sopapilla* (traditional fried breads), she had made for the journey.

19 FEBRUARY

In the early afternoon, we find Orlando in the captain's house on the waterfront to say our goodbyes. We are invited into the lower house to find most of the family in the small front room, where a wood stove burns and pink baby bootees, baggy underwear and a baseball cap dry on wires strung along the walls beside the chimney. The men are drinking a concoction of liquidised pineapple with rum and something else besides. The family is warm, friendly, amused by our poor Spanish and interested in our work. When the final glass of sweet pineapple poison has been drained, we bid our goodbyes and walk the long stairs to Orlando's office. He is in a cheery mood and not only provides all the statistics and specific information we require, but also gives us a wall map of the parks under his jurisdiction. I feel impoverished in that I am unable to offer anything in return. In our short time together we have become friends, for we are kindred spirits, and our goodbyes are heartfelt and our thanks sincere. His greatest wish is only to have one of the books that we produce donated to the library in Caleta Tortel, so that the youth may see that the wild land about them is considered important on an international scale.

As we drive away I remember Orlando's indigenous garden along the main walkway into the town, with each tree and shrub bearing its own carved wooden name board. It was only the dullest visitors who did not pause to study the names of the trees or ferns that they had seen so often along the road. Orlando's work was a cheerful, free and surprisingly effective open-air classroom, and as we drive the beautiful riverside road towards Cochrane I am determined to add my contribution to his list of understated achievements.

20 FEBRUARY

Cochrane is in a quandary. There is rioting in Coyhaique and the protesters have blockaded all roads south to commercial traffic. They want to get the attention of the faraway political decision-makers in Santiago, who are planning the construction of five hydroelectric dams on the Río Baker and Río Pascua, the lifeblood of local agriculture and tourism. Added to this, the hydroelectric schemes are to produce electricity for copper-mining operations in the wealthy far north and will have no benefit to the rural people of Chilean Patagonia.

There is no fuel and the mercado is filled with shoppers stocking up in anticipation of the crisis. In the street, we are accosted by backpackers and travellers seeking a lift out of town. Our most difficult moment is to turn down a charming German couple, in their seventies, who have already been stuck in town for several days. Although we have enough fuel in our reserve tanks to get to the Argentinean border, our plans are vague and we will definitely not be making the drive in one leg. We still have much to explore and so cannot take them along. A frowning, middle-aged American woman is less understanding of our explanations, and hangs around sullenly until we drive off.

We drive out to the campsite at Reserva Nacional Tamango, on the banks of the crystal-clear and meandering Río Cochrane. The place, however, is crowded with stranded visitors and we opt to head out of town to Estancia Valle Chacabuco, a recent acquisition of the Tompkinses, who are in the process of rehabilitating it into Parque Patagonia. The dream is to create the Parque Nacional Patagonia, joining the tiny Reserva Nacional Tamango to the south to the much larger and wilder but inaccessible Reserva Nacional Lago Jeinimeni to the north. It will be one of the few valleys of Patagonian steppe grassland within a national park and critical to species diversity and protection of the endangered huemul, Austral viscacha, Darwin's rhea and puma.

As we turn off the main Carretera Austral and find ourselves on a road through a landscape unrestrained or bisected by fences, my spirit is lifted to lightness. It is country of semi-arid grassland and lenga forests, set in tumbled hills of crumbling black rock. A herd of guanaco grazes on the valley floor, seen from the neck of a short rise, and to be with animals again triggers something within me in which I feel returned.

At the administration building we are greeted by a vivacious, diminutive woman in a plaid skirt and top, and gumboots, who comes up the stairs behind us. Beverly recognises her immediately as Kris Tompkins. She invites us to join her at the lodge in a few hours and shows us the way to the campsite. The park centre is a mass of construction work, where gabled buildings of dark stone with spired, copper-sheeted roofs rise in all stages of completion. In our get-together with Kris Tompkins, we are joined by seven young Americans on her staff, and what we had hoped would be a time of insight into the wilderness of Patagonia becomes an intense and somewhat overwhelming discussion on ideas of principle, of cause and effect, and of the fundamental philosophies behind wilderness.

22 FEBRUARY

We meet with Dagoberto Guzmán, the Parque Patagonia administration manager. Despite frequent interruptions, he is surprisingly relaxed and unhurried, and at his suggestion we make plans to spend the evening and the next morning with Daniel Velasquez, who monitors the huemul herds to the south and west of the park.

After loading five litres of oil and a set of jumper leads, we set off late in the afternoon to Daniel's puesto. On the way, we stop at the top house. Down a passage, I enter a vaulted, open-roofed room, the entire north wall of which is glazed, with a view of the far mountains across the Río Chacabuco. Kris is busy in the spacious kitchen off the rear of the space, while her assistant, Nadine, sits at an expansive rustic wooden table tapping energetically at a laptop. Despite Kris's easy-going manner and the studied, rustic warmth of the furnishings, it feels a world away from where we have been the past few months. Kris tells me that Doug, her husband, will be coming tomorrow and says that he would like to meet us too. She proposes we get together then. We are happy to agree, as we have long wished to spend time with the Tompkinses.

As we drive away, the sun comes out, and we look forward to seeing huemul in the warm, evening light. The track to the puesto passes several small, reed-fringed lakes and then descends steeply, even by Land Rover standards, through lenga forests towards the shores of Lago Cochrane. Daniel Velasquez's simple, shingle-walled home stands in the rear of a flat, cleared field, punctuated by a single giant lenga tree. The whole family is gathered on the short grass to greet us. We drink tea and maté around the small family table and eat the best sopapillas we have had in Chile. Daniel's wife, Veja, is clearly a talented cook, and the fried breads are light, fluffy and not greasy. She offers us homemade strawberry jam to eat with them. There is no milk for our tea as they have not had a cow on the property since the place became a reserve, and they have no refrigeration. I look around and note that there is no electricity either. The house is square, six or seven paces across, with two bedrooms leading off the main room. An enamelled cast-iron stove warms the room, and when I notice that it is not the model with a water heater, Daniel grinningly admits to cold showers. They have two sons: the older one, Christían, is tall, gangly and blond, like his mother, while Daniel resembles his father in cheerful, rounded squatness and jet-black hair. They are a soft-spoken and gentle family, and there is a unity and contentment here to which I am drawn. We quickly relax, and despite our poor Spanish the conversation flows. We head down towards the lake, to look for huemul in the sunset light, but at a promontory overlooking a grand vista of the lake Daniel's radio telemetry gives no answering beep. We are all disappointed and make our way back to the puesto, and Beverly and I set up camp in the bottom corner of the field. They have made a fire for us and I cook spare ribs. Behind me, a sagging, low shed of corrugated iron bulges outward under a press of hay, and rawhide bridles and halters hang in neat order from horseshoe hooks. Just before we eat, Daniel comes by, flushed from walking hard. He has picked up a signal from one of the nearby valleys. We agree to try there first thing in the morning. The night is soundless and warm.

23 FEBRUARY

We walk out into a thin, cold dawn. Daniel stops where he found the signal the previous evening. Forty minutes and a steep climb later, we find three huemul, silent, in the open under-storey of the forest. A radio-collared female and two males: one young with small and misshapen horns, and the other a buck in his prime, with a most impressive rack. The forest, although open, is consistently thicketed and we struggle for a clear image of these endangered deer. They move slowly, browsing on low leaf growth, and rest often. After three hours, they have grown calm in our presence and will walk closely past us if we sit still. They eventually emerge into a clearing but seem to almost consciously avoid our desire for them to step into one of the scattered patches of sunlight filtering through the trees.

We spot a small Austral pygmy owl that flits in the branches above our heads. It so resembles the barred owl of home that without reference I would have thought them the same. We have found it more than fascinating that many of the birds of South America so resemble their counterparts in southern Africa that it requires only a hop of the imagination to see their common ancestry. By contrast, the mammals are disparate in species, type and number. When I put this question to Orlando, on our way to Bernardo O'Higgins Glacier, his answer was immediate and simple: 'The cold.' I sensed without hesitation that he was right. South America has survived extreme periods of glaciation and is still cold. Africa, by contrast, is hot. Birds move more readily than mammals and can bypass land passages blocked by ice.

By midday the huemul are resting again and the light is too bleached of warmth for worthwhile images and we make the hike home. Veja invites us to a lunch of lamb-bone soup and, learning that Beverly is a vegetarian, prepares a fresh salad from the garden. We linger over the table and take photographs of ourselves with the family before we leave.

Back at the administration buildings, Kris Tompkins is in meetings and we meet instead with Christían Galvez, who is head of the conservation pro-gramme for the park. His English is flawless and our interests the same, and some hours later we are still in his office in animated conversation. We agree to spend the next night together, patrolling for puma along what Christían calls Guanaco Valley.

We drive out to a point to which Christían has directed us to photograph viscacha. The Austral viscacha is one of those mammals that seem to be the sum of parts of others: it appears mostly rabbit-like in its body and face, but it has a long-haired tail and short rounded ears, and its paws are like those of a hyrax. It dwells in the tumbled scree of boulders on hillsides, teetering on the edge of the vulnerable list. Viscacha are inquisitive by nature, and although they flee into the rocky crevasses at our approach, they soon reappear, peering over edges for our whereabouts. We scramble about the sharp-edged scree and its vegetation of dense, tightly rounded, pale shrubs with spines for leaves until the sun sinks behind the mountains and the cold settles its blanket on the land. For the first day in months, we have been warm and have spent the entire day working with animals, so we pour ourselves a glass of wine in celebration. We stand beside the Land Rover in the yellow grass and watch a guanaco run, calling, towards us and then on past and into the mountains beyond.

24 FEBRUARY

Although the day starts cold, with high cloud, by mid-afternoon the rain starts to fall. When we return from rehanging our washing under shelter, Kris Tompkins pulls up beside the Land Rover with an invitation for dinner at their home.

The Tompkinses are high achievers who have made radical and alternative decisions about their lives. Decisions I admire. Philanthropic decisions that have a deep regard for the wild. I know, too, by reputation, that they see their projects through. By any standards, this requires a fierce energy and an unwavering will, and I have already noticed that, despite Kris's disarmingly relaxed and informal nature, she has the capacity to receive the focus of people without demanding it. It is an attractive, magnetic quality and its power to action is infectious. Doug, by contrast, is reserved, although not shy, and stands to the side of the spotlight. He regales us with self-effacing stories of some of his early escapades in Africa. I am surprised by his clear reticence in company, which almost registers as a discomfort. For someone with such an impressive list of achievements, not least of which includes several major land acquisitions that have been donated back to Chile and Argentina as national parks, I had expected a more obvious, extrovert dynamism. This is not the case. But when, at midnight, we are still sitting hunched in animated conversation, I realise that Doug Tompkins, like other of my close friends, is a man set alive by wild land. Set alive by mountains and walking, by sitting beside a campfire and the vitality that fills you when you turn around on a hilltop and feel utterly alone, distant from the clamour of people.

We are alike in our regard for the land, and I am disconsolate when I see how much of what their foundation has preserved has escaped us. Perhaps, the Tompkins' passion for the conservation of wild land will draw Chile into a new approach to their national parks. The western sector of Parque Pumalín that we saw through the rain represents perhaps 2 or 3% of a total of 3 250 square kilometres that stretches, range upon range, one undisturbed valley after another, all the way east to the Argentinean border. The 2 940-square-kilometre Parque Corcovado, a short skip south, which we could only look upon peripherally from the boundary, is a wonderland of mist-shrouded rivers, giant granite buttresses, forest and lakes in high craters or reed-fringed at the edge of the sea. 'This is wild land,' Doug says, turning from the pictures to me, 'wild as anything you can find.' He is right and we have missed it because it was beyond our means to access it. Although I am sad, perhaps jealous of the seeing of it, I feel that it is in its way right that it should be so. That what is wild should be beyond us, free.

25 FEBRUARY

The mist lies in veils across the valleys, clouds obscure all the heights, and rain falls intermittently all day. We follow a track down to the river, where the fishing is excellent and I catch two trout, fat with gold-and-black caterpillars, to bolster our food supplies. I walk far upstream, sheltering from the rain beneath the lenga trees. For hours, my world is private and content. I search the ground for tracks, but of the big elusive cat there is nothing, only fox and guanaco. In the evening, we photograph the waterbirds in the lakes and camp on a ridge, where we have a high view of Chilean flamingos in the shallows below us.

At 9:30 pm Christían and Luigi Soles arrive to take us night spotting. During the afternoon of our discussions with Christían a few days previously, I had asked him how they went about collaring a puma with GPS-plotting collars, and he explained that the puma is chased on horseback and treed or cornered by a pack of hounds. Once Christían has assessed that the puma is suitable for collaring, he darts it with a small dose of sedative, sufficient to sedate it without it losing its ability to stay in the tree. One of the gauchos then climbs the tree and ropes the puma's legs so that it can be lowered to the ground without injury.

Tonight, Christían tells me that Luigi is the tree climber. I am glad that the night is comparatively warm. Luigi has good eyes and at one point there are a few moments' excitement when an animal is glimpsed close to the road, but none of us gets a certain view before we lose it. 'Puma escondido,' says Luigi. Seen yet unseen.

29 FEBRUARY

In the early hours of the morning a wind picks up. It has the sharp bite of ice. Cloud drifts around the base of the jagged peaks to the north, and as the sun rises, trailings form like pennants from the highest spires, bright-gold in the first light. We are quickly processed by the Carabineros de Chile at La Frontera and then cross a flat valley of no-man's land for more kilometres than feels comfortable, with the Argentinean border post nowhere in sight. We stop several times on the crossing to photograph birds on a silted, shallow lagoon, and several large raptors feeding on the meagre remains of a hare beside the road, feeling all the time the uncomfortable sensation of being watched. Paso Roballos, when we finally come across the Gendarmeria de Argentina post hidden round a sharp bend in the road, is a small green oasis of poplar and willow trees about a short swathe of cropped green grass where a canal of clear water runs quick and silent. A large flock of upland geese waddles off at the approach of our vehicle. The youth attending to the processing formalities is uncertain and keeps disappearing into a back office for confirmation of procedures. The entire process takes more than an hour, while we sit outside in the welcome warmth of the sun. When I bring out the map he tells us the road to Los Antiguos is closed due to the heavy rains of the past week. It is an earth road, as opposed to gravel, and according to him it is '*Muy malo!*', an expression I have heard throughout Patagonia when conditions are ugly-rough.

We turn onto it anyway, I am not entirely sure why, partly because we had seen two condors flying over as we sat in the sun at Paso Roballos, but mostly because it seems right, that by driving past we would be forfeiting something. The Argentinean border guard is right: the road is muddy and quite hazardous in places, with the deep drainage trenches typical of Argentinean rural roads. But the worst areas occur in the lower sections of the valley, and as we climb towards the peaks the road becomes drier. We stop for lunch beside a small lagoon high in the cusp of the hills, just below the peaks, where several hundred waterbirds, including swans and flamingos, feed around the densely weeded fringes. A black-chested buzzard eagle comes to land just ahead of us, prey in its talons. Snowy peaks are everywhere, and through the binoculars condors cruise the updrafts in groups of three or four. There are no trees, and in the early afternoon the road passes beyond the snowline into an arid, stony landscape where the only traces of green lie in the sharp crevices of the streambeds. The landscape is vast, lunar, with the earth in faded hues of burnt orange, copper-blue, grey and pale-yellow. It is presided over by columns of rock: solitary obelisks of eroded sandstone and giant, sharp-edged peaks, too sheer for snow to cling to, that stand stark, black and orange against the sky. The ground is littered with rock and the soil is soft underfoot. Climbing to photograph two singular sandstone columns, I find the tracks of a condor in the fine soft scree of the steep hillside. The tracks are as big as my hand splayed upon the ground and are sunk as deep as my boot tracks. I follow them for 120 metres, steeply upward, to where the bird had found a ledge and taken off. It is a landscape more varied, bizarre and beautiful than virtually any we have seen. It is as if we have come upon the scorched earth of some ancient place of reverence, where cathedrals and castles surround a ground of ritual and burial, and long-ago totems and monuments are weathered to decay. We are transfixed by it. By evening, we have only travelled 11 kilometres and decide to return to the lake of birds to camp so that we may see it again at sunrise. Because of the altitude, the evening is bitter cold, even though the sun stays with us till the last. I am amazed at the capacity of the birds to withstand it. In the night, when the wind grows strong out of a clear, moon-blue sky, I listen to their pretty, liquid piping calls and am warmed by their resilience and close presence.

1 MARCH

The wind is strong and the sky thickly overcast, but the sun breaches it from below and for a time the peaks and the cloud and the curtains of thin rain, falling in the west, are cast in gold. A new world emerges that yesterday had gone unnoticed: the ramparts of the High Andes to the west that crowd the horizon right to the clouds. The cloud is thinning, and where the sun breaks through, it spotlights cordons of high rock and snow against a deep, dark, purple backdrop. The road descends gradually until the day is warm. On the streets of Los Antiguos, people are wearing shorts and sandals and we must shed the layers in which we started the morning.

Los Antiguos is a quaint little town, its main street divided by a wide, tree-lined island. It is spruce and clean, in contrast to the drab and bedraggled towns in Chile. The spirit of the people is the same, and there is a joie de vivre on the streets that may be just the warmth of the sunshine but feels somehow more than that. The people we talk to are forthcoming, and I am intrigued to find the women looking me in the eye. In Chile, we had grown accustomed to a reticence that bordered on the dour. The shops are well stocked and the produce fresh, and we buy more than we should in sheer exuberance before heading back to the mountain road to Paso Roballos. Clouds are gathered over the heights of the Andes to the west, and the sun, breaking through their frayed eastern edge, delivers shafts of light onto the land that is already extravagant in its scenery, with the embellishment of what a pilot we had flown with in Alaska called 'the Jesus light'.

2 MARCH

At the Paso Roballos junction to Bajo Caracoles, we turn east into a wetland valley where a lake stands blue between the hills in the distance. We stop for lunch where pintails and widgeon are feeding in the pool beside the road. Driving over a causeway in the overflow of a small lake that spills from between two tall, narrow cliffs, I see the fin of a fish breaking the surface. Closer inspection reveals a huge trout in ridiculously shallow water. And although I suspect this behaviour to be as a result of spawning and that the fish will have no interest in feeding, I decide to throw a fly, regardless. I am wrong, and I spend the rest of the afternoon in fly-fishing heaven, sight-casting to fish feeding in the shallows, around dense weed banks. I lose count after nine fish, none of which weigh less than six pounds, and which, bursting from the shallows for the safety of deep water, thrill my fisherman's heart with blistering, reel-screeching runs. I return all of them to the water, for they are too big to kill, and I linger, lying on my belly on the bank, with my last fish of the day cradled in my hands in the water as I ensure it is properly revived from the fight. It is a huge fish, just longer than the full stretch of my arm, and the bright pink of its flank is in pretty contrast to its silvered side and smooth, green back. When I release the fish, it cruises slowly along the channel, the wind-driven ripple casting bright wavering lines along its back.

3 MARCH

The day is overcast and windy, but the wetlands around us are so densely patronised by waterbirds that we are still working furiously when, at mid-morning, the wind dies and the sun comes out. The road leaving the valley climbs out beneath a rise of cliffs looking down on the stream-laced, overgrazed pastures of the brightly roofed Estancia Sol de Mayo. Beverly spots a condor on a far cliff, and we are walking cautiously closer when the sky fills with condors. They fly close above our heads and land on the cliffs above, showing scant concern at our presence. After three months of stalking this keen-eyed bird with little success, we are suddenly presented with them willingly on our doorstep. Up close, they are ugly, the dominant males with large, flapping, pink-skinned wattles, but are extraordinarily graceful in flight. I cannot find a comfortable place for them in my mind. I think I preferred them as a high-flying, elusive quarry of the wild heights, but taking the camera from my eye and watching three of them thermal and cruise on the updraft of the cliffs and then stoop close-winged, in an audible rush of whistling feathers, I am bewitched. As much as they are undisturbed by our presence, they, to my surprise, show clear signs of discomfort at the wind that pushes through in occasional gusts. By the time we settle down to lunch we are alone beneath cliffs scoured by winds grown to gale force.

The wind lifts sheets of pale dust as we travel the deserted road east, dropping ever lower. The land becomes arid scrubland, with dry, mineral-encrusted pans lying between flat-topped hills of pale-yellow and faded oxblood. It seems another country entirely, forsaken and deserted, and we pass only two estancias. We find the fresh roadkill of a hairy armadillo and stop because we have yet to see one, although they are supposedly common.

4 MARCH

At the tiny settlement of Bajo Caracoles there is no one in sight. The square façade of the Hotel Bajo Caracoles is pierced with white windows set in blocks of russet sandstone. The centres of the lintels are carved with flower motifs. Blue plastic chairs stand against the wall as if blown there, and a hosepipe overflows onto the dusty gravel towards Ruta 40. A curious resident drives out to where we are parked beside the fuel pumps, circles once slowly, and then drives off.

The door to the hotel is unlocked and I step into a cheerful, warm room with tables at its centre and a counter running around the rear walls. I call into the silence and a man appears from down the passage with a dishcloth in his hand. They have no diesel or petrol. He answers my other queries from behind the counter while drying coffee cups that seem already dry. Heading back to the Land Rover, I find the curious resident doing another slow circle.

Ruta 40 is tar and then recent gravel and we make good time. By late afternoon we are photographing guanaco against the dramatic backdrop of the Andes and Parque Nacional Perito Moreno. The *guardaparque* (park ranger) tells me that we are the only visitors in the park. There are notices at the information office regarding puma, so I ask him if he has ever seen one. He replies that puma can be seen anywhere in the park.

5 MARCH

Wind-driven snow issues in flurries from low cloud. On the drive to the Guardaparque Estación El Rincón, we pass through a landscape segmented by fences and see more cattle and horses than guanaco. Outside an estancia building, dogs run out to greet us. I find myself growing upset, for this is not a national park; it is a farm with guanaco wandering its plains. I simply cannot reconcile the Argentinean definition of what constitutes a national park with the idea formed of my own experience. I expect wild land with wild animals, both unrestrained, and the presence and influence of humans kept to a minimum. Without exception, in every national park I have entered in Argentina, I have come across livestock and the buildings of private enterprise, whether these be estancias or cottages, or fishing and other lodges, and the sense of pristine wild land has dissipated. I am relieved when we pass through an open wooden gate into a country free of fences.

The ranger at Estación El Rincón is bow-legged and wide-faced, with teeth that have rotted to stumps. He speaks in such rapid bursts that it seems that the words rush from him in pent-up relief from his life of quiet isolation. He assures us that although the cloud has broken up and now clusters only around the mountains, obscuring their heights, it will continue to rain and snow intermittently all day. '*E viento fuerte,*' he adds almost as an afterthought, for strong winds it seems are a given on the eastern escarpment of the Andes.

We motor slowly towards Lago Vulcan and stop to watch two condors flying almost in tandem in the updraft of some low stone hills. It is a few minutes before I realise that they are not preparing to soar but are watching something on the ground and want to land. I scan the surrounding ground and see two caranchos, southern crested caracara, also perched on a rock nearby. I am certain there is something dead in the vicinity. Beverly opts to stay behind, as one will be less disturbance to whatever is out there, and I walk the kilometre across the plains towards the hills, tracking downwind from the birds so my approach cannot be heard or smelt. Climbing into the stony hills I walk more slowly, and with the binoculars scan what I can of the ground ahead, puma on my mind. I curse when one of the caracara sees me and takes off, and talk to my imagined quarry under my breath, encouraging it to ignore the warning. Circling, footstep by cautious footstep, around a scrub-jumbled slope of sand I am overcome by the sense of being watched, and the hairs on my neck prickle erect.

Halfway up the slope, circumventing the dense shrub, I come across the first puma track, and looking up, 10 metres ahead, I see a jumble of tracks. My breath becomes laboured with adrenaline and I keep looking out over the land as I approach the place. Fresh tracks, yet to be marred by the snow or rain. I am right on them, a female and at least one cub by the look of things, and they are close. I stand still and test for any hint on the pushing wind, but there is none, and my methodical inspection of the 360 degrees of my surround is also without reward. I work my way forward and in the short grass of a depression below me I see the kill. It is a Darwin's rhea, and although much of it has been eaten, meat still remains. I pick up the partially consumed thighbone and turn it over and there is blood on my fingers. I wipe them on the grass, and drop to my haunches to study the area for the most obvious route to choose to flee. There is a shallow gully running up and around the hill, passing through copses of thicket and rock. I choose that and scout out from its radius for tracks, moving further and wider with each arc of my search. Beverly joins me there, and I tell her of my find and show her the kill and the place the puma had rested and played in the sand. She tells me that while I was gone she had seen two animals running across a short stretch of open space, downwind of my approach. Without binoculars, she could not say what they were. But I like to think that that is the way of the wild cat, wily enough to outsmart my move to outsmart it.

7 MARCH

In the evening, we drive out to Parque Nacional Los Glaciares and camp just short of the gate in a dense stand of willow trees, above a wide wetland where geese are gathered in their hundreds.

Just as evening settles, I notice a small creature passing the front of the Land Rover. Taking out a zoom lens, I am delighted to find a Patagonian skunk. It seems unaware of my presence, and I use the wind to disguise the sound of my walking and keep my scent away while I make images of a creature I have only ever seen as roadkill. When it does become alert to my presence, it lifts its head to the wind and I back away smartly, wary of the skunk's reputation. It is, however, remarkably tolerant of my presence. An hour and a half later, the skunk is busy sniffing out bugs and larvae just below the surface while I walk beside it. When it digs, I lie on my belly and it continues without pause in its enterprise, with me no more than 50 centimetres from its nose in the bright light of a full moon.

9 MARCH

At the entrance gate of Parque Nacional Torres del Paine an enthusiastic man takes our money while the official on duty surfs soft-porn images on a computer in the corner. We drive a circuit around the mountain towers and beside the lakes for the afternoon and decide to return to the extreme eastern end to be able to photograph the highest columns in first light.

We have just begun retracing our steps when we find a vehicle, with a big lens protruding from the window, stopped beside a small reed-covered depression beside the road. To our query they answer that there is a puma in the reeds. We pull in behind them and, carefully scanning the reed bed, which is perhaps 80 by 30 metres at its widest, we see nothing. We wait and scan and spot the movement of small birds, but not a puma. We wait longer, the silence of the place complete in the intervals between passing cars. It is only when a vehicle stops and, in typical Chilean fashion, the exuberant occupants spill out onto the road, talking and slamming doors, that the puma lifts its head. It is right in front of us, fewer than 20 metres away. The reeds are green and yellow, and its eye dark-amber in the quick catch of sunlight. Its face and head are broad but not as heavy as the big cats we know in Africa. Its look and its manner are not what I had expected. It has about it the reticence of the cheetah, an attitude of reserved and cautious regard. As if at the outset its choice is one of evasive concealment rather than confrontation. It moves off, its body slim, lithe and muscled, and its walk stealthy and soundless, low to the ground. It takes shelter in the shrubs of a low depression hidden from view by a high bank.

Beverly and I climb from the Land Rover and head away from the puma and climb instead the opposite bank, which is high enough that we can look down on its hiding place. We stop short of the top so as not to reveal ourselves in silhouette and sit down on the steep shale, finding meagre shelter from the wind in the lee of a bush. I scan the bushes opposite, and in the deep shadow of where I think the puma to be I notice an ear twitch, and slowly in the gloaming I make out the rest of the head. It is watching us intently. It remains hidden and then I am completely taken aback as the puma lies down to sleep.

After half an hour only ourselves and one other vehicle remain on the scene. It is occupied by two photographers and a third person, who uses a small video camera. I call them to me and reveal the puma hiding, and we move our position slightly so as to get a clearer view. It watches us, but shows no alarm or concern. It turns out that the person with the video camera is Heriberto Yaeger, the local CONAF guardaparque and a puma specialist. The five of us keep our vigil in the thin wind on the high bank while the puma sleeps, occasionally lifting its head. At 7:30 pm Heriberto suggests that we move closer because that will offer better photographic opportunities. I have deep misgivings, sure that we will provoke the puma to flight, but he seems certain that it can be approached more closely without disturbance, so I acquiesce. He says we should move in twos, and I opt to go last so that if he is wrong and the puma does move, I will be able to photograph it clearly from above. Heriberto is entirely correct, though, and with a beating heart loud in my ears I join the others sitting out on open ground 12 metres from the wild cat. I am completely at a loss as to its tolerance of our presence; although alert, its manner borders on indifference. I had expected an animal with such a history of persecution to have long fled. With the sun low and the light the warm yellow of evening, the puma stands and stretches. Its behaviour, thus far, has been quite leopard-like, although it lacks the heavy, solid build of the leopard. It arches its back into a high C and then stretches full length, extending each leg alternately to its maximum reach. Its tail is long and thick, right to the end. It moves off, and when it disappears from view, Heriberto stands slowly and follows. A short way on he finds it again and beckons us to follow. It is sniffing at the trunk of a burnt lenga tree, and then walks slowly on up a wide gully until it crests a low shadowed saddle, where it stands in silhouette against the distant sunlit hills. Again, Heriberto follows, instructing us to stay as a single group and not spread out. In the saddle, it is Heriberto who spots the puma first. I am impressed by his quick eye, but learn, too, that whereas I am searching the entire country for the animal, he restricts his attention to the shadows. It is in the shadows that the puma moves. Watching it, I become enthralled by its natural sense of the land, for it keeps to the lowest parts, to the valleys, gullies and depressions, never breaking a horizon except briefly to cross a hill and then always where its form is disguised by rock or bush. In a particularly narrow cleft, between two steep, bare hills, where the scree slopes provide little cover, the puma climbs the side to a point above us. Heriberto stops us as it comes into view, and 30 metres from where we stand the puma, seeing us, stops, sits and then lies down, peering occasionally from around the black basalt edge of its perch. I keep shaking my head at its behaviour. Heriberto, sensing my incredulous disbelief, smilingly acknowledges, '*Muy tranquilo, eh!*' When it moves off I can see the black underside of its paws as it nonchalantly traverses a narrow ledge. We wait until it has disappeared from view across the steep scree slope and we must walk fast to catch up. For a time, as we cross the ridge and look down on the wide valley below us, the puma remains invisible, and I follow the tracks in the direction in which Heriberto has intuitively headed. A small lake appears in the ford ahead of us where three valleys connect, and we see the puma in the trees skirting its edge. Heriberto leads us into the trees where we last saw the puma. The cover here is thick and none of us finds any movement to give us a clue to the cat's whereabouts. We edge slowly forward, and Heriberto is scanning the bank to our left when the puma lifts its head. It is right in front of us, barely 12 metres away, exactly where we last saw it. Clearly, it has been drinking, and it peers at us over the rim of the bank, with just its ears and eyes revealed. We freeze, but the puma is remarkably unalarmed and remains watching for a time before getting up and walking off through the trees, pausing to sniff at the trunks as it goes. We stand watching a grey form, lithe and supple, moving through the static graphic of the still reeds, whose tips all curve towards the water.

We have seen a puma. After months of searching, on the penultimate day of our time in Patagonia, we have found its most renowned resident. As we make our way back to the Land Rover in the gathering twilight, our mood is buoyant and excited. At some stage on the walk back I find myself alone. A few stars are showing in the deep blue of the darkening sky and I recall walking out of a restaurant in Los Antiguos called Viva La Viento. I had thought that the owner, a grumpy, self-important individual, would fittingly be the wind referred to. By the door of his restaurant, draped over a gaucho saddle standing on a trestle, was the skin of a puma, a hardened parchment with holes where the eyes had been and the legs stiff and awkwardly splayed. It saddened me, and made me regret giving his business my custom. I think as I walk now that the wind will always be in Patagonia as long as the Sierra de los Andes stand. The puma we just saw seemed as ethereal as the wind, but it was a living breath, a breeze, a shadow filled with the heat of life. I wish it now '*Viva La Viento*' ('long life, grey wind'), for as long as it is there in the shadow and the night not only will the mountains still be there, they will be wild.

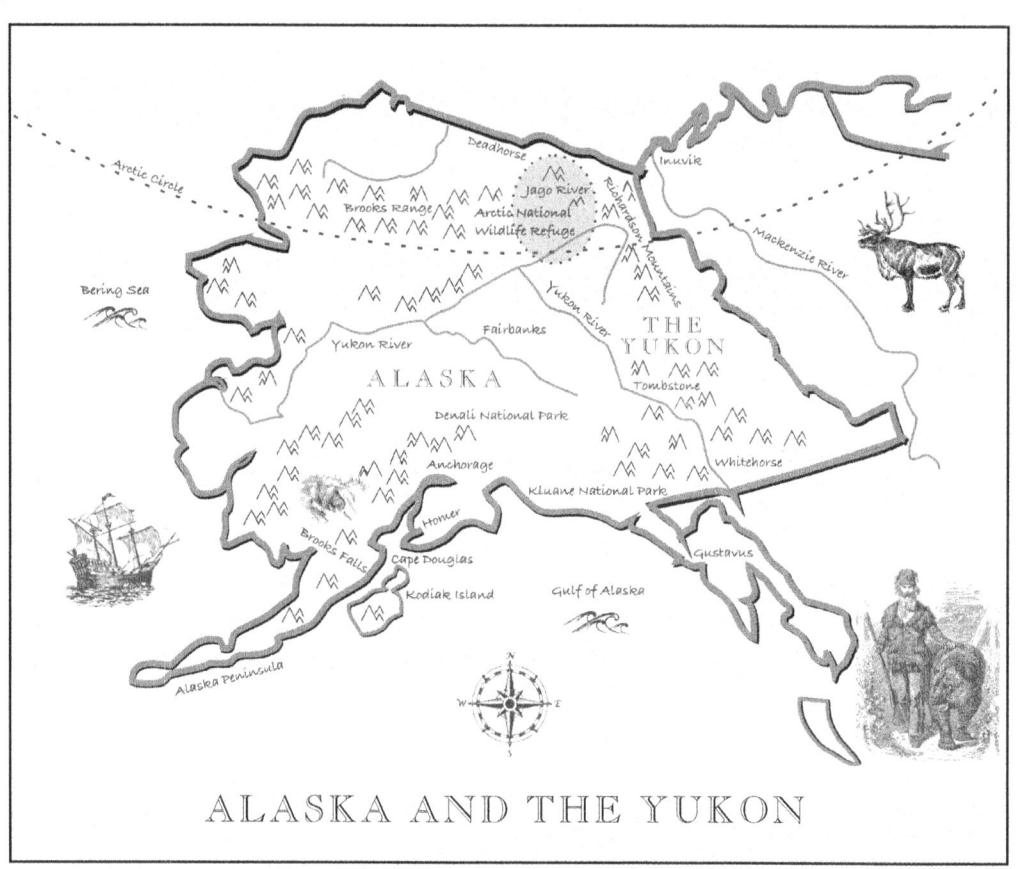

ALASKA AND THE YUKON

III

NORTH AMERICA
Alaska and the Yukon

SUMMER 2012

*Climbing above the land-bound cloud, the mountains of the Brooks Range
rear suddenly before us. We climb to 7 500 feet, high enough to fly through
the saddles, but the peaks still soar high above our heads on every side.
There is no clear path through …*

MAY 2012

We have spent the past month, at the kindness of our friends the Rinearsons,
in a lovely seaside home on Vashon Island, in Puget Sound, across the bay from
Seattle. We are awaiting the arrival of our Land Rover, which we shipped from
Santiago, Chile, to Vancouver in Canada.

It has been interesting to live in a house again: large windows and glass sliding
doors with a private veranda beyond. We have looked out over Quartermaster
Harbor, where a bald eagle perches high in the pines, its presence making foraging crows and ducks nervous. In the early hours of the morning, deer have fed
on the lawn, and, the tide being at a sufficiently low ebb, a raccoon, with its stiff
old-man's gait, has emerged from the level of shoreline trees to dig between the
stones for invertebrates and clams.

I have stood with my mug of tea and looked out at the thin rain dripping
from the eaves, or glanced up from my work when the sun has broken out from

behind drifting cloud to flood the interior with its sudden warm light. But, over the month's passage, as the apple trees have burst into bloom and the budding leaves have turned the forest from bare to green, I have yearned more and more to get up from my desk and stand in the brisk air outside.

On our last evening, I stand on the veranda at twilight, watching the water reflect the rich steel-blue of the sky, when a bird I cannot see calls a plaintive song from the trees. It raises in me a yearning for the open road.

23–28 MAY

Collecting our Land Rover from the shippers in Vancouver, we travel directly north, heading for Whitehorse, capital of the Yukon Territory, where we stay with our outdoor-enthusiast friends, Jill Pangman and Bruce McLean.

We attend the open-studio evening of a local artist, and as we stand beside the wood-fired stove, spooning gingerly at steaming bowls of moose stew, we are collared by one of the doomsayers who seem to haunt every community. Learning of our intention to spend the summer in the Yukon and Alaskan wilderness, she regales us with horror stories of bear attacks. Although our narrator seems never to have personally been subjected to the interest of a bear, nor to have spent much time in the wild, she is on intimate terms with all those who have. 'Play dead' is her final advice, which virtually every good authority I have consulted since would regard as the least-wise option. I am somewhat cheered when our host, joining the conversation, remarks quietly and almost offhandedly that she has guided more than fifty wilderness expeditions and has never experienced a serious incident with a bear.

29 MAY

In pursuit of the Dall's wild mountain sheep, we climb the slopes of the mountains that rise immediately from the Alaska Highway into Kluane National Park and Reserve. From the base of the mountain, with binoculars we had spotted a group, their pure-white coats starkly visible against the grey shale of the mountainside. The mountain is steeper than it appears from below, and with our heavy camera packs, our going is laboured and slow. We stop finally on a rounded knoll where a herd of Dall's sheep are resting while they chew the cud. Each paws a small space in the dirt before lying down, some on their side with their legs straight out, as if in repose. We stop close by, where the edge of the short knoll joins the trees and we can shelter somewhat from the wind. In the lee of the trees the warmth of the sun is welcome. The sheep stand up to graze away, but return quite shortly to lie closer still to us. Encouraged, we move uphill towards them and begin to work. It is a group of females and their horns are short, straight and goat-like. On the ground, I find tufts of their shedding coats. The hair is coarse and strong.

Later in the afternoon another herd appears on the shale scree above us. Glassing them with the binoculars, I discover that this group contains several rams, with their distinctive curl of horn. The horn is heavy and thick at the base, tapering to a point as it describes a full circle around the face. They are compellingly handsome and so we abandon the ewes and, shouldering the packs, climb towards them. The loose shale is steep and with each step slides away a little beneath one's feet. It takes time to reach their height, and because, with the camera packs on, it is impossible to do anything but stand, we leave them in a small depression where a spread of low heath kinnikinnick grows. Carrying the cameras in our hands, we climb the last reach towards the sheep. The footing is precarious and from time to time I use one hand to steady myself on the sheer side. As one who suffers from vertigo, I do not look down and instead will myself forward, focusing on the big ram that is feeding towards where I climb. I find a shallow foothold and stop there, waiting for the sheep. A young male is heading fast down the scree toward me and I lie down against the slope to brace my heavy lens in my hands and immediately start to slide. Instinctively, I raise the camera out of harm's way. This saves me, for my elbow digs into the slope to form a painful but effective anchor. I realise then that for purchase on this steep, loose mountainside, one needs to have weight pushing down on your feet or hands so that they dig in. It is after 9 pm when we finally return to the valley. My legs are stiff and unwilling and my chapped hands covered in small lacerations from the sharp shale.

30 MAY

Beverly notices some sheep moving rapidly uphill. Scanning the slopes, I see two wolves approaching them from below. The sheep climb higher in short bursts and then stop to watch the predators below, allowing them to within 30 metres before moving again. The wolves' pace is almost leisurely and they turn constantly to sniff the vegetation that clings to the slope or grows in the crevices of rocks and small cliffs. One is pure-black, the other a silver-grey and paler on the chest. The grey wolf is a third larger than its companion and the colour of its coat is such an effective camouflage that even with the binoculars I constantly lose sight of it and must rely on the black wolf to reveal their whereabouts. When they lie down in the low-growing vegetation, the grey wolf vanishes entirely. They follow the climbing sheep until the hunted finally seek refuge on a high cliff. They have climbed three-quarters of the way up the face of the mountain and I am glad I am not following them on foot. The wolves traverse the high slopes towards two other groups coming at them from above and to the side, but the steep scree of the upper slopes offers virtually no cover and the sheep see them well in advance. Again, they move away and upward, manoeuvring so that they will have the height advantage on the wolves. The hunters follow without urgency, driving their quarry towards the edge of a sheer ravine. At times, the sheep allow the wolves so close that I am sure that they will break into a charge, but they do not. As they draw closer to the edge of the ravine, the two groups of sheep are forced together, the wolves ever closer on their heels. When two young rams turn before the wolves' advance and break into a run, I am sure that the inevitable has arrived. But the wolves just stand and watch. It dawns on me that the wolves are specifically hunting lambs. It is lambing season for Dall's sheep and the wolves' behaviour is identical to that of a cheetah hunting in the Kgalagadi during the springbok lambing season. The advance in the open, without seeking cover, drives the adults forward, but the predator, as opposed to chasing them, scours the ground behind for where they have secreted their young, which will lie flat and still until their parents return. We have seen very few lambs, only just into the double figures, a grim reminder of how effective is the wolves' unpressurised but thorough sweep of the ground. The sheep must also face the attentions of bears and lynx, as well as the golden eagles that we have watched soaring upwards along the face of the mountain. The sheep climb to the safety of the cliffs that line the ravine edge and the wolves follow a path that carries them beyond our sight into the trees and scrub of the ravine. Although we watch for a long time, we do not see them again.

31 MAY

Rising early, we climb towards where we can see two separate groups of ewes with young lambs. They are taking refuge on either side of a narrow, cliff-sided valley, walking out on short feeding sorties into the open sage meadows of the mountainside and scampering back to the safety of the cliffs at the first sign of danger. We are two-thirds of our way up the mountain when, 30 metres to our left, the black wolf appears in a shallow saddle between two rocky knolls. It is only for the briefest moment that we see it, but the scene will remain etched in my memory. The thick fur, touched with the warm light of the first sun, is blown by the wind from behind. The wolf stands outlined against the sheer white of the distant mountaintops, cradled in the green and rocky cup of the saddle as it sniffs the ground. As I crouch to free my camera pack, it lifts its head to look right at me, unsure for a moment as it stares directly into the sun. Then it turns unhurriedly and is gone.

Returning to our climb to the sheep, I notice that the wolves have used the same path to descend. As I fear, when we arrive at the cliffs there are no sheep to be found, all of them having fled the wolves. We slump against our packs. Far below, the lake is still half-frozen, and every horizon about us is cut short by tall black peaks whose pinnacles are under deep snow. While we drink a cup of tea, we see what we assume is the first of the group of ewes crossing far below us, across the gully to the north. We have climbed straight up for more than two and a half hours and now the only way for us to cross to the north side of the gully is to descend, walk around its base and start the ascent yet again. An icy wind has started to blow hard. We are almost at the bottom, resting in a sun-soaked opening between the spruce trees that shelter us from the wind, when above us the ewes and their lambs appear again on the skyline of the cliffs. They had not fled, merely hidden themselves in the sheer folds of the cliffs where neither the wolves nor we could find them. It seems too much to climb again to where we were, so we opt to attempt a crossing of the scree slope that separates us from the cliffs. From where we sit among the trees it seems quite plausible, but halfway across it becomes apparent how reckless it is. The scree crumbles beneath us, and the slope becomes increasingly sheer and unstable. Finally, my nerves shrieking in protest with vertigo, we abandon the attempt. By the time we reach solid ground again, we have been on the mountain for five and a half hours and are no nearer our goal.

We opt to try walking around the base of the cliffs in the hope of finding a way up a slope of vegetation. Halfway around, I spot a single female with a very small lamb in a cleft almost at the base of the cliffs. To get to her again requires crossing a scree slope, but this one is short. Beyond it, although steeper, is a narrow, shallow gully where a few spruce grow, and the ground is covered by a carpet of kinnikinnick that reaches right to the base of the cliffs. As I get close to the top of the gully next to the cliffs, I am climbing on all fours. At the top, I rest the camera pack against the base of a spruce to prevent it rolling down. Edging out below the cleft,

I find not one but three ewes with tiny lambs looking down on me. We are careful to give them their space and they quickly relax to our presence. For the next few hours we are able to work with the animals coming to us rather than pursuing them. Brief periods of ten to fifteen minutes of feeding are followed by longer resting times as they chew the cud. The lambs stay close to their mothers, and when the ewes cannot be badgered into allowing them to suckle, they scramble onto their prone backs in a game of king-of-the-castle. I am glad for the periods of cud-chewing when the females retreat to the sheer sides of the cleft, for it gives me a chance to calm my anguish at the dizzying height. While we photograph the ewes and lambs, they constantly dislodge small avalanches of stones that tumble down the slope beside me, gathering momentum as they go and setting others free with their speed, which does nothing to appease my discomfort.

Suddenly, Beverly, crouching slightly above me and to my right, exclaims, 'I'm going!' I lift my head to see her sliding down the slope, dislodging a shower of stones. As if in slow motion, I see that she is lifting the camera out to me as though I should take if from her. She scrambles with her leg for purchase and as she passes I catch her behind her bent knee and grab hard on the cloth that comes to hand. For a moment, my arm gives in to her momentum. As I slew towards her in my effort to hold on, I can feel my own precarious footing shift. It is only a second and then we are still, and as I catch my breath I can hear the shower of dislodged stones far below coming to me on the strong updraft of the valley wind. I hold Beverly's camera while she clambers to a more secure footing. It is only by a gross effort of will that I stay, distracting myself by concentrating on what I see through the lens.

In mid-afternoon, the ewes finally abandon the safety of the cleft and, walking a few metres from where Beverly and I share an apple, cross the scree slope below us and climb towards the cliffs to the north, where we cannot follow. By the time we return to the Land Rover our bodies are bone-weary and we collapse into our seats, happy to be out of the wind, which is now raising an opaque grey cloud from the silt of the river delta that feeds into the lake.

At the Arctic Institute of North America, Kluane Lake Research Station, we enjoy a hot shower and then head out to camp at one of the secret spots of Lance Goodwin and Sian Williams, who run the station. We cross three small rivers to get there, and in a country of folded hills, with quaking aspen and spruce-filled valleys, we camp beside a small lake. When all is quiet a fat beaver waddles from the water to gnaw at the budding trees, and diminutive spotted sandpipers utter their plaintive call as they patrol in their bouncing bob-walk along the shore. The evening is still and sun-drenched. In my shirtsleeves, I study the saucer-sized hoof prints of bison that form sharp depressions in the soft peat of the lake surround.

1 JUNE

The country is soft green in the first flush of spring, and the rounded crests of the short hills that encompass a myriad clear and tiny lakes make for lovely walking. A lone bison bull appears on a hillside and we make our way towards him. It is not difficult to get close, but the wind and the sunlight stand opposed and we retreat, before he becomes aware of us, to seek a better alternative. We decide to try our approach out in the open. At 250 metres away, we step out from the trees and into the open of the hillside, walking at the pace of a grazing animal. The bison sees us and watches for a while and then, to our delight, resumes grazing. We work our way slowly closer. At 100 metres his bulk is massive in the lens and when he raises his shaggy head to regard us I can see the snot that hangs in long spider-thread tendrils from his nose and his strangely small black eye. We work our way around the slope to skyline him on the ridge where he stands. He stops grazing and walks along the ridge towards us, stopping every few paces to watch. I can hear his breathing. We move closer to the trees as he advances, until our backs are against them. His hump-shouldered bulk looms as high as the peaks behind, and when he stops he is panting more from the heat of the sun than from the climb. At the apex of the hill he stops and for a long while regards us without moving, then takes a short step down into a hollow we cannot see and has a dust bath. He moves off when he is done and we follow him onto an open hillside with a lake below, where he resumes grazing. For a time, he grazes away from us, but then turns and grazes right in our direction. We back away until once again we are against the trees, but this time he is below. On the steep hillside, the water of the lake makes a dramatic backdrop to the scene. He keeps coming slowly forward as he grazes, and only when he is 30 metres from us and Beverly stands to move deeper into the trees does he stop feeding and raise his head to regard us. He seems prehistoric. His horns are small and black, and point upwards, but his head is broad and heavy, the hair curly black over his forehead, with a thin mane of long straight black hairs at the base of the skull between the horns. The head is carried low before the huge hump of his shoulder that is his bulk, with its shaggy covering of wavy, brown hair that fringes the short, close hair of his hide. He trots briefly down the hill and then breaks into a run, bouncing on all four feet in short little bursts. It leaves both of us laughing.

3 JUNE

Finding a repair shop open on a Sunday in Tok, we decide to make use of the otherwise unproductive day to do the few mechanical repairs that have become necessary. 'Willard's' is a tall, unpainted shed with two wide doors standing open towards the highway, across a mud-puddled yard, where an assortment of tow trucks and a great variety of other vehicles in various stages of being dismantled crowd so closely to the entrance as to make access to the workshop a cautious affair. The interior is a chaos of tools, pipes and cables, and the gravel-strewn floor is thick with dust and the soiled brown of old oil. I find a robust, well-bellied older man searching in the back of a pick-up laden with toolboxes and oxyacetylene tanks.

I greet him, and indicate that our Land Rover requires a few minor repairs. He does not look up, but after a time grunts and comes up with the tool he has obviously been looking for. Over the next few hours, as I work on the vehicle and Willard repairs the parts I strip off myself, there is a steady trickle of visitors. Most have grey hair, which they wear long, and the men have beards of equal shaggy length or they at least, as Willard does, sport three-day-old stubble. Time is a luxury and visits are not marked by ceremony. Conversation is, however, unhurried and punctuated by such long pauses that I sometimes wonder if there are two people involved.

From beneath the Land Rover, I see a pair of stained, baggy blue trousers walk slowly past towards the rear. The walk is almost a shuffle.

'So's everyone gone home then?' I hear the visitor say.

There is no reply, just the pop of gas being lit and the clang of metal. For a time, there is just the hiss of the gas.

'Who's everyone?'

The clang of metal again and then someone walking past the vehicle to the front. Silence. The person returns to the workbench at the back.

'Well, Willard, just you're here.'

A grunt and the pop of the gas being lit again and this time it hisses for several minutes.

'Bruce's wife called, said his steak was cold. He said he doesn't eat cold steak, he'd have to feed it to the dog and cook himself another one. Cousin stood on a nail while we were moving that stuff.'

There is a sound of loud hammering, and I can see Willard's feet moving back and forth around the workbench. The visitor's feet come round to the side where I am working and stop. A face peers at me round the wheel arch.

'Where're those plates from?'

'South Africa.'

'South Africa, aye! Had a chap here from South Africa a few years back. When he went home they held him hostage. Least that's the way he tells it – probably his own family did it. When he come back he had the biggest, ugliest, smokiest diamond you ever saw on his finger. He was proud of it though.'

He pats the side of the Land Rover and walks around it to stand back from the workbench again.

'Wasn't a rusty nail. Galvanised. So, I guess that's lucky.'

When we leave Tok it is still raining, but the rain thins and then stops as we travel towards Valdez. We camp on a grassed, open place beside a crystal-clear creek that flows wide and shallow over grey gravel and rounded river stones. It comes around a low curve through the spruce towards us and, just below our camp, passes under a narrow log bridge into Lake Mentasta. Beside the bridge is a sign that reads:

MENTASTA FISH HABITAT PROTECTION PLAN

TAIL'AHWT'AENE, THE HEADWATER PEOPLE, ARE THE STEWARDS OF THIS LAND. WE LIVE IN THE PLACE WHERE OUR ANCESTORS WALKED, WHERE THE WATER RUNS CLEAN, AND THE FISH PEOPLE RETURN EVERY YEAR TO PROVIDE US WITH NOURISHMENT AND A CONNECTION TO THE EARTH. WE ARE ALL GUESTS HERE, EVEN IF IT IS OUR HOME. THANK YOU FOR BEING RESPECTFUL OF THIS PLACE.

THANK YOU FOR KEEPING THIS LAND CLEAN

TRASH DOES NOT BELONG ON THE GROUND

PLEASE DO NOT DISTURB THE FISH

DO NOT GO IN THE CREEK UNLESS

GRANDMA KATIE SAYS IT'S OK

TSIN'AEN

It is signed 'Robert John, son of Katie John, elder Mentasta Tribal member, legal owner and steward of the land beneath your feet'.

The mist hanging low over the trees parts at some point to reveal a slice of high, snow-capped peaks close behind, cut softly by a thin, late-evening sun. In the silent twilight of midnight, we are woken by a shot close by. It is followed, a few minutes later, by four more in quick succession. Silence, then another shot, vicious in the calm. Far off I hear two voices whooping.

5 JUNE

Stopping on the banks of the Copper River, we investigate a strange series of paddlewheels, which turn out to be salmon traps. I approach a young man in a pick-up stopped beside one to find out more about them. He is talking on his mobile phone. His arms are tattooed in green and red. On the seat beside him, tucked beside the gear stick, lies a rifle with a telescopic sight.

He finishes his call and, without introducing himself, walks me out to the home-made paddlewheel turning slowly through the turbid water of the river. 'Don't fall in, you won't come up,' he warns as we cross the narrow gangplank over the fast flow of the water. A wind is blowing strongly upriver, driving a stinging cold rain before it. The wheel is four-bladed, two paddles and two nets, each standing opposed, turning on a central axle mounted on two pontoons anchored in the river. The mechanism of nets and paddles, turning between the pontoons, is driven by the swift current. The nets are rigid structures 1.5 metres across, a metre deep and reach probably two metres down into the river.

The salmon swimming upriver to spawn, unseeing in the discoloured water, are trapped in the net moving against their direction and are lifted free of the water as the wheel turns. As the net rises vertically, the fish drop into a chute that slides them out to the side to fall into holding cages beside the pontoons. The trap is simple and effective, and the holding cage seethes with the blue, silver-grey backs of salmon.

Later, in talking to a fisherman launching his boat below the traps, I learn that he has a free subsistence permit, only available to Alaskan residents. He can take two hundred fish from the river, which amount may include no more than five chinook (or king) salmon, as he is a household of one person. He tells me that a household of two or more may take five hundred salmon, with the same limit of chinook. He uses dip nets, essentially just oversized fish landing nets with a mouth 1.5 metres across and a 3-metre handling pole. Motoring upriver, he drifts downstream, holding the nets below the surface, and literally scoops the salmon from the water when he feels them swim into the net. 'I'll be happy if we get a hundred,' he tells me. And it dawns on me that he may be speaking of just today.

I stare at the surface of the wide river and wonder at the magnitude of life that surges upstream, unseen, a few metres beyond where I stand. The sheer number of fish I have seen taken from the river leaves my heart on the side of the salmon. Out to sea they must escape the deepwater purse-seine netters as they draw closer to the shore and begin forming shoals before they drive towards the fresh water they can sense and recognise as the place they were conceived. Close to the river mouths, pods of orcas hunt them with ruthless efficiency, and seals grow fat on their numbers. Entering the river, they face men and their nets, and then, as they turn from the main river into the smaller tributaries, they find the men there again, this time with hooks. Where the tributaries grow shallow or form waterfalls that channel the fish into narrower passages, they are hunted by bears. In their final spawning beds, they are still not left alone as bald eagles swoop to snatch them from the shallows. The journey is often hundreds of kilometres, against fast-flowing water, cascades and low waterfalls, and yet those remaining fish succeed in their thousands. Although I will eat the salmon, I would as soon not if I were in any way to jeopardise the opportunity to stand as a spectator to its wonderful, incredible race.

11 JUNE

At the Kenai and Russian rivers the salmon derby has become known as 'combat fishing'. On every approachable bank of the river are fishermen, only a metre or two apart, standing in waders in the shallows casting into the current. They are not fishing in the true sense, for the salmon that run up the river have stopped feeding, and to catch them one must jig or snag them with the hook. I have noticed that the regulations stipulate that this is a fly-fishing-only river, but what most of the riverside salmon seekers are using is a regular casting rod with a weighted lure superficially dressed with feathers. To attempt to proper fly-cast in the press of anglers along the bank would be almost impossible, let alone dangerous. The lure, or so-called fly, seems to be of little consequence other than to lend sufficient weight to the cast to get the hook as far as possible into the river. Where space allows, it is retrieved fast with a jerking rod action as the angler attempts to impale a passing salmon with the point of the hook. These are spirited fish, and one loosely played once it is hooked will wreak havoc over several metres of the river, tangling lines and fishermen as it goes.

Steering clear of the fishing section of the river, we shoulder our heavy camera packs and head for the Russian River Falls, where those fish that have survived the deadly gauntlet downstream can be seen performing aerobatics to clear the fast water. As we leave the car park, two fishermen pass on bicycles. The first has a short-barrelled pump-action shotgun slung across his chest. We come upon a sign that states, 'This is Bear Country', and lists precautionary measures to be taken to avoid attracting bears. We pass a few more fishermen along the path and then an older man who, like us, seems to be out for the pleasure of the walk as he carries his gear. He is walking towards us in a slow, stooped gait with one arm straight down by his side, the other, crosswise across his waist, clamped out of sight at his belt. It is a strange posture and when I turn to look after him, once he has passed, I see that he holds a rifle concealed behind his arm, and a belt of cartridges at his back.

It is only an hour's walk to the falls, but in that short space of time I see more guns carried openly than I have seen anywhere, except perhaps in the Omo Valley in Ethiopia, where every male from age fourteen carries a weapon. There is a man carrying a baby in a papoose. Beside him are two women and a dog. The women push a pram, and the man has a six-shooter strapped below his left shoulder alongside the papoose. Two girls walk slowly beneath the heavy weight of their backpacks, with black-handled knives strapped to the outside. One of the girls holds a short-barrelled revolver at her hip. There is a man swathed in gear and sharp shades, riding fast on a bicycle, his dog running with its tongue lolling out as it follows behind, a rod and a rifle protruding from the man's backpack. Later, at the overlook of the falls, I photograph a Mennonite youth who seems in his mid-teens and carries a Colt .45 Peacemaker in a holster strapped brazenly across his slight chest, with the brass cartridges lining the shoulder strap. He is protecting three young girls who are wearing plain, ankle-length dresses, and have white head coverings pinned to their hair, which is drawn back into a tight bun.

Despite the many walkers, we find ourselves alone at the falls when we arrive at the sturdy wooden overlooks. I am still shrugging my pack to the ground when Beverly issues a stifled cry. Unable to speak, she just points towards the base of a series of cascades that constitute the falls. A large female brown bear is clambering hurriedly down the steep ground of the far side, with two cubs following her. Walking without hesitation into the water, leaving her two cubs perched in anticipation on a rock, the sow navigates the swift current with practised ease. Using the eddies to carry her forward, she strokes powerfully across the short channel of strong surge to bring herself to a central eddy where the waters boil with bubbles from the turbulence above. Then, rising on her haunches on a shallow rock, she launches herself bodily at the white mass of bubbles with her forepaws held wide, disappearing in a surging splash. She surfaces with a shake of her shaggy head and for a time paddles around in the cauldron of frothing water where one can see by her posture that she is probing with outstretched forepaws for fish. Back towards the calmer downstream end of the pool, and with her back towards us, she suddenly snags a salmon, pinning it to her chest with both paws. Lifting its struggling body clamped in her jaws, she crosses the river, carrying it back to her cubs, who jostle back and forth on the bank as she approaches. She holds it pinned to a rock until one of the cubs has a firm grip on it, and she then releases it to them. The two youngsters, with their prize, climb quickly up the bank into the sanctuary of the trees while the sow has a good shake and stands for a moment looking over the river. This time she tries a different tactic, and entering the river, wades with her forepaws probing wide through the deep eddy right beside the bank. From my high vantage point I can see the salmon scatter through the clear water at her approach. Trying the same place, she moves in the opposite direction, but finding nothing, moves to the bottom of the pool and crosses with nonchalance to the centre through the relative shallows where the current is so swift as to sweep any man off his feet. Again, she finds her rock in the shallows and launches her bulk in a blind leap into the central cauldron of the pool. Emerging empty-handed, this time she returns directly to the submerged rock and launches herself again. I am not certain if by this method she actually anticipates catching a fish, or merely wishes to spook them to the shallower, more accessible perimeter of the pool, for she turns and, burying her face in the white bubbles, courses the edge. She is soon successful, and although her snout is submerged I can see the quick, urgent movement of her head as she snatches to secure the salmon in her jaws. Lifting her dripping head from the water, the fish arching its body as it fights to be free, the sow crosses quickly to the bank and into the trees to join the cubs.

In the hope that the bears may reappear, we spend the next five hours focusing our energies on trying to photograph leaping salmon in mid-air. Looking down from above on the race of white water, I am astonished as I watch the fish fight their way upstream. Some seem lucky, or savvy, or both, and pick a route that avoids the rush of the water, swimming to the shelter of small eddies where they recuperate before their next push. Others take the current head on. It is remarkable to see a creature fight its way forward through a shallow rush of white water. Its back stands clear of the current as it enters a column as deep as a man's waist, and falling as a solid, two-metre-high chute, then it powers its way upward against this unrelenting resistance. Those who miss a route, or who tire, are swept and bounced unceremoniously downstream into the eddies at the base, where they can be seen jockeying for position as they gather their strength to make their run again.

Ecstatic with our day, we camp that night at a pretty site beside Lake Skilak where we look across the far fetch of the turquoise water to the Harding Icefield.

14 JUNE

We return to the Russian River to fish. The sky is broken cloud, and when the sun is on one's back it is warm. I am using a ten-weight fly rod, fast sinking line and a strong thirty-pound leader, with a hefty split shot of lead tied about 60 centimetres back from the fly. I am told that the fly makes little difference, but notice that most people use either red or neon-green and so try something I find in my fly-box that has a bit of both. The method is a roll-cast across the river, as the trees stand close along the bank and the current is swift any distance out from the shore. The lead shot lands with an ungracious plunk. With a quick mend upstream to correct the drag, I keep light tension on the line as it is swept downstream waiting for the snag.

My first few hours are disillusioning as cast after cast goes by and my only snag is the bottom. We are making our way slowly upstream and away from the crowded river in the vicinity of the boardwalk. There are still fishermen, but they are fewer and I am able to find stretches of river entirely to myself. The water is unsullied, clear and cold. I cast across the river where boulders provide eddies in the swift current, but by midday I have still only made contact with the bottom and the trees.

Upriver, we come across a female black bear with two cubs patrolling the opposite bank. The cubs cavort in the thicket and climb a thick aspen with remarkable speed and dexterity for an animal that seems to lumber along the ground. The mother patrols the edge of the water and fishes out the filleted carcasses of salmon discarded by fishermen, pulling them free from the rocks with a dabbing motion of her paw. She seems to ignore our presence, except that she never looks directly at us. She fishes downstream, leaving her cubs far behind. Eventually they pause in their climbing and log balancing and scramble off downstream to disappear into the woods.

For the first time, we are entirely alone on the river. Finding a good, deep eddy, my third cast is rewarded with a solid take. The short stretch of free line is snapped from the water as the fish careens down the current, jumping twice, high, as it goes. This is the fish I watched with amazement a few days ago swim up the falling white cascade of a waterfall, and now I feel its strength and speed as I manage to do little but hang on, with the heavy rod bent in a deep C as the salmon vanishes into the fast water, headed towards a bend 50 metres downstream. I am frantically contemplating the mad scramble I will have to make through the fallen debris that clogs the riverbank when, to my great luck, the fish turns and charges back upstream. Even swimming against the current, it keeps the rod bent to its trembling extreme. When it is opposite me, it leaps clear of the water again with a great shaking of its head. Its body is bright-silver, tinged with a slight tone of blood-red, torpedo-shaped with a small triangular head and a wide tail. For the rest of the fight it stays close to the pool, but when I work it to the shallows it is still too lively to handle and, sprinting away as it sees me, it works itself into a fall of deadwood and snags the line. Beverly runs from the bank with the net while I hold the rod high and head into the deep water to try to free the line with my right hand. The tangle of branches slows us down and the salmon breaks free. I sit disappointed, but animated, on the bank, and rig a new fly. This is a quarry among quarries.

The next place we try is a fast, deep race above a long eddy and my third cast connects hard, tearing the line from my left hand to snap taut against the rod. The fish vaults upstream in a quick series of spectacular jumps, then turns to the current. The water is faster here and I am forced to follow. The bank is steep, but fortunately not wooded, and working my way around a tipped-over stump I am able to play the fish into the quieter current of an eddy. As the fish tires, I call urgently to Beverly to bring the net as I struggle to keep the salmon away from a fallen tree trunk along the shore. The fish is lively. Scrambling through the shallows, Beverly slips and trips as she turns, trying to snare its fast form, and falls sidelong into the icy water. She stands up without hesitation and, leaning out, makes a sweep at the passing fish. For a moment, I forget to play the fish and stand in wonder as to how I was ever so blessed to find such a companion. The fish is too big for the net, but together we finally get its head into the scoop and lift it, as it kicks hard, from the water. I look it in the eye and say my thanks and apologies before I kill it. The flesh is orange and I cut the carcass and the gut into small strips and feed it to the herring gulls that wheel overhead.

17 JUNE

On our first morning on Kodiak Island we meet Chris Flinchman, who runs cattle on the southeastern spit, near Pasagshak Bay. He likes my truck, and I like his land. He has two dogs and a rifle with him and he is out hunting bears. They kill his cattle, he says. I tell him of a cow carcass I had passed earlier on the hillside opposite. The bald eagles had drawn my attention to it, standing clustered around it and calling in their thin, high-pitched cry, which does not suit their size. He shakes his head. He shot the last bear that was killing cattle not a month ago. 'You shoot 'em, then another one comes,' he says, expressing a lament that I have heard often in Africa.

Chris is the first person we have come across in Alaska who is farming live-stock. He runs a herd of shaggy, stocky cattle reminiscent of animals we have seen in Scotland. The country here is not forested, except for copses and gullies of low alder trees. On the peaty ground the grass grows in runnels and humps that appear rippled from above. It is clear that running stock here is a fringe exercise, compounded by the presence of bears. Perhaps this is why Alaska is such a 'huntin', shootin', fishin'' place. With an abundance of land in a natural state, wildlife exists in comparative plenty in relation to the low human density. Hunting permits are inexpensive and Alaska is one of the easiest places in the world to buy a gun. A man walked into Mack's Sport Shop in Kodiak and bought a Concealed Carry .45 over the counter and was done and gone before I had finished selecting and paying for a few salmon and trout flies. People here take their meat from the land. It is not seen as a luxury, but a part of life.

At the neighbouring ranch house, in Bill Burton's living room, bearskins hang from the walls and antlers are tied to the balustrade of the upstairs balcony. 'The problem with shooting the bears is that they just get more: way more today than when I bought here in '67,' Bill claims. His theory is that the bear population is self-regulating, with the big boars keeping other males out and killing cubs. He reckons that when they killed all the big bears they just made more space available for the lesser ones. But, still, he is only half-joking when he says he would prefer I took one of his rifles and shot any bear that I saw with that rather than with my camera. He farms with bison because their herd and protective instincts are stronger than that of cattle and they survive the bears better.

His land falls on a gentle coastal plain caught between the steep rise of mountains and the sea. It has a long aspect in every direction and flows in low, rolling undulations between blue, clear creeks where trout and salmon run. There is a lake at the base of the mountains, and the scalloped black-beach coast, littered with giant, bleached logs, is punctuated at its points by small islands and spires of rock where seabirds roost.

In the twilight of evening we watch a coastal Arctic fox feeding on a bear-kill carcass. Its longhaired coat is a charcoal grey-black, grizzled with cream at the extremities. Its tail is bushy and thick, more cream than the body, and it holds it out straight behind as it watches me. Its eyes are amber. From the hill behind another fox begins calling, and the eyes turn away only briefly before returning to me.

18 JUNE

Heading north toward Cape Chiniak, I am fascinated to see a grid placed across the Pasagshak Estuary. It is for a salmon census and I am surprised by the simplicity of the enterprise. It is in essence a barred fence through the water with a narrow gate allowing access upstream. Two men in waders keep it clear of drifting sea grass and count the salmon through the gate with hand click counters. As yet, only forty-four have swum upstream.

The northern coast of the Marin Range that forms the Chiniak Cape is forested, and logging trucks carrying giant logs, which my heart aches to see, pass us hurriedly in billows of dust. The gravel road wends through a landscape of cliffs, horseshoe bays of clear, green-blue sea and rippling, gravel-bedded creeks. We stop for lunch at a bay where a grey whale feeds close inshore, to the gentle sigh of the wavelets and call of the gulls.

At Cape Chiniak we branch off onto a narrow two-wheel track and find a bald eagle's nest, with a single chick, on a spire of rock that stands 20 metres out from the cliffs. Further on, we climb a hillock at the end of the road to an old Second World War bunker. From the height of its vantage point, we look out onto an island around which seabirds wheel in flocks of thousands. It is too far to see what species they are, but we camp at the end of an isthmus of sand between the sea and a freshwater lake behind, determined to try to get out to the island tomorrow.

19 JUNE

We inflate the kayak and paddle the kilometre across to Chiniak Island. Close by the leeward beach we find a small group of harbour seals on the rocks. They take to the water, despite our measured approach, and then come up behind us as we paddle on. Nearing the island, we pass a raft of sea otters that rise half up out of the water with raised heads and then dive to pass close beneath the kayak, their bodies appearing a creamy, silver-white as we watch them through the greenly transparent sea.

The birds we had seen the evening before are kittiwakes and puffins. On a stretch of calm water between Chiniak and another small island, there is a raft of thousands of puffins on the water. They are tufted puffins and their black bodies and orange bills cast a quilt of patchwork colour over the sea that rises and falls with the gentle swell. Thousands fly about the cliffs, all turning so that their black silhouettes appear spun on the same pivot. On the eastern tip of the island, where a solitary spire of rock stands 150 metres out in the sea, the circling birds, their flight directed into the easterly wind as they fly along the cliffs, are moving in two opposite rotating circles. Driven side by side, where they meet at the point, it is like the meshing cogs of a giant machine.

20 JUNE

Aboard the M/V *Tustumena* ferry to Homer, a group of Russian Orthodox Christians camp out with us in the rear of the observation lounge. They are led by a rumple-featured man whose hair and long beard are silver-grey, his bushy eyebrows jet-black. His eyes are deep-set and he wears a dark, felt fedora low on his head, which he turns slowly side to side as he takes in his surroundings, giving the impression of a walrus blinking at the light of day. He is accompanied by four women and a flock of children. The older women wear dark knee-length skirts and stout square shoes, but the younger women sport pretty, ankle-length flowing dresses of a silky, shiny material with bold floral print. All the women have their hair tied back beneath a headscarf that is folded in on itself to form a long pouch from which no hair escapes. The youngest has a flower by her right ear. Their leader is clearly a spirited young woman who bustles and organises with a brusque but gentle efficiency. As the lights are put out the children are unheard.

The sea is as calm as if we were in harbour, and when I lift my head from the bench seat of the table booth where we sleep, the far-off mountains are slipping past between a grey sea and an identical sky.

21 JUNE

After a brief stop to replenish at Kenai, we decide to use the fine, sunny day for a final hike up the Russian River to photograph salmon underwater where they pool beneath the falls. With the spell of warm weather melting the snow, the river is even higher than previously. I cross the river upstream of the falls where three large salmon, just out from the bank in a shallow run, are beginning to show their red spawning colour. Beverly stays behind on the viewing platform to warn me of approaching bears.

Alone on the pathless far side of the river, carrying the bulky yellow underwater camera housing, I feel suddenly exposed and peer more closely into a bush that I approach, and begin singing softly so as not to surprise any unwary bear. I find my way to the salmon pool, signal through the trees to Beverly on the far side, and then, taking a long look in both directions, head down to the pool. The salmon are stacked side by side in the slower water of the eddy at the edge. I can see their fins and grey backs rise through the turbulent reflection of the surface, but when I lower my camera housing into the water they move away. They do not return close enough to photograph and I work my way cautiously around the edge of the pool towards the falls and the current where my presence will be disguised by the turmoil of bubbles and current.

The water is bitterly cold and the current tugs at the housing and my legs, but after standing still for a few minutes I can feel salmon bump into me and I take the first few exploratory photographs. When I stand up from the water I pause to survey downriver and the steep descent of the bank that rises in a semicircle above me, making time to listen before I turn to the photographs. I work for forty minutes on the edge of the cascade. My arms are bright pink from the cold and my hands claw-like in their stiff function. When the camera becomes too cold to operate properly, I climb out and find a place to sit in the sun. As I work my way up the bank I can see the paw prints of bear cubs etched clear in the mud, and the smell of fish on the ground is strong. From across the river Beverly waves the all-clear. I warm the camera in the sun, then return to the river and work for another hour before the cold claims me again. Believing that I should not push my luck with the bears for another session, I retreat back upstream, singing, through the trees.

24 JUNE

At the Eielson Visitor Center in Denali National Park, Beverly walks the trail down to the river while I find a quiet corner to write. Glancing up, I see a bear in the distance cross onto the near hillside. It moves surprisingly swiftly towards the visitor centre. Others see it too, and a general cry goes up as people crowd the windows, balcony and trails for a better view of the still-distant bear. The genial information ranger is forced to become an authoritarian to control those whose excitement is spurring them on to recklessness. Far below on the path I see Beverly approaching, on a collision course with the bear. The ranger gesticulates in her direction, but she is too far off to respond. I give our contact whistle and point out the bear when she looks up. She continues upward, which produces further gesticulations from the ranger, but at the crest of the rise she switches to a longer route, moving away from the bear. When she finally arrives, breathless, at the centre, I am amused that, for all her fearful anticipation of meeting a bear face to face on foot, her only reaction is frustration at not having taken a camera with her.

On the bus, returning to the park entrance, I note that virtually all the other passengers have fallen into a dulled soporific reverie and some are asleep. There is something in their numbed, listless apathy that gnaws at me. I feel as if the singular wild country through which we travel is no more to them than a passing entertainment. Why do I know without asking that it is the highlights that the people about me seek, not the power of the land itself? If I were to shout wolf, they would all start awake. But it is the wolf in the context of this wild land that ignites us.

Outside, the multihued hills and their cast of rivers through low, wide plains of green pass unnoticed. It is only when a young couple standing on the roadside with their heavy backpacks flags down our bus and collapse on the seat opposite us that my faith is a little restored. They have hiked for three days through the mountains and their faces are radiant with their adventure. If someone asked me later what my enduring image of Denali would be, it is of their young and animated faces huddled close, side by side as they reviewed their pictures of a Dall's sheep that had appeared without warning over the rim right before them. Behind their heads the bus window was opaque with dust, cast yellow by the low sun, and framed through the open top half stood the sharp cone of a snow-streaked peak of orange earth and black rock.

26 JUNE

In Fairbanks, we spend the evening with Lon and Nora Kelly, who invite us in out of the rain. Lon works for the Bureau of Land Management and has a large portion of the central and western North Slope under his jurisdiction. The North Slope is the expansive reach of flat land that stretches between the Brooks Range, Alaska's northernmost mountain range, and the Arctic Ocean. When the conversation turns to Native Alaskans, as it often does when one is talking of land, wilderness and wildlife in Alaska, I am interested that Lon comes from a standpoint not shared by many others.

There is, in the northern realms of North America, both in Alaska and the Yukon, a decided sense of 'them and us' between settlers and the Natives, and it cuts both ways. Strangely, I never hear envy expressed for the wealth that has come to some Natives from land-settlement deals. The gripes I hear most often have to do with the utilisation of wildlife. Not all northern Native North Americans live on mineral-rich land, and many rely entirely on subsistence to live. This was recognised and incorporated into their rights to the land. Natives today may hunt and fish when and where they please, regardless of season or the land's designation, including in parks. It is when this right, perceived by many as a privilege, is abused, that it brings about a disgruntled reaction from the settlers who share and hunt the same land.

In Alaska this year, the king salmon run has been one of the weakest on record, and all king salmon fishing is accordingly prohibited in an effort to protect the few that arrive to spawn. In protest of what they blamed as improper management, concerned principally with the lucrative commercial exploitation of the king salmon resource, some Native people ignored the ban and caught salmon. It is their right to do so and it brought attention to their protest that they have been too long ignored in the sharing of this natural wealth. But, to exploit a resource that hangs in precarious balance is seen by many as rash, and devoid of the empathy the protestors claim to have for the environment. It has won them no friends. I did not speak to Lon of these things, but instead asked him about conflicts that have developed in my own mind with regard to subsistence.

Subsistence has, I feel, as part of its definition, an inclination to tradition and history, and an adherence to cultural ways. These constraints in method and action are in their own way limiting, but the contemporary North American Native uses every modern convenience, including snowmobiles, rifles and exploding harpoon heads, to pursue their traditional quarry. What was a traditional harvest for subsistence becomes entirely altered and can, if abused, lead to wasteful slaughter.

Lon is a slow speaker, but his answer is unhesitating and brims with a suppressed passion. It is at the outset wrong, he points out, to attempt to make argument with, or understand, another culture from our own standpoint. If we are to debate an issue, then we must first stand in the shoes of those concerned, abandon our own cultural rationale, and adopt the point of view of those whose rights are at stake. Many, if not all, Native North Americans were to some degree historically nomadic in their efforts to secure enough food during the brief summer to endure the protracted winter. When the state parcelled out rights to some, but not all, land by asking people what land they historically used, Native history was dramatically altered and it cut deep into traditional ways. Lon explains that we lose sight of how recent this change has been. He has held meetings with people who were part of communities where many perished of starvation in a winter, following a summer too wet to dry the salmon. He argues that all humans adapt to change, that we embrace what eases the burden of living, and that the twenty-year-old hunter who heads out on a snowmobile dragging a sled and kills ten caribou with a rifle is not making a bloodthirsty slaughter; he is a critical provider of food for a community where many are too old to endure the extreme outdoors.

Lon empathises with a change come too fast upon a people to adapt without pain and loss. He sees people weep at meetings with regard to what road to take into the future, where communities eroded by drugs and alcohol are still tight with love and the bond of their community. I hear in his soft talking that he understands their aloneness, feels the fear that makes them wary, defensive of their land and their right to it. It was here before we came. By what authority do we now claim to judge fairly, and to whose end do we know what is best?

28 JUNE

In the early-morning waiting room of Wright Air, a sleepy-looking pilot is talking to Steve Springer, our guide from the Arctic Wild expedition company. They stand below a big wall map of the Brooks Range and the North Slope. I see Steve stab with his finger at the map on the three potential campsites, but the pilot shakes his head. A cold claw grips my chest when I hear him say that nobody has seen any caribou in the entire area.

Yesterday Steve told us that our quarry, the Porcupine Caribou Herd, which numbers some one hundred and sixty-nine thousand animals, is unseasonally still in Canada, and has not as yet crossed the Alaskan border to their traditional calving grounds in the Arctic National Wildlife Refuge. Seeing our dismay, he quickly consoled us with the news that a herd of about thirty or forty thousand caribou from the Central Arctic Herd have recently been recorded by a friend who researches caribou movement in the area.

We have committed everything to an expedition into the ANWR, which will require a four- or five-hour flight in a small aircraft into the heart of this vast, wild place. We have done exhaustive research into caribou movement and timed our efforts to coincide with what has historically been the peak of their concentrated numbers, as they pass through the narrowest area, from the Kongakut River west, before they once again disperse. We are aware that the place they usually move through is hundreds of kilometres across but hope that the flights of other expeditions and data from researchers will narrow our margin to a certainty. We stand on the precipice, filled with built-up anticipation, for this plunge into one of the remotest places on the planet and realise we have not bargained for what we have just heard. We do not have the budget of a BBC documentary team, we cannot afford maybes, there will be no second chance for us. I do not want to believe what the pilot says, I want to believe what Steve told us yesterday. I am, as yet, unfamiliar with this place and do not have a grasp of its physiology, its vastness, its sharply folded landscapes, as big as an entire European country, where not a road or a village exists. I am relying on, wanting to believe, those I have come to for their knowledge of the land. I do not yet understand how unpromising a promise it is to find a herd of even forty thousand caribou in a place that is hundreds of kilometres long and as much wide, with more steep-sided, hidden valleys than one can count radiating from a single river.

Beverly suggests to Steve that we contact the head of Arctic Wild and request that, since our entire journey here is to photograph migrating caribou, we postpone the trip until the caribou have at least crossed into the ANWR from Canada. Steve says it will not be possible to change our flights at this late stage. He seems bullish about our chances, and thinks that we may do best to base ourselves around the Aichilik River, where the herd was last seen. There will be several pilots flying over the area, and we accept his persuasion that one of them could well spot the herd and lead us to them.

Outside, we are led by our pilot, Daniel Hayden, to the Helio aircraft standing on the apron. The Helio is no longer manufactured, but several of those singular aircraft are kept in service in Alaska because their extraordinary take-off capabilities make them ideal bush aircraft where no formal runways exist. It is a tail-dragger, and has been fitted with outsize, balloon-like tyres that reach as high as my thigh for landing on rough, uneven terrain or rocky river-bars and beaches. To provide extra lift at low speed, it also has ailerons on the front of the wings that are gravity- or air-speed-activated. When we are cleared for take-off, I am amazed at how the Helio 'wants' to fly. We are at maximum capacity load, but at 50 metres the tail is well off the ground and by about 130 metres we are airborne.

The White Mountains are the first striking geographic feature we encounter: bare, pale grey-white, limestone peaks around which the clear Beaver River meanders a convoluted course. As Daniel dips a wing towards them for our photography, he spots a herd of Dall's sheep on the upper slopes, and we count fifteen lonely white specks moving across the high green pasture below us. Ten minutes later, Daniel excuses himself on the headset and mine goes dead as he has a long exchange over the radio. When he comes back on we are turning around and headed back to base because he has felt an intermittent vibration he is not happy with.

Two hours later, after the aircraft has received a thorough going-over and been taken for a test flight with Wright Air's chief engineer, the Helio is declared mechanically sound. The vibration is attributed to dents in the propeller causing slight balance issues when under power, but still within allowable tolerances. Just before take-off Steve asks me where I would like to head. I am dumbfounded by the question. It has from the outset been clear that we have one prerogative for this expedition: to find the caribou migration. He explains that the initial idea to camp near Caribou Pass, on the Kongakut River, in the far east of the ANWR, would be a long shot. We have two, possibly three, alternatives to encounter the only herd that has been seen, and he rolls off the names: Jago, Hulahula and Aichilik. I leave the decision to him, for he understands its import, I hope, far more clearly and fully than I do, but a leaden weight of doubt sits heavy in my belly.

On the flight out, I learn that Daniel grew up in the mountains of the Brooks Range, living in a tiny cabin with his parents and his sister, making their living from trapping. His sister was born on a split-pole bed in the dark of winter, his father tying off the umbilical cord in the smoky, yellow light of the kerosene lamp that, together with the wood-burning stove, was their only convenience. They were home-schooled, and Daniel worked with his father setting and checking the two hundred-odd traps that made their trapping life. Marten was their main quarry, and they trapped an average of one hundred and fifty, perhaps two hundred, in a good winter. It paid for their staples and for the gasoline they needed for the outboard motor, and later for the snowmobile and the chainsaw. They lived only 35 miles from Arctic Village, 'but I never went there', he says.

'Never, ever? Not once?' I ask.

'Never.' And then, after a pause, 'My father is a white guy.' I can see from his features that his mother must have been Native. 'The people of Arctic Village would throw stones at our dogs.' So, they left and never went back. When I ask him if it is still like that, a brief wistful look crosses his face. 'No,' he says, 'not any more, not like it was.'

Near the village of Venetie, we cross the Arctic Circle. We are flying over the Yukon Flats and the land below us is level, sparsely wooded and pockmarked by myriad lakes. Daniel talks of the millions of waterbirds that come to this haven to breed. Although it is yet too early for most of the chicks to have fledged, he points out the occasional small lake where flights of a few birds can be seen winging low over the water. At Arctic Village, the cloud is down and we fly at a few hundred feet following the course of the river, with the hills starting to rise from the plain, their height hidden by the cloud. There is no one to meet us at the airstrip. Daniel tops off the wing tanks himself and soon we are back in the air again.

Climbing above the land-bound cloud, the mountains of the Brooks Range rear suddenly before us. Billowing cumulus clings to the higher peaks, whose grey-black bases shed curtains of rain that obscure the valleys. Sunlight breaks through like spotlights on a stage. We climb to 7 500 feet, high enough to fly through the saddles, but the peaks still soar high above our heads on every side. There is no clear path through, and it is only Daniel's practised ease that calms my apprehension at the forbidding frontier that looms constantly in the windscreen. We twist left through a saddle, fly towards two peaks that block our way, and turn sharp right into a monochrome valley where sunlight makes the fast tumble of water a bright, thin thread embroidering a twisting, convoluted white between sheer, bare-rock slopes. We rise up at its end, scraping the clouds with our heads, to pass over its saw-toothed rim into a caldera-like valley where unbroken sheets of blue ice coat the high sides, like icing clinging to a bowl. Again we turn away, this time dropping down towards the U of a saddle that is the only outlet that can be seen. We pass between two peaks, beyond which we can see patches of sunlight, again at the end of a high, black-stone valley. The sky beckons, a ribbon of blue to the north.

Daniel is talking on the radio with another pilot. I can hear their conversation through the headset. My heart is heavy, for it is clear that the caribou herd we seek is lost to us by the sheer scope of the land, and I reluctantly accept what he suggests as our best option: to make camp on the Jago River. The other voice proposes it as the most scenic of our three options, with good hiking. Nothing could be further from my mind than hiking and, as if reading my thoughts, the voice intrudes to say that the chance of caribou would be the same for all our choices: a crapshoot at best.

Dropping below a cloth of cloud strung between the heights, Daniel flies us down the Jago Valley and lands on a short stretch of tundra beside the river. Into the silence, as the engine dies, Beverly requests that we try to find any sign of caribou from the air before we choose a place to be dropped, as this is what the itinerary proposed. Daniel is uncomfortable, and in a flat voice explains that he is late for his next pick-up so cannot give us any more of his time.

An Arctic ground squirrel watches us unload, twitching its tail in alarm beside the entrance to its burrow. Then Daniel flies away and we are left with the rush of the river. Dwarfed by a still and silent landscape, I feel the desperate weight of our mistake. Despite the magnificence of what surrounds us, this is not what we came for. We have used 80% of our funding and been brought to an empty landscape. A landscape no different to that we could walk into from any point off the Alaska Highway by simply stepping away from our Land Rover. I feel sick, and have to will myself to action to avoid infecting the others with my despondency.

With the tundra soft and damp underfoot, we make camp on a rise overlooking the broad, braided riverbed. Just before dinner I spot a single caribou bull coming upstream. The sight of him gives me a rush of hope and I watch fixedly through the binoculars. He is being driven mad by the mosquitoes and the flies. He bucks suddenly sideways and, racing through a shallow braid, plunges bodily into the icy, grey, silt-laden current of the main stream. Emerging on the far side, he shakes, and then plunges back in again, swimming strongly. He stands still for moment on the bank, his nose and heavy-antlered head hung towards the ground, water streaming from the chest and belly hair of his coat. Then he is running again and does not stop. I watch him until his movement is swallowed by the blur of heat shimmering off the far-distant riverbed. Later, sitting in the warm evening light, I search again upstream for sight of the caribou and find instead a grizzly bear heading downstream on the opposite bank. It is still a kilometre distant and I call Beverly and Steve and we watch its ambling but surprisingly rapid search for food for the time it takes to pass. From time to time it raises its head to test the wind, following scent trails up the hillside to a reward we cannot make out. In a field of white-headed bear flowers, it walks swinging its head rhythmically side to side, mowing the stalked heads with a casual chomp of its jaws. The sun does not set, and in the north the midnight sun stands well clear of the low hills, bathing our tent with warmth.

29 JUNE

A blanket of mist finally burns off at midday, and we take a seven-kilometre hike downriver towards where the mountainous land flattens into hills. Our hope is unspoken, but in each it is the same: to find the herds of caribou on the move. The river valley widens and narrows and where a knoll rises at its edge above the plain, we find the ground littered with wolf dung: dry, white turds in which shards of bone show between matted, compressed hair. We see no sign of caribou, but the tundra is bright with flowers, and from the stony gravel bars to the heights of the hills we come across nesting birds, or those with young chicks.

A semipalmated plover feigns a broken wing and, fluttering, leads us away from unseen chicks between the gravel stones. A pair of wandering tattlers fuss as we collect water from a stream, and their downy grey chick runs out from the riverside caribou path beneath our feet. A female willow ptarmigan, the colour of orange-brown sandstone, and her four yellow chicks just able to fly, all move only a metre out of our path and watch us pass with unblinking eyes. On the plains, there are savannah sparrows and a pair of Smith's longspurs, a rarity that some come here especially to see.

For a better view downvalley, we climb to the tops of the hills and put a pair of rock ptarmigans to flight from right beneath our feet. Their call is coarse and guttural. Near the heights of the hills are American pipits and the strikingly marked American golden plover.

At the top of the hills the land is flat. Cresting the ridge first, I walk out ahead with a sudden desire to be alone. I am finding my disappointment and the banter of conversation distracting me from where I am. I am not open to it and it feels strange, for that is my customary place. Alert more to the reaction of my being, of my inner voice, than to the busy mind that now distracts me. I find a horned lark and, remembering that Barry Lopez records bowing to this bird in his masterwork, *Arctic Dreams*, I stop and sit in the hope of approaching that same instance of the place.

There are no caribou herds to be seen from the heights, and we trek back to our camp with the wind at our backs and the mosquitoes forming clouds around us. While Steve prepares dinner, Beverly notices a bear far off on our side of the river. It is walking close along the riverbank and not stopping to feed, and it is approaching fast. We are downwind. Dinner is put on hold and Steve refreshes us in the etiquette of a bear encounter, as we screw bear bangers onto their small pen-like launchers and remove our cans of pepper spray from their sheaths. Steve loads his revolver and we gather pots to bang together. The breeze is growing steadily lighter, and when, 200 metres from camp, the bear disappears behind a hummock of land, the wind dies altogether. The minutes tick by, and I check and recheck the cameras. Then, as suddenly as it died, the wind rises again, but in the exact opposite direction, blowing our scent towards the bear.

I utter a quiet curse, and five minutes later my fears are confirmed when Steve spots the bear lumbering fast up the steep side of a gully into the mountains. It stops periodically to raise its nose in our direction and then continues on its high, steep climb to be away.

Well after midnight, after I finish writing, I come out of the mosquito-net tent to find another bear in almost the same place as we saw the first. This one, however, is feeding avidly, and its progress towards us is slow as it meanders back and forth across the rise of the foothills that fall to the river. I wake Steve, and this time he suggests that we walk out to the bear. At 3 am we sit in a low depression of the floodplain, with the bear approaching through a cast of yellow-headed flowers. At a little less than 60 metres, it sees us and pauses, unsure. It rises, and what was big becomes twice so. It looks down on us with one paw held out to the side. It turns quickly and shuffles away fast, stops to look back and, suddenly certain, breaks into an ambling run towards the river. It crosses the water without pausing, and on the far side heads quickly downstream. We walk parallel with it until we are in camp and then watch as it climbs away into the mountainside.

30 JUNE

It seems certain that we are not to see caribou where we are camped. Their coming up or down the valley, we now realise, is reliant on chance, and the landscape is on a scale that reduces chance to, quite literally, a needle in a haystack. We ask Steve to contact Bill Mohrwinkel, the co-owner of the expedition company, and in a satellite conversation we decide to glean as much information as we can from pilots and other expeditions and then, if possible, be moved to an area into which any groups of caribou are headed.

Even this, however, is no certainty. The caribou, I am learning, are entirely whimsical and their movements are more dictated by weather than any instinct. To call it migration is a misnomer, for that suggests regular and predictable patterns. This is not the case. The caribou move in response to bugs. Their food is abundant everywhere in the ANWR, and only the ice fields, coastal wind, water and foul weather in the highlands of the Brooks Range provide relief from the mosquitoes and flies, and it is this that they seek out in their movement and not the known, fresh food sources, or safe raising grounds for their young, that drive other animal migrations. Although the pregnant caribou are drawn to the coastal plains to drop their calves, this area is vast, and prevailing weather conditions are the only certainty that will determine where and when this will occur. And once they have calved they will be on the move once again.

I feel the desperate helplessness of our search: our constant, frequent and pathetic hope, our eyes cast imploringly skyward. We are overwhelmed by the scope of this vast, wild place. Its expanse is far too great for searching. In the afternoon, we walk into the sheer valleys to the east, and although every far mountainside and valley floor is striated with the pathways of caribou, they are all years old and the valleys lie silent and still. It is only after dinner, when I am catching up with my notes, that we see a second caribou. A single male trotting upriver in the shallows of the wide side braids, raising a splash of sun-brightened water. Although his coming gives rise to fresh hope, he is soon gone, and no more follow on his heels.

1 JULY

The early-morning satellite call to Bill Mohrwinkel does not go well. The connection keeps dropping, and it is only after repeated attempts that Steve is able to glean that no caribou have been seen for the past four days, and no pilots are free to move us to the Hulahula River where the herd of thirty or forty thousand caribou were the last to be seen.

I cannot bear the thought of another day walking the same path, of deepening my already-saturated desperation, incrementally, with each heavy step. I do not see the landscape. The arrangement that we had with the outfitter was to locate the caribou migration from the air, then be dropped in the path of the migration and set up a base camp. We do not even have backpacks, but rather trunks and sealed containers that we cannot carry to another site. This means we are not even able to use the time to walk a trail to the Hulahula, as we have to return to the same base each night.

I am blinded by my heavy heart, my desperate sense of futility. To distract myself, I spend the day writing, coming out only to glass the valley, eat a snack lunch, and go for a late-evening walk when I have exhausted myself beyond feeling.

2 JULY

The ground-to-air radio crackles to life. Daniel is flying overhead, on his way to collect another expedition further east. He has not seen anything yet, but will call again with any news on his way back. Despite our hopes, his news is not good: they too have not seen a single herd of caribou. The expedition leader, Don, who has extensive experience of the caribou, suggests we move to the Hulahula, as there is no sign of the caribou towards the mountains. Daniel is enthusiastic about the move and, after dropping Don and his group at Arctic Village, anticipates having two and a half hours free, which would give him time to move us. Steve arranges to make satellite telephone contact in forty minutes.

For the first time since Daniel delivered us into the Arctic, we permit ourselves to feel excited. We retreat to our tent to pack our few things. Daniel's call is late and we pace the tundra, waiting. When we do finally patch through, after several aborted attempts, the news is better than we dared hope. At Arctic Village, another expedition has also landed, having completed a raft trip down the Hulahula River and bumping into the caribou on two occasions. They estimate the herd they encountered to be between thirty and fifty thousand animals and last saw them yesterday crossing the Hulahula to the west, right near the coast. Daniel reckons he can easily pick us up and move us down to the Hulahula Delta in the time he has, but must first verify the change and the cost with the Wright Air office in Fairbanks. He'll be back in ten minutes.

We are overjoyed and start breaking down our tent and packing up the supply tent, half finishing everything, jumping to the next matter. The satellite phone rings and we listen with half an ear to Steve's conversation about the cost. But when the tone of his voice sinks and he offers only yes and no, we do not need to hear his words: we can see his face. Wright Air has recalled Daniel; he won't be allowed to make the flight. I turn and walk out alone onto the gravel bars of the river. My mind will not focus, and I search among the river-smoothed, flat rocks for one with a cross of white quartz in the black stone to give to my sister-in-law, who is an ordained priest.

3 JULY

Persistent rain, a grey sky and a cold wind from the north keep us close about the camp for the morning, but towards midday the sun breaks through and a rainbow appears across the valley. Its downriver arc is pale against the sunlit hillside, but as a cloud's shadow crosses the valley, I see that its end is touching our tent. It makes me realise just how much I have been counting my cost and not my blessings.

The Jago River is as wild as a place can be. It is the earth untouched, accessible only during the brief window of the Arctic summer. I walk away from camp to feel it more fully, for I have found that it is mostly, or perhaps only, when I am by myself that there rises in me the context of the land about. As if it were solitude that opens some inner ear. It brings about in me some slow release, like sucking chocolate on my tongue. I feel the spongy give of the tundra beneath my step and, looking down, see its tiny complex form: delicate, intricate and splashed with small dots of red. There are no trees here that reach anywhere beyond the height of a shrub and so the landscape is open, its sharp contours softened by this thick pliant mat of green. The ground is damp, often wet, and water is never far away. It is the season of flowers, and there are vivid splashes of white where the fine, whorled heads of cotton grass grow in profusion on the wetter, tussocked seeps. Along the river's edge, and bright against the grey stone of the gravel bars, the rich saturated purple-pink flower of the Eskimo potato is as bright as splashed paint. In lesser stands, scattered more widely across the green undulations, are heads of pale pink and faded lilac and purple-blue. Stooping to examine them, I see that they are exquisite in their tight and complex detail. When I lift my head to regard a high, rounded, stony hill that sheds the cling of vegetation until its sides become sheer, sharp ridges of unadorned rock, my eye is arrested by movement. A caribou cow, the first we have seen, is trotting upriver. She is moving surprisingly fast and covers in minutes a distance we would measure in terms of hours of walking. Directly opposite camp she stops, and looking for a moment hard upriver, suddenly springs away to the side, then with her tail raised, erect, runs in a straight-legged gait towards the far bank. I look upriver searching for the source of her agitation and find a pale grey wolf. It trots out onto the open riverbed and watches the caribou without altering its pace. It has a pale, almost white head that at its tall shoulder falls to a grey, flecked thinly with black. I can see the long hair of the coat flow to the wind as it runs. It pays no heed to our camp, nor to the caribou that is now well up the opposite hillside. I would howl if I knew how, not only because I have been told that wolves respond to calling, but also because that is what the wolf's presence evokes in me. It makes the land primal, and is gone too soon into the distance. The caribou is no longer in sight and I am left alone on ground that, though silent, stands frangible in anticipation.

4 JULY

Kirk Sweetsir lands his big-wheeled Cessna 185 at mid-morning on the short stretch of dry, level, hard riverbank that serves as a runway. We watch it bounce once, twice, and then it is down and he brings it to the end of the level ground and swings it around to a stop with practised ease, placing the tailwheel a mere two metres from the river's edge. At Bill Mohrwinkel's request, he has brought a drum of extra fuel to fly us in search of the caribou. Kirk is tall, lean and soft-spoken, with an easy, disarming demeanour. He is one of the most respected bush pilots in the Arctic, with the quiet reticence of an intellectual rather than the swagger of cowboy bravado.

Discussing what little we know of the caribou whereabouts and movements, we readily concur with Kirk's suggestion that we fly a grid: west to the Hulahula, then north to the coast, returning east to the Jago. We take off into what Steve calls a bluebird day, headed north, downriver, with the mountains soon left behind. We turn west over the low foothills, the plains stretching to flatness on our right, the mountains a sharp rise to our left. Looking down, I realise the futility of our hope that the caribou would pass by our camp. I have heard people breathe deeply when confronted by a big landscape, but this place takes one's breath away. Its sense is too strong, and I feel overwhelmed.

Kirk has keen eyes, and is familiar with the signatures of the wild from above, but we find no animals, except a lone bear. At the Hulahula River we turn north towards the coast. Halfway we find the tracks of caribou, intersecting lines that show clear in the gravel bars of the riverbed. They are not those of a large herd, however, and we fly on.

Near the mouth we find tracks: a vast spider's web of wandering black lines through the tundra and gravel bars that veer to the side of the myriad small lakes to come together beyond. Kirk flies lower, but where the ocean meets the land we must lift away from our pursuit of the trails, for a band of coastal fog sits upon the sea, reducing visibility. For a time, we fly east with the fog at our wingtip, but then, on a hunch, Kirk swings the Cessna around and for ten minutes we head west. The final juncture of the continent falls in crescent bays and low headlands to the ocean. The beaches are strewn with bleached driftwood. Caribou tracks trail the shore into the water, but where they lead is hidden beneath the blanket of fog.

Caribou will readily wade into the cold sea and the bitter bite of the fog to rid themselves of the insects that hound them on the land. Now the fog puts them beyond our reach too. We turn east again and fly higher, so that more of the land is accessible. On a patch of tundra, we find a group of caribou resting in a close circle, but our hope for the great herds that lend this place its reputation is fading. The land is too extensive, its scale such that even a herd of thirty thousand animals becomes a pinhead. Between the distinctive, widely braided course of the rivers the boundless arena of the land is punctuated by small lakes, so numerous that they appear like raindrops beaded to a flat surface.

Turning south back towards the mountains, Beverly and I become resigned to our defeat. We should in a way be glad. Glad that we have found a wilderness too immense to know completely. Glad that there is still such a place that is beyond the sure knowledge of men, that is wild enough to defeat us, to humble the certain conquest with which we reduce the earth. It is what we have come in search of. But to say that defeat does not rest heavy on our shoulders would be a lie.

7 JULY

After waiting a day in Homer for suitable weather to coincide with the low tide required for a beach landing on Cape Douglas on the Katmai Peninsula, we take off in a small plane with Hugh Rose, our Arctic Wild guide. Barbara and Jessie May, and Patrick Enders, a wildlife photographer and friend of Hugh, are in a second plane behind us. Although the day is grey, the cloud base is high enough for the pilots to drop us off and return to Homer. On the far side of Cook Inlet, the cloud is broken and sunshine plays on the heights of the Alaska Range and the volcanic cone of Mount Augustine off our starboard wing. Cape Douglas is a spectacular, spade-shaped jut of land that stretches out from the mountains into the sea. It is threaded by a river and small creeks, which fall in white cascades on the uniquely stepped terrace of the foothills. We land on a wide, sloping beach of black gravel and rounded pebbles, and humping our gear through the last half-mile of soft sand, make our camp below a rise of grass and lupin-covered dunes. We are still pegging out the electric fences that will protect our camp when a bear comes wandering down the beach. It passes close between us and the sea, ignoring us completely and leaving plate-size tracks in the black sand. In the evening, we walk to the estuary of the river, where grazing bears are highlighted against the lime-green of the sedge meadows.

In the cocoon of our tent, when a thin dark has descended, a brief patter of rain arrests my falling asleep. In the wash of quiet after its passing, I hear the waves at the height of the eight-metre tide. Beneath their lulling drone the soft pad of a wolf that steals by within metres of our camp goes unheard, and we are left only with clear, sharp impressions in the wet sand to tell of its secret late-night visit.

9 JULY

Barbara comes breathlessly into the mess tent, saying there is a bear right behind our trench latrine. We abandon breakfast and find the bear, who shows us less regard than the sparrows that fly up startled from under its nose. In the dense, riotous growth of lupins, grass, cow parsnip and wild celery, it is feeding not 50 metres from camp. Hugh leaves us briefly to turn on the electric fence. In a wind-blown hollow at the base of the dune, not 20 metres from where we stand, the bear lies down to sleep, and almost simultaneously another bear raises its head over the ridge of the dune. All of these are brown bears, known colloquially as grizzlies. These coastal bears are markedly larger and generally darker than those we saw in the Arctic National Wildlife Refuge, and the one that comes now over the crest of the dune walks wide-legged in its bulk.

Its legs seem short for its size, which is emphasised by the low carriage of the heavy, wide, round head. In its direct, head-on advance, the dark pit of the unseen eyes at the base of the long snout, it has the aspect of a tank: something brute, bulldozing and unstoppable. Only the small, furry, rounded ears suggest anthropomorphic cuddliness. There is little about the bear that is soft. It is craggy and strong, and seems fitting to the land: the cold, the loneliness, the sheer rock and the ice.

The approaching bear encounters the sleeping bear, but the latter simply rises and gives way. Towards the mountains, a pure-white she-wolf comes into view, but she is travelling fast and is soon gone in the undulations of the plain. For the first time the cloud is thinning and for a moment a single beam of sunlight breaks through to shine almost exactly where the wolf had stood, but the stage is now bare.

In the afternoon, we can see patches of the sky, and the amphitheatre of peaks that surrounds us is revealed by slow and jealous increment. The peaks soar beyond any height that we could have guessed from their base, and thin trailings of cloud cling to their entirely snow-shrouded summits. They dwarf the land. Their presence adds dimension to this place, like music to the scene of a play.

11 JULY

It starts to rain, and at the first heavy, squall-driven gust Hugh raises his eyes to the roof of the tent and grimaces. The rain pelts down. Even the short walk from our tent to the food tent leaves one drenched. The wind is gusting at 30 knots and the cloud is almost to the ground. No flight will be able to get in today, and Hugh takes stock of our food and works out an informal ration for a three- to four-day wait. All our tents are small, two-person structures, which, shaken by the wind and stung by the constant rush of the rain, make the storm an overbearing presence. The sky is black, and through the opaque gauze of the rain the sea is slate-grey and cold. In the food tent, we must raise our voices above the din of the storm to be heard. Some sleep, some play cards or read, and at every momentary lapse in the storm all eyes turn upward, but there is no respite and I am glad when it becomes late enough to sleep.

12 JULY

I wake to a storm that seems, if anything, more violent still, and turn over and sleep some more. By mid-morning, however, the wind has shifted more to the north and the volume of the rain begins to lessen.

By late afternoon, the rain has slackened and the cloud has lifted enough for us to see down the beach. When it reduces to a sparse drizzle, we don our wet gear and go for a walk to the mouth of the estuary. The beach has been swept clean by the storm, and a fresh set of wolf tracks is etched sharp in its surface. They run almost its entire length. The wolf has come twice, on both occasions passing not more than 10 metres from the main tent. I look hard up and down the shoreline dunes, for there is a sense of it that remains on the air.

13 JULY

The cloud hanging at the height of the peaks is too thin for rain and lets a brightness through that bathes the rain-washed landscape in a summery fresh-ness. Before I am fully dressed comes Hugh's shout of 'Bear in camp!' Standing in my bare feet in the wet grass, I see a young sow walking circumspectly around the food tent, testing the air with her upraised snout. The electric fence is off and Patrick and Hugh stand between her and our food cache, separated only by a few metres. They are talking loudly, holding their arms akimbo. When I join their thin defence the bear moves off behind a clump of high grass, then saunters past the sleeping tents and onto the dune behind and begins to graze the wild celery. Finally, she swims the river.

The news is that we will be collected this afternoon, as long as the weather holds. We break camp and, between alternate bursts of sunshine and fog that rolls in off the sea, lug our gear in relays to the landing site. On the last run to the landing site Beverly and I are in the lead, dragging two cooler boxes and shouldering a bulky but light drum, when I notice a bear approaching from the far end of the beach. It is still a dot, and our unattended gear piled at the landing site sits almost halfway between us. We pick up our pace and for a time it seems that we will arrive at the cache first, with a minute or so to spare. The bear is closing fast, however, and despite dropping the drum and walking as hastily as the dragging coolers allow me, I realise at 300 metres from our cache that the bear will beat me to it. Unsheathing the can of bear spray from my belt, I begin to run. Fifty metres from our cache of gear the bear pauses to my shouting approach, then raises its head. For a few seconds, it holds its ground and lifts its head higher, standing on the balls of its forepaws. I feel the hairs on the nape of my neck prickle erect. The bear at close quarters is huge, just short of half a ton, and I feel puny and vulnerable. I shout louder and, glancing over my shoulder, find Hugh coming at a staggering run beneath a load of shoulder-slung dry bags, 200 metres behind. The bear drops to its feet and turns side-on, which I know from my time with wild animals is a gesture to emphasize its size, but it hesitates, unsure in its stance, and at my continued running charge it gives way and lopes up the beach and over the dune. I stand by our luggage, breathless, with a racing heart and the taste of adrenaline dry in my mouth. We fly out into a clear sky, and looking down I trace the familiar contours of the land with my fingertip on the windowpane, longing already to return.

15 JULY

From Anchorage, we fly to the small ex-military town of King Salmon and, on the river that flows through town, board a floatplane to the Katmai National Park. There are two bears on the beach as we coast to a stop before Brooks Camp, and the helpers on shore, standing in hip-waders to catch and turn the plane, must beat a hasty retreat, leaving us to glide slowly onto the gently shelving gravel. Out in the lake another bear is wading about on its hind legs, searching for salmon by sight. A heavy drizzle is falling and the light on the lake and in the sky is pale powder-blue.

It is a half-hour walk to the falls on the Brooks River from the campsite, and the trail leads past the front of the lodge and the floatplane-landing beach to a pontoon-supported footbridge across the mouth of the Brooks River, where the river issues into Lake Naknek. Beneath the footbridge I can see schools of salmon in the clear, shallow water. On the far side is the first bear-viewing platform, with its raised view of the bridge and the river mouth. Salmon are jumping and a few fishermen stand in the water casting unsuccessfully into their midst. There are two other bear-viewing platforms, reached by a trail that follows the dirt road for a time before turning off into the riverside trees.

From the Riffles Platform, the bears at the falls are quite a distant view, but after a thirty-five minute wait a place becomes available at the Falls Platform. It is crowded, and it takes me some time to adjust to where I can bunker myself and see only the bears. There are three fishing immediately in the deep water below the falls, and their habituation with the raised platform allows one the extraordinary privilege of seeing entirely natural behaviour at very close quarters.

The bears wait patiently with all but their heads submerged in the cold white effervescence of the deep pool. They catch salmon by pinning them against their chests and then carrying them in their jaws to the shallows, where they hold them fast in their long, robust claws and strip first the skin and then the bright-orange flesh. Glaucous gulls wheel and fuss about their feeding, but the bears discard very little, even neatly biting away the skull of the salmon above the eyes where the rich brain and sweet cheek flesh are situated. At 6 pm it starts to rain and the crowd on the platform thins until, by 7:30 pm, we are only four. The rain grows heavier and eventually we abandon our uncomfortable vigil in favour of sleep, with a dawn start in the morning.

16 JULY

Alone at the Falls Platform in a rainless dawn, under low cloud, we are treated to a young bear's frenetic plunging about the river below the falls as he dives repeatedly after salmon he can see but not catch. It is the first time we have seen a bear snorkelling, where he walks along with his muzzle and eyes below the surface, only his ears showing above. We can see the fast dash of salmon out of his path by the quick twist of his head as he watches them. At just before 10 am we hear the first floatplanes arriving with day visitors, and we return to camp for breakfast and sleep.

Part of the etiquette we have been taught in Alaska is to make sufficient noise while walking so as not to alarm a bear as you come upon it. This is emphasised by the park staff and our sleep is so interrupted by calls of 'Hey, Bear! Ho, Bear!' and 'Val-deri, Val-dera. Ha, ha, ha, ha, ha. Val-deri, Val-dera,' by returning campers that I wonder how the bears themselves manage, for all the bears I have encountered like to nap.

Returning to the falls in the afternoon, we recognise some of the bears gathered around as those we had seen the previous night, and I glean from a ranger who is accompanying a special-interest group that all the bears here are resident, with a distinct social hierarchy. I learn some of the names they have for the individuals: 747, a huge-girthed bear that is an expert in catching salmon in the deep water of the pool below the falls; Lunge, a bully of both bears and people and an out-and-out thief; and Scaredy Bear, who stands in quiet seclusion on the far side of the falls beneath a short rock face, where the water runs fast and shallow, snatching salmon that accelerate up the race with their backs exposed, and giving way with a look of apprehension to virtually every bear that enters the river from the steep path behind.

During the Brooks Lodge dinner hours, when the crowd on the platform is only a handful, 747, catching a fish in his favourite spot, is still turning toward shore when he is charged by Lunge, who has been hanging back in the shallows. Lunge stops just short of physical contact in a welter of spray, and the air is filled with the two combatants' loud and threatening growls. They stand head to head with their jaws agape, their threat rolling deep in their throats. 747 holds the fish to his chest. And then suddenly they are fighting. Standing over 2.5 metres tall, they rear out of the water and in a swift exchange lambast each other with broadside blows of their giant paws, each lunging with their jaws for the neck of the other. 747 is knocked off his feet and, as the fish falls to the water, Lunge breaks off his attack. After a few last cursory growls, he backs away and, snorkelling, searches for the dropped fish. He finds it in an eddy at the downstream end of the pool and, lifting his head, streaming with water, he carries the fish to the shore, where he eats it with his back to 747. We see two more challenges between these two contestants, of nearly a ton each, before 747 resorts to cunning and wades fast across the river with his catch to where he is given enough time to eat most of his prize before Lunge catches up.

18 JULY

Bears walking along the lakeshore with high, sun-streaked cloud behind them cut into coffee time, and we follow them to the river mouth. The camp and lodge are quiet and we are alone with the bears that, after swimming the channel, warm themselves in the sun. Vapour rises from their coats, that are made golden by the soft warm light. A sow with three very young cubs comes out of the thicket and leads her brood onto the spit. She is cautious and wary of other bears, and the cubs mimic her, standing and looking around. When she too lies down at the edge of the lake to bask in the rare spell of the sun, the cubs climb and loll on her back or play catch in the shallows. Beverly leaves me to cross via the bridge back to the campground side of the river in case the sow decides to swim across. It is a good decision, for at the arrival of another bear she takes to the water. Only one cub follows, however, and halfway across, the calling of the two cubs left on the shore turns the sow around and she swims back and retreats quickly though the shallows into the thicket.

I am keen to see the images Beverly had the opportunity to make of the swimming bear and its cub, but instead of beaming joy, I find my wife in tears: she has been subjected to a rude and bullying rebuke by a ranger. I have no idea of the reason, for at no time have Beverly or I breached the Katmai National Park 50-yard distance limit for a bear, or the 100-yard limit one is required to be from a sow with cubs. We walk out in search of bears to distract us; however, our pace is listless and our only sunny morning has been spoiled beyond repair. At midday, we fly out.

20 JULY

Under a heavy, rain-threatening sky we spend the first part of the day in Denali National Park, then continue on to Fairbanks, where we stay with Lon and Nora Kelly. Nora is more distressed by the rain than we are. She relies on July to provide her with the dose of sunshine and warm outdoor time in her well-tended garden that will succour her through the long winter. 'You cannot know how hard the winters are here,' she says wistfully, looking out onto the rain-wet porch.

Later into the evening, we get to talking of hunting and the usually reticent Nora tells us that most Alaskans will defend to the death their right to hunt. This serves to endorse what Beverly has aptly coined the gold-rush mentality of Alaska: the land is there for the taking and the people of Alaska are accustomed to its harvest. Any threat or suggested compromise to this bounty has almost without exception raised the ire of even the most mild-mannered individual with whom we have discussed it. Alaskans eat wild meat: moose, sheep, caribou, whale and salmon. In the ten years after the turn of the millennium, Alaska's population increased by 16.5%. That much of the land is wild is without question, but for the creatures that inhabit this place to remain as wild and healthy populations requires that the land remains so. Many Alaskans consider the wildness of the land to be beyond taint, to be too inhospitable for humankind to exercise any permanent or consequential change. I wonder that the cowboy who rode west to hunt gold and bison might have argued the same, because in sleeping under the stars or a sheet of canvas he could not imagine highways, cars, aeroplanes or refrigeration. It has historically fallen to visionaries to protect what later seems so obvious. Climbing into bed late at the Kellys', I find myself wishing for such a visionary for Alaska.

24 JULY

Central is a small, unusually neat Athabaskan settlement of wooden cabins set back from the road between the trees. One can only wonder as to what incident lends the main road its name of 'Runamuck'. All the forest here has been burnt and the hills stand pink with fireweed between the crisscross of the black, thin, leaning thicket of dead trees. In the lone stand of unburnt trees beside the road, a bear scrambles for cover at our approach.

At Circle, another Athabaskan village, the road runs to its end on the muddy banks of the Yukon River. There are a few wooden buildings, a Russian Orthodox church and a handwritten sign that brags, '5000 Watts of power broadcasting to the State'. The small house with an unkempt yard beside it has used a faded 7-Up sign as a patch beside the front door. Beyond is a closely clustered group of small cabins.

It seems that families here call themselves 'outfits'. A happy term to my mind that smacks of unity and a blood bond that cannot be intruded on. There is the hand-painted, US-flag sign of the Carroll Outfit tacked below its entrance porch, and across the road on the river frontage, the Niemann Outfit has its small, stencilled sign directly over the front door. I recall seeing on the steel span of the bridge over Birch Creek 'Carroll Outfit Bitch', in an untidy scrawl. I wonder if she might reveal herself, but there are only children moving in the street.

25 JULY

Yesterday evening I noticed two women in an old-model Ford 4x4 pick-up who spent an unusually long time around the public facility at Birch Creek. This morning I glance up into the exposed rafters and see a small bottle secreted there. It has a press cap with what seems a wick, and when I draw it open my nose is stung by the vapour of almost pure ether. I have discovered someone's stash, and I close it and put it back, standing on the toilet seat to reach the rafters. During our time at Brooks Falls, I had spoken with a young woman who had spent much time photographing Native communities in North America where, she said, 'huffing' was common. 'Huffing' is the concentrated inhalation of fumes from any source that has a volatile, high-proof alcohol base and produces a quick high, with some dangerous side effects. Some isolated villages have declared themselves 'dry' towns in an attempt to curb the excesses that the long months of winter seem to exacerbate. In these places, the desperate will turn to any source – Aqua Velva cologne is the most common.

26 JULY

The road north to Deadhorse runs straight up and down the hills beside the Trans-Alaska Pipeline and is populated mostly with large articulated trucks going at speed. The official name is the Dalton Highway, but everyone refers to it as the Haul Road. It runs alternately dirt and tar with no clear reason for the changes in surface. The truckers have their own names for particular sections, such as 'Beaver Slide' and 'Oh Shit! Corner'.

At Finger Mountain, the dense spruce forests suddenly end and an open, rolling country falls upon the eye. It is a pleasing view of wide valleys that hold a scatter of ponds, some the size of small lakes, their blue in clear contrast to the lime-green of the tundra bogs. I take off my shirt and walk bare-chested into the close hills where even near the road all sign of human presence is soon left behind. It feels strange to know that it was near here on 28 January of this year that a temperature of -62°c, equalling the coldest temperature ever recorded for the whole of the United States, was recorded at the Jim River Maintenance Station.

At Coldfoot we visit Dirk Nickisch and Danielle Tirrell, who own and operate Coyote Air. Dirk flies and Danielle, a certified aircraft mechanic, tends the family and the aircraft on the ground. They are slaughtering their chickens in preparation for winter, and drink beer while The Eagles play from a portable CD player on a camp chair in the vegetable patch. The children herd the chickens, placing them head down in snouts below which buckets spattered with blood collect the spill of their beheading. We are hoping yet to find caribou, but although Dirk has not seen many big concentrations, he promises to bear us in mind if he does. We will have to call him on the satellite phone from time to time to check, as there is no mobile phone network this far north.

A Coldfoot local tells us of a pack of wolves that has been frequenting the gravel bars of the Middle Fork Koyukuk River, just south of Wiseman. Stopping there in the evening to listen, we are almost immediately rewarded with the first wolf howl we have ever heard. After dinner, we walk downriver, and although the banks hold plenty of evidence of their presence, we do not see any animals. We camp just off the airstrip. As the night that remains as bright as a late afternoon grows quiet, the wolves call into the stillness. By midnight they have moved nearer and we are rewarded at 12:45 am by two wolves appearing on the fringe of the trees beside the airstrip. They are wary of the unfamiliar presence of the Land Rover and circle circumspectly through the trees as they move past, emerging periodically to examine our presence. They call several times an hour. Around 4 am, when they are very close again but unseen, I can make out four distinct voices in their chorus.

27 JULY

Shannon Kieselbach lets us set up on her land in preparation for an all-night wait for the wolves. We rig the Land Rover as a blind, with the side windows open and the cameras standing out beneath the drape of olive-green curtains that will hide our presence. We darken the entire vehicle so that no movement can be seen within, for even at midnight it will still be light enough to see. Our plan is for Beverly and I to take shifts to watch for the wolves to arrive and wake the other when they do. However, at 10:15 pm, while leaving the vehicle to relieve myself, I come face to face with a wolf not 20 metres away. Both of us pause in mid-stride, and for three or four seconds I stand with my foot still coming off the ground as it studies me before turning back into the trees. At midnight, we briefly see two more that we think are pups on the edge of the bush, but they are far off and difficult to see in the gloaming. We must wait another two hours and forty minutes before we see them again. It is two adults that finally come trotting down the track in full view. The female is pale, almost white, and is quickly gone as she veers off onto a path towards where I have been hearing the pups, about 25 metres from where we watch. The male is perhaps a quarter larger and is a big animal, standing at least as high as my thigh. His coat is a grizzled, grey-black, turning to white on his belly, and the ruff about his neck flows to his easy, light-footed trot. He pauses to sniff the side of the track, and when he stops to scent-mark the bush, where their path branches from the track, he is looking straight into my lens. The eyes are yellow and offset by kohl-like markings of the fur around them. I am drawn by their intensity. The eyes are narrow-set at the base of the long snout, behind which the head flares wedge-shaped to the listening, alert ears. The wolf vigorously scratches the ground after marking, as all dogs do, and then turns to sniff his work, pausing once more to look back up the track before following down the path towards where the female has gone. We hear their greeting yipped and whined on the still night, then one raises its voice to a howl and all follow. The call is high-pitched and drawn out, winding down to a whine, before, with a few yelps, it begins again. The chorus fills the valley and I count five voices, when suddenly they are answered by another pack and the night brims with haunted, wild sound.

28 JULY

It is early evening when we arrive at the rise to the Chandalar Shelf. At its base stands the most northerly black spruce tree, a landmark final soldier of the boreal forest. It is two hundred and thirty years old, of neither great height nor of substantial girth, but what is beyond belief is the fact that it has been killed by a vandal. It has been ring-barked, and in its sad, brown skeleton rising out of a thicket of low willow I have a sense of foreboding about the future of human-kind. A conviction that in this senseless act lies the prophecy of our final human destiny, which will not fall to any intellectual authority or cognisance, but to stunted intellect in which brutality and selfishness supersede all regard.

29 JULY

The Atigun Pass is as lovely as its name. Even the silver millipede of the Trans-Alaska Pipeline, on its raised steel stilts, cannot detract from the spectacle of the land. It compels one ever northward with its vast vista, through a fall of wide green valleys, where slow rivers meander like fallen thread, and sharp-edged mountains, where drifting white cloud adds crisp definition to high black ridges. On the first plains of the North Slope, where we have stopped to photograph long-tailed jaegers in their swooping, angle-winged flight, Beverly spots a herd of nearly a thousand caribou far off the road. They are moving steadily west, and we drive back towards the hills, overlooking a small river, where we imagine we could intercept them. We pack our cameras into backpacks and provision ourselves with energy bars, water and a precautionary can of bear spray before setting out into the hills.

Tundra is an entirely uneven, humped and wet surface, and the going is awkward and hard. The ground is for the most part not ground at all, but a sometimes metres-thick mat of tight vegetation that gives way anywhere between seven and twenty centimetres to one's step. It grows in tightly knotted tussocks that fall sideways beneath one's tread. It is the worst walking terrain we have ever attempted. The only places where it is somewhat hard and level is where it is wet, and we follow these edges where we can.

It takes us three hours to walk the seven or eight kilometres into the hills that will bring us to an intersection with the caribou herd. We are breathless with effort and every joint in my legs is sore. When we are a kilometre short of where I think we should position ourselves to photograph the herd crossing the river, the caribou unexpectedly change direction and turn south towards the mountains directly away from us. A straggler, passing not 30 metres away from where I stand slouched against a rock in dismay, trots so fast and easily over the wicked terrain that in six or seven minutes it is crossing the ridge to the south over which the herd has disappeared. It would take us at least forty-five minutes to reach the same place. Defeated, we eat our lunch in silence.

30 JULY

Snowy owls hunt beside the road, their rounded wings carrying them in a light drifting flight low over the tundra. Occasionally, they pause in a brief hover before plunging down. Walking out to photograph the owls, we find the mosquitoes prolific, and along the river's edge, black flies, known locally as white socks, add to the buzzing persistence.

Beside the Franklin Bluffs that rise out of the now entirely flat tableau of the country, we find musk oxen on the gravel bars of the Sagavanirktok River. The afternoon grows cloudy, and pretty patches of light drift across the bluffs as we wade across shallow bars to get close enough to work. Ancient beasts, their long coats hang in a straight-edged skirt just above their hooves and, caught in the wind, flow backward in liquid movement. The humped rise of the shoulder of the bulls is covered with a coarser mat of hair, blonder than that of the body. Their low-slung, heavy-bossed horns, with only the tips curving away from the head, make the animal seem reticent despite its size, as if retreated into some resignation. Perhaps, like that of elephants, a cautious disposition gained from the wisdom of prehistoric aeons. There is the occasional challenge between the bulls, but it must be too early for the rut as the clashes are quickly over.

To the southwest the sky is burgeoning to a livid storm and we are forced to retreat from the gravel bars as dark curtains of rain approach and the river begins to rise.

31 JULY

In patchy fog, we arrive in Deadhorse and go in search of supplies. However, the Prudhoe Bay Supply Company has nothing that we seek – no milk in any form, nor any beer or eggs. Deadhorse itself is barely a town; it is a staging point at the end of the road for the oil fields in the restricted area beyond and lacks any sense of permanence or attachment.

I am fascinated by the bizarre and unusual technology of oil exploration and extraction visible beside the roads. Each drilling company has laid a gravel plat-form onto the wet bog of the tundra to accommodate its operations. Some are a few hundred metres square, others, where tall drilling platforms stand in wait or for repair, much larger. There are peculiar, bright-orange vehicles, half the size of locomotive engines, with wide rubber caterpillar tracks, crowded in rows three deep and forty wide. Long, six-wheeled buggies with soft balloon tyres wider than my outstretched reach and taller than my waist. A single, long-snouted trench digger with a sharp-toothed proboscis like that of a swordfish, and two sets of caterpillar tracks on each side. Heavy machinery, piping, and articulated trucks are everywhere, and on the wet dirt roads, giant yellow earth-moving trucks, with wheels nearly as large as our Land Rover, rumble slowly carrying gravel for yet more platforms or to use in repairing those that are sinking.

We stop and climb down to photograph three caribou bulls grazing in front of a tall blue drilling platform. When we return to the Land Rover a kind-faced, middle-aged woman with a wind-burnt complexion is waiting beside the vehicle. She uses a false, high-pitched sing-song voice, the kind that grandmothers use when they want to reprimand a young child without being harsh, to tell me that it is against regulations to walk on the tundra. 'It's called disturbing the tundra,' she says. I turn and look back at the narrow strip of green on which the caribou graze. It is perhaps 60 metres wide and is surrounded on all sides by roads, gravel platforms, trucks, drilling rigs, square box buildings and a small army of pick-up trucks plying in all directions over what was once a pristine landscape.

1 AUGUST

Beside the abandoned remnants of a small fire, I collect one hundred and seventeen spent .45 auto brass cartridge casings and three handfuls of 12-bore shotgun shells, one of which is still live. The shinbones and hooves of two caribou legs lie beside the river. To the north and south I can see the winking lights of drilling platforms, before the wind rises, bringing with it a fog that closes the world out.

2 AUGUST

At Happy Valley, a helicopter pilot comes over to chat while we fill up with drinking water. He is here flying an archaeological team out to survey the proposed route of a new road out to the west, to Nome. 'More roads, more area to hunt,' he laments, the first Alaskan I have heard express such sentiments. 'Look at the airstrip,' he gestures with his chin, where a few RVs and some tents are arranged along the edge. 'Twenty-five hunters. I don't even walk there any more. I used to walk there in the evenings. Next week there'll be a hundred and fifty.' Next week is when the caribou are expected to begin crossing the highway as part of the annual cycle of movement.

At many of the turnouts along the highway, hunters are already camped out in anticipation of their arrival. They range from extremely basic operations out of the backs of station wagons to highly sophisticated outfits with luxury RVs, freezer trailers, big generator units and large-wheeled aircraft drawn up on old highway work depots. On a long, shallow pond right beside the highway, a float-plane is tied up. In 2010, some twenty-two thousand caribou were shot in Alaska. Because of the easy accessibility, it would make sense that many of those were shot along this highway. There is a five-mile-wide restriction to 'bow hunting only' from the highway, but hunters may travel on foot, boat or aircraft beyond that limit to hunt the caribou with a rifle. In this area, Alaskans are permitted to take a staggering five caribou per day, per hunter. All other hunters are restricted to a total of five caribou per season.

Crossing the Toolik River, in almost exactly the same place we had seen the herd a few days before, we find a small group of caribou trotting parallel with the highway up the far valley through the grey sheen of the rain. At the top of the hill, there are more. We stop a short way down the road, beside the Trans-Alaska Pipeline, and make camp on a small promontory overlooking the Kuparuk River Valley. It is a pleasing scene across the meander of the small river, made complete by our longed-for view of caribou crossing the hills and spread ant-like across the valley floor.

3 AUGUST

The caribou are crossing the valley below us in a constant stream of varying numbers, but the intermittent heavy rain keeps us in our camp. We read the 2012–2013 Alaska Hunting Regulations, put out by the Alaska Department of Fish and Game, pausing periodically to glass the valley when the rain stops for a time. For the purpose of hunting, the State of Alaska has been subdivided into 26 units, each with its own set of conditions regarding big-game hunting. In Alaska, the term 'big game' includes brown or grizzly bears, black bears, bison, caribou, Dall's sheep, deer, elk, moose, mountain goat, musk ox, wolf and wolverine. I read that, in Unit 24, Koyukuk, and Unit 26, Arctic Slope, which between them cover much of the Dalton Highway to Deadhorse, aside from caribou a hunter may take one grizzly bear, one moose, three black bears and ten wolves. Coyotes are also on the list but are not considered big game. I have yet to see a coyote in Alaska, but for hunters along the northern parts of the Dalton Highway, there is no limit to the number of coyote they may kill and there is no season closed to coyote hunting. While they are out there, hunters are not restricted to those predators either; they may in the course of their sojourn in the wild bring to an end the lives of two Arctic fox, ten red fox and two lynx. One other unexpected fact strikes us: in the Arctic Slope Unit, an Alaskan hunter may kill as many snowy owls as they see fit at any time of year. Who would kill an owl, and what for? We can only imagine that their feathers may have been used as decorative clothing items in the past.

In the night when I wake, a three-quarter moon is showing through patchy cloud and I realise that I have not seen the moon in weeks and do not know whether it is waxing or waning.

4 AUGUST

Beyond the reach of most of the vehicle-bound hunters along the highway, we hike far upriver. On a steep hillside, looking down on the river, we set ourselves up to wait for the caribou in the slight depression at the base of a small pingo, or palsa, which is a mound forced above ground level by a core of ice beneath the tundra. We do not have to wait long. The first group runs across the valley towards us, between the small tussock-fringed ponds on the river floodplain, and plunges across the river less than a hundred metres from our set. Climbing out on our steep hillside, they pause to raise their heads with what I now recognise as a characteristic wide roll of the eye as they try to catch our scent. We can hear the distinct click-click of their tendons as they walk. The group continues past us without alarm and some of the inquisitive even approach us more closely to stand briefly skylined, against the deep steel-blue and silver of the clouds.

For the whole day, they come out of the east, cresting the far ridge of the river valley under heavy cloud for almost as far as we can see. They pass over the land in spates as irregular as the rain. Some groups comprise just a few cows and their calves, others are hundreds strong. While some pause to graze in a leisurely diagonal into the valley, others come in urgency, running to cover, in ten minutes or less, what would be a forty-minute walk for us. There is little time to eat, but we could not be more content in the middle of this life pulse of the land. The coming of the caribou is a spark that ignites the country. Like a taut telephone line over which the wind starts to blow, it produces a dimension that is not an exaggeration of its physical composition but an entirely different attribute; it begins to hum. Finally, after a particularly heavy downpour in the evening, their numbers begin to lessen. An hour later, cold, wet and hungry, we retreat as another squall crawls dark and threatening towards us.

Driving the short way to our chosen vantage point to camp, we notice a hunter running across the downstream hilltop. Far below him we spot a bull caribou limping heavily as it enters the river's edge. As it emerges onto the opposite hill I can see through the binoculars the hunter's arrow protruding from the left shoulder of the bull. Around the wound its fur is haloed dark with blood, and the caribou is labouring so heavily up the steep, uneven ground that I can see its tongue in silhouette in its open mouth. The hunter's partner is in a vehicle on the highway. Their luck is in when, half a kilometre further on, the wounded bull turns back towards the road. The partner positions the vehicle in the animal's path and, standing on the obscured side, waits for the bull, which to my amazement is still running. It passes close by the back of the parked vehicle, and through binoculars I catch the glint of light on the arrow the partner shoots, which lands well behind the bull. For another ten minutes the animal runs across and downhill until, utterly exhausted, it stumbles twice, heavily, as it crosses a stream and, limping out at the far side, stands still for the first time. It walks a few paces and lies down, 50 metres from the pipeline.

On the road below, the hunter's partner is standing watching with binoculars. As the caribou bull gives in to exhaustion, I am troubled that he does not take the opportunity to close, but instead stays watching. After a few minutes, the bull rises again and hobbles in a jarring hop beneath the pipeline and onto the hill beyond, where 50 metres further it lies down again, and after a brief look behind for pursuit, lays its head outstretched on the ground. The hunters have lost sight of it from below, and, coming across to our side of the valley, I point it out to them. Again, I am stunned by their inaction as they content themselves with watching it with binoculars and a spotting scope. I must assume that they are waiting for it to bleed to death, for three hours after we first saw the wounded bull the hunters have still made no move to close with their quarry.

Beverly and I can find little to say in the face of such suffering. After dinner, I find that the hunters have retired to sleep in the back of their pick-up. Forty-five minutes later the wounded animal stands and, stopping every four or so paces, makes its way in a stiff awkward hop up the valley. We lie in bed awake and sore right through.

5 AUGUST

There is frost and snow on the peaks of the Brooks Range, of which we have a clear view, and a cold wind is blowing out of the south, bringing a chill to the plains. It makes my eyes water as we hike out to our post above the river. It is apparent as we walk that the caribou numbers are down – just a few small groups scattered over the entire reach of the valley. By midday, when only two lots of a few females and calves have crossed below us, we pack up and head back.

The blueberries are out in profusion and we stop to collect a small bag of them from the low-growing bushes. The leaves here are already starting to turn and their flare of red is bright between the green. I notice a similar but blacker, shinier berry on a spiky-leafed plant. Later, when we pause to photograph a covey of rock ptarmigan, I notice that this berry is their preference. These are crowberries, and I have noticed the long-tailed jaegers eating berries too, which seems out of the ordinary for a predatory bird. Their preference is for cloudberries, which are like small raspberries held upward. Perhaps it is that they look like a small knot of salmon eggs, for the birds prefer the pinky-orange to the red. Flights of whimbrels are calling as they weave in erratic, swerving flights along the valley before settling to feed where surface water gathers in small pools and seeps in the tundra bog. They seem on the cusp of their southern migration.

6 AUGUST

We return to Fairbanks, the road once more passing through boreal forest, which limits any view of the country except from the heights or where fire has laid it waste. Fireweed is bright on the burnt hillsides, and a bush with a pale yellow-green leaf adds contrast to its vivid pink palette. At a section of road under construction outside Coldfoot we are led through the works by a pilot car. At a pause while we wait for heavy vehicles to pass, the driver of the pilot car tosses her cigarette out the window onto the road. On impulse, I climb out, pick it up, and give it back to her with a word about littering in so pristine a place. She looks at me in surprise and then tosses the wet butt back out the window.

The Alaskans I have known from past visits were passionate outdoorsmen with a deep-seated love and concern for the wildness of the land. I worry for the new. They are as passionate about this vast, undeveloped state, but the new ethic is utilisation, and some seem blinded to the fragility of the frontier that they enjoy. History has shown us that the rampant exploitation of a wild frontier country eventually leads not only to its impoverishment but also, ultimately, to changes in the essential character of the place, so that what was once is changed forever. Gone. Alaska, I feel, is ignoring history.

II AUGUST

A 5 am check-in for the ferry to Gustavus leaves us blowsy and red-eyed with lack of sleep. The freezing wind keeps us to the warm cocoon of the ferry's forward observation lounge. The solitude of the fjord landscape, cast in faded blue and grey by the rain, is drowned out by the thrum of the vessel's engines. At Gustavus we disembark in the rain onto a flat land sandwiched between the sea and a close damp sky. The tide is out and sand flats are fringed with a wheat-like sea grass. A sea otter paddles slowly in the shallows.

When I ask the portly Gustavus ferry official what the weather prediction is, he holds out his cupped hand for a few seconds, then looks up, grinning, 'No, not raining yet.' The drizzle stands in droplets on his short grey hair and beard. An older man who walks with a stick approaches across the pier. 'Ask him,' says the ferry official. 'He's the local oracle.'

'What?' asks the man, pausing to turn towards us.

'Fella wants to know what the weather's gonna be for the week.'

'Look up,' says the man, 'look up,' and walks on.

'I read the weather report for today,' says the jovial official, 'but it can't be right 'cause it has us at 3 000 feet. I know a couple of Gustavus residents get quite high but I don't think the weather report takes account of that. Seems most likely they got it wrong.'

We make our camp down a track towards a tidal estuary where small craft make their way cautiously to the sea and a bald eagle perches atop a tall channel marker pole. Our view is out across Icy Strait where Point Adolphus shows occasionally as a dark smudge through the obscuring rain.

On the ferry jetty, we chat with Todd Sebens, who runs twice-daily whale-watching trips to Point Adolphus. His engine hatches are open and his hands greasy to above his wrists, but his genuine lack of agitation at our intrusion and his easy manner and willingness to sit and discuss our needs disarms us. Unfortunately, his prices are beyond our budget, however, there is an undertone of sincerity here that makes us want to extend our stay.

It is with Peggy McDonald and Dennis Montgomery, who own South Passage Outfitters, that we finally find a boat to suit our needs and our pocket. They own a small lodge in Gull Cove on the other side of Icy Strait, southwest of Gustavus. We make arrangements for Dennis to collect us at the ferry dock the next morning.

12 AUGUST

Rain. I am fed up with the weather and missing home. Missing being able to swim in the sea, the company of my friends, comfortable conversations with family and the feel of my dog's head in my hands. The long spells of heavy rain spoil any plans we have to be on the water with the whales, but Dennis suggests that we fetch the skiff regardless and bring it back to Gustavus, which is a shorter run in the boat each day to Point Adolphus than from Gull Cove.

On the far side of the strait, the cloud lifts, giving pause to the rain, and I feel my spirits lighten and then quicken as three whale spouts rise from the water close to the steep and shadowed land. We see many more on the hour-or-so run to Gull Cove, which appears around a low headland. Tendrils of mist hang like silver scarves draped over the trees, and a large motor yacht stands bright-white against the dark forest, where greying wooden cabins stand spaced between the trees. The dock is a mismatch of listing wooden gangplanks, and is littered with oddments of fishing equipment, tools, pipes and other paraphernalia. We have arrived with the groceries and clients for a full camp, but neither Dennis nor Peggy seem moved to any urgency. They stand and chat while we find our way about the kitchen making tea.

Our boat is beside the dock but the motors do not want to start. Two hours later, with unsuccessful spare parts scattered on the deck and one tool glinting at us from the shallow bottom beneath the boat, Dennis utters a heartfelt curse and retires for a Jameson-fuelled reflective pause. But the motor stubbornly refuses to respond to our attentions. As evening approaches and the margin of time for a safe return journey to Gustavus narrows, Dennis abandons the work and gives us his own personal fishing skiff to use until he can sort out the problem.

There is a flat sea that reflects a lightless grey sky, and I drive close beside the coast so I can familiarise myself with its landmarks. Off Point Adolphus we come upon a large pod of humpback whales feeding in the chop caused by the tidal ebb. We stop to photograph them, and in the silence we are suddenly aware of whales all around us. We can hear the puff and suck of their breath, and when one makes a drawn-out whistle as it dives, I smile to its unusual overture. They move beyond us. When I turn the ignition to carry on I am greeted with a flat click and the smell of burning from the battery wire. I glance towards the shore: the outgoing tide is carrying us backwards at four or five knots, the speed of a slow river. I bite my lip, twist the battery terminals until they give a fraction and try again. The motor starts and I let out my pent-up breath and turn out across Icy Strait towards Gustavus and our camp on the beach.

13 AUGUST

When I peer out from behind our curtains, dawn is bleak and grey, but an hour later, when I clamber stiffly out of the vehicle, I can see a patch of clear sky and sunshine low on the horizon to the west, over the Inian Islands. Slowly the sky clears, revealing a land hemmed by ranks of mountains that recede to blue. Snow lies cold in their hollows, and driving the skiff out into the strait we can feel its hold on the air, despite the welcome sun.

Several pods of whales are concentrated around Point Adolphus, and we motor against the tide and switch off at the junction of currents that ripple a small chop with the pace of their flow. The whales feed past us and we drift towards another group that is feeding towards the land. Only when the tide has carried us well beyond the point do their numbers thin and we start the motor and head to our starting point, to begin again.

Even as a relatively small member of the whale family, humpbacks are huge, and a group at close quarters is a sight to quicken the beating of one's heart. The pods here feed synchronously, and at the sight of seven or eight abreast, swimming directly towards our small skiff, I feel the flutter of hesitation in my chest even though I trust their benign demeanour. They move without haste and swim so close side by side that I am sure they must touch, although I cannot see beneath the reflection of the calm sea. From time to time one turns on its side beside another and briefly lifts its elongated fin into the air. The trailing edge is serrated by a strange series of small bumps, like the stumps of severed fingers on a hand. There are bumps too along the line of the jaw, which is a smooth, lustred black beneath the snout-like blowhole. The dorsal fin, raised on a hump in the middle of the back, is small, more akin to that of a dolphin or porpoise. The finely curved tail is broad, wide and thick and notched in the middle of its rear edge. I love the view of the tails raised above the ocean against the dark, shadowed land, trailing runnels of water that shine silver and bright in the sunlight. When they breathe great plumes of fine spray into the air through the mirror stillness of the sea, it feels as if I am witness each time to some creation or birth, and it elicits in me a sense of discovery, as if I am the first to come upon life in a still place.

However, it is their voices that charm me. They have a wide repertoire of drawn-out sounds that can only be likened to song. There are whistles and squeaks, and deep moaning groans, and a rolling guttural boom, but most impressive of all is a trumpet so loud that I wonder whether they too can hear its echo off the hills. They sound like elephants, like several herds coming together through the trees, and it awakes in me a nostalgia powerful enough to raise my longing until it comes as a lump in my throat. They are the elephants of the sea. As big, as gregarious, as long-lived and caring of their young, with an identical voice. And I am left wondering again why it is that these greatest of beasts seem so possessed of a wisdom, and why, so unlike humans, their unopposable size and great strength make them gentle and benign.

Later, in the afternoon, when Dennis arrives with our boat, having found the faulty part and repaired the motor, we are rafted together transferring the gear when a pod surfaces close behind us. Turning towards us, they swim leisurely but directly beneath the boats, diving below only a few metres before they would have made contact and so close that their plumes douse us. It pleases me to see that even after fourteen years of living on Icy Strait, Dennis is as moved by their proximity as we are. He bangs on the boat and shouts, 'Thank you! Thank you!' at the now-empty surface of the sea.

By late afternoon all the whale-watching boats that flood the area in waves are finally gone and we are left absolutely alone upon the sea. I turn fully around. The sun is dropping to the sharp, black outline of the mountains to the west, the ranges before it etched as silhouettes of gold and blue. We stand in a small boat upon a sea as calm as glass that sheens the smoky opal of the sky. The hills are as silent as the forest that shrouds the land and I can hear the whales breathing around me. It makes me want to sing out. Not words, just sound, and to lift my arms out wide to the sky. In the middle of the strait as the day ends, two whales approach us where we have stopped, and swim beneath the boat. They surface beyond us and, turning in a wide circle, approach us again. One swims directly at the side of the boat, only submerging when it is 10 or 15 metres away, the other comes obliquely past behind us. It is fascinating behaviour, and it is only when they have repeated it twice more that I realise they are feeding on a school of baitfish seeking shelter beneath the boat. Their blows catch the colours of the sunset, but their breath has the fetid reek of fish. It is the closing darkness that drives us finally to the shore. There are stars in the sky, the first stars we have seen since we came to Alaska.

14 AUGUST

A thin fog, turned pink by an early sun, burns off quickly to reveal Icy Strait as shiny calm as mercury. It seems every boat in the small, exposed anchorage of Gustavus is headed to sea, and at Point Adolphus the whale-watching boats are more numerous than the whale pods in the slack tide. I wonder if it is this pressure or some other reason that has dispersed the whales, for their numbers are only a fraction of what we saw yesterday.

It is only in the late afternoon, when we have drifted in our waiting, well beyond Point Adolphus towards the Inian Islands and the open sea, that the whales come about us once more. Todd Sebens comes to drift close by, and for an hour whales mill so close about our boats that I can see them a few metres below the skiff. Their calls fill the still air, and when one particularly large whale rests still on the surface, only 20 metres from where I watch, I can hear its soft low rumble that among elephants would be a reassuring contact call. In the evening, the sky darkens to the west and fog comes pouring over the Inian Islands, bringing a chill wind and a chop to the water. Unsure of how quickly the fog will descend on the strait, I abandon our last forty minutes of rare sunset light and head towards the dock. The fog flows behind, coming down in thin tendrils out of the mountains onto the sea like the searching flicker of a serpent tongue.

15 AUGUST

A dense fog driven by a strong westerly wind makes the dawn clammy with damp. I cannot see as far as the ferry jetty, only the close shore slapped by the wavelets of the choppy sea. A black raven flies silently through our curtailed world and seems a harbinger of something that does not come.

After a few hours, as the fog lifts, we make the trip across to Point Adolphus, which is easier than I had expected, but on the far side of the strait the wind is still blowing, and we anchor close by the shore in the lee of a small bay. The sea is choppy. The ebbing tide running against the wind has caused short, deep-troughed waves to form, and although the whales are only 400 metres off the beach, where the water is a little calmer, it is still difficult to hold a camera still enough to photograph. We realise now how lucky we were to have had the first few days of good weather. For, with a rough sea, the scene is not the same, the whales difficult to distinguish against the broken surface. When the fog appears again on the western horizon we pack up and head out towards it, making for Gull Cove. The wind soon picks up, and for a time we both wince as our skiff slams down on the back of the waves, sending up a spray that the wind drives back onto us. Soon the chop becomes too vicious to motor at speed, and I am reduced to manhandling the boat at dead slow through the steep troughs. We must hold on to the sides to stop being thrown about. We make little progress, moving only as fast as the outgoing tide, and we have far to go. As the fog to the east builds, the wind increases. I consider turning back, but that is just as unattractive an option. Beverly has tied the camera packs to the boat and implores me, in earnest, to head closer to the shore where the water is calmer. For a while I batter straight into the chop. But it is both frightening and horrible, and although driving the shore will double our distance and be littered with rocks and reefs, it still seems a better option. I turn the boat south towards the land. In the lee of the shore, the water is flat and we are able once more to drive the skiff on a plane. It is assuring to have the shore as a measure of our progress and a close safety.

Twice we must head out to sea and away from the shelter of the land. The wind has risen in excess of 25 knots and I avoid, as best I can, the crumbling wave tops that rush white onto our low bow. On the final point, with South Passage Lodge in sight, the sun breaks through the cracked clouds and we stop to photograph the drama of the fog-shrouded hills and sea. The outgoing tide is running like a river and I let the motor idle with no desire to be carried back to what we have just come through. Later, back at the lodge, in the camaraderie of a stove-warmed room, with a hot halibut dinner, the hostility of the outdoors recedes to a distant view beyond the wide windows.

19 AUGUST

Taking the ferry to Haines, we return to the mainland, then cross the border into Canada. Searching for an access to Klukshu Creek, we pass right by the turnoff to Klukshu Village as it is not signposted and the road is nothing more than a nondescript opening in the trees. We have already missed the turnoff to Dalton Post because of its anonymity. Both are settlements of First Nations Canadians, and I wonder if it is anonymity they seek, because when we drive slowly through the small settlement of Klukshu, our tyres crunching loud on the gravel, of the few people on the road not one acknowledges our presence.

It is a tiny cluster of small, attractive wooden cabins that sit squat against the land, as if they are hunkered down, sheltering against the elements. At the turn-around at the end of the village, we find that the road does come to the banks of Klukshu Creek, and I step hard on the brakes when I notice the scarlet-red backs of sockeye salmon in the shallows. We spend the day captivated by these bright fish that tussle and chase each other across the shallow gravel. The females flutter every now and then, on their sides, digging their spawning beds with their tails. The humpbacked males swim in close attendance and yawn their exaggerated hook-jaws, charging down and occasionally biting intruders that do not heed their warning. The river is little more than a stream, not twelve paces across, and so shallow in places that the fishes' backs show through the surface.

The life history of the salmon is well documented and so broadly known that is loses the edge of its incredible process. To squat on one's haunches, however, on the bank of a stream with a fish with a yellow eye in a green head and a fire-engine-red body not two metres away, and to realise as you watch the lazy way it holds against the current that this creature has swum more than 200 kilometres upstream and climbed more than 600 metres through the mountains, is like watching the final lap of a great marathon runner who comes to breast the tape with no competition in sight. Here, however, there is no staggering, stumbling last steps, no cheering, clapping crowd brought to its feet, just a tranquil village where children skip stones on the water and a woman opens a cabin window to let the smell of fresh bannock eddy through the still day.

It is so reminiscent of the rural villages of Africa that we feel quite comfortable here. There is no rush, the greatest clamour two youths playing tag on their bicycles. People drift by. Some stop to chat, others simply nod a greeting as they join us, standing and watching the scarlet bodies through the smooth surface that reflects the green of the forest.

In the afternoon, a group of people saunter down to the stream and a youth takes a long pole with a large gaff fixed to its end and, standing on the river's edge, tries to snag the passing salmon. It is not easy. The fish are lithesome and the gaff heavy. His young companions taunt, 'So many fish and you can't catch one.' The young gaff handler moves slowly upstream until he is finally successful on the lower spawning beds, where the fish are less inclined to flee. The youth seems completely surprised by his success. 'I got one, I got one,' he repeats, lifting the heavy, flapping fish from the water, the hook high through its dorsal hump. His younger companion holds a plastic priest and swipes inaccurately at the wriggling head.

21 AUGUST

Wandering out from our camp on the shores of Mush Lake in the early morning, I find bear, moose and wolf tracks etched in the sand. The day is utterly windless, no ripple mars the reflection of the mountains with their gatherings of cloud, white and grey against the blue. A few trout rise here and there across the lake, and I listen for animals and scan the shore because everything feels closer in the stillness. There is an anticipation in the hush, as if the creatures there may be drawn out, like myself, into the open by the welcome warmth of the morning – free for a time from the wind and driving cold that makes life hunker down and hide away. And, even though nothing shows, there is a sense of them that makes the land seem fuller, that draws me to stand on the shore with my hands in my pockets and wait.

Three loons squabble far out on the lake but I can hear more than them. I can hear the land. It may seem strange if you have not heard it, when I say that a land's voice lies in its stillness, that silence has a language of its own. I could not hear it in the wind, but I hear it today, and the quiet water at my feet reflects it back at me, until the place is loud with the presence of the country: of wild, unpeopled space.

22 AUGUST

As we drive, the close, dull sky makes the landscape a monotony of lifeless colour and, robbed of the spectacle of its hills and mountains, it becomes a flittering repetition of trees. We listen to music and I fall into a faraway reverie. My thoughts turn to fathers. I am not sure why. Perhaps because Dennis Montgomery had said so curtly that he did not like his father, while Peggy, in another conversation, had shaken her fist at the ceiling when talking of her father. 'The old bugger,' she exclaimed. 'Sometimes I get so mad that he died and left me to deal with all this by myself. I miss him so. His glass of whiskey in one hand, cursing into the telephone with the other.' There is a photograph of him on their fridge doing just that. In the image, he is staring like a bull, ready to charge into the lens. He is wearing only his underwear.

I miss my father, too, far away in South Africa looking after our dog. We were not always friends. In my youth, I knew him as a distant if somewhat authoritarian figure, and in the years of my rebellion we were frequently at loggerheads, with me often disappointing him. It was only years later that each of us let go of our history and accepted one another for who we are. It was only then that I saw him finally for the man he was, for the man I like and now regard as a close friend. He is old now and I fear his dying while I am away.

We lost Beverly's father like that. We were in Bhutan. The face I saw in the casket when we flew home was, strangely, not the man I had known. Without life, there was a hold on his features that was missing, a vital ingredient that had moulded the clay to his form. He was an eccentric. Not the kind of eccentric we fondly call our friends, or vaguely wish upon ourselves, but the genuine sort, the kind that is different, without regard for convention, steered by an offbeat star. Although never father and son, I had cherished his friendship and unique individuality. I was the first of Beverly's suitors that he accepted, and in doing so he unwittingly charged me with a responsibility to meet his expectation. I still feel it today and, in trying to meet it, it has made me a better man than I would have been.

I am not a father, but in the small army of our godchildren and siblings' progeny I find in myself a fierce love and it makes me wonder what it might be that could ever drive this away. I can feel its root right through me, and no matter what storm and tempest has shaken its tree it has remained fast.

In Whitehorse, Jill Pangman and Bruce McLean welcome us into their home once again, and after my day's reverie I watch with special interest their friendship with their son.

24 AUGUST

The sun streams warm through the window, waking me in our garden yurt and beckoning me out into the day. By the time we are finally gone from Whitehorse, however, the clouds are back and rain threatens with thin, faint droplets that continue to grow larger.

As the road northward winds through a picturesque country of lakes set in wide treed valleys, between low hills, I wonder at the clear distinction of our having come into Yukon country. The people are soft-spoken here as if, like their cars, they perceive that less will achieve as much as more. Gone is the sprawl of smallholdings that spreads like fingers from the arteries of the roads. The land is wild right to the edge.

In the evening, we camp where the Tatchun Creek enters the Yukon River, by pure chance bumping into fellow wildlife photographer Brys Bonnal. I talk with him late into the night and he expresses an idea that I feel stands at the core of the defence of wilderness. He is talking of an argument he had with a woman who asked him: what is the point in saving the cheetah? With his subdued but brimming French passion, the steepled fingers of one hand held shaking before his face as he speaks, he says to me, 'Peter, it is not just the cheetah that we fight these blind people for. Behind the cheetah stands the panda, the rhinoceros, the owls, the squirrels and the mice, everything! It is all one, and these people cannot see that even they are included. For each thing that we kill for good, that we let slip away, we take one more chip out of the foundation of our own house. If we cannot stop now, one day we will kill ourselves.' I believe he is right.

The Yukon River runs almost silent in the gloaming, its fractured, dark surface fragmented with the silver-grey reflection of the sky. It is too perfect a place for man to leave alone. At some time in the future – fifty, one hundred or five hundred years from now – humankind will find its way here and desire it. Even if there is nothing of gain to be had in the water or the ground or the trees, it will start with a cabin just for the solitude and the view.

25 AUGUST

The road is hemmed in by forest, but along the Yukon, Pelly and Stewart rivers one finds breaks that in spite of the rain give a view over a wide, undulating country, through which determined and desperate men have waded and trudged in search of fortune. The Tintina Trench, whose tectonic upheavals caused gold to be brought to the surface and resulted in the Klondike Gold Rush, is a wide valley where fire has killed the trees and the second growth is not yet a metre tall.

I remember Lon Kelly's definition of where wilderness exists, here in the far north of the American continent. When I asked him, he paused with his head back. 'One day's walk beyond the ATV and skidoo tracks,' he said, looking at me again. In our flights over the country I had noted the narrow, innocuous-seeming paths – made by quad-bikes and snowmobiles – between the trees and across the tundra and that radiated out from the roads, the rivers and the isolated villages. Arctic Village, for example, is the centre of a network of expanding threads that reach more than 20 kilometres in every direction, like the strands of a spider-web.

The rain stops, and at the turn-off to the Dempster Highway, which will carry us 730 kilometres north to Inuvik, the sun comes out. I walk about outside with my face towards it, my eyes closed. As we drive into the Ogilvie Mountains, the drifts of sunlight highlight patches of neon-bright-yellow where the leaves of the aspen and willow are turning amid the green of the spruce, and the mountainsides are a patchwork of reds and orange close to the ground.

At Tombstone Territorial Park the road climbs beyond the tree line, and in the high valleys open to the view there is an extraordinary palette. In the crease of the hills, gathered at the base of the tall peaks, the shallow dip of each small valley or drainage runs a dark maroon-red, seemingly leaching to concentration the orange and rust colours of the hills. Scattered, single spruce trees divide its ebb with their colour, and along the valley floor the shiny silver braids of the wide riverbed run between splashes of yellow and dull gold. When touched by the isolated beams of sunlight they glow as if an electric switch has been flipped. A storm musters over the far Tombstone monolith and the drama of its advance down the valley has us riveted, the narrow beams of light shining silver in the dark grey curtains of the rain, the foreground a riot of saturated colour. The storm recedes before it reaches the road and we work until dark, our hands stiff with the cold.

26 AUGUST

With the dawn light from the east, the Tombstone valleys are presented in an entirely different guise. And although the morning lacks the drama of weather, the turning autumn colours of the alpine vegetation are saturated to richness by the butter-warm light. A covey of ptarmigan, their colour turning to the white of winter, feed beside the road. When I sit still and let them come to me, they gradually accept my presence until they feed less than two metres from my feet. I love the deep-alarm chirr that the male gives when an unseen bird passes high overhead, the twitch they give their tails when uncertain, and the strange behaviour of the male when he crouches low to the ground and sways his head with quick jerks from side to side, the reason for which I do not understand.

There is a moose in a pond far off the road, but it takes fright at our yet-distant approach and trots hastily over the hills. Tombstone Territorial Park is a multi-use park, a bland terminology that disguises the fact that hunting is permitted here. Ultimately, all that is actually protected here is the land.

28 AUGUST

The dawn is a grey swirl of mist and rain. I am tempted to roll over and go back to sleep, but I do not. Later, I will wish I had. The road is wet and slippery and we drive slowly beside the Richardson Mountains, where breaks in the mist reveal shallow valleys, vivid with the turning colours of the low bushes that hug the contour of the land. At the top of Wright Pass, where the Yukon gives way to the Northwest Territories, Beverly spots a group of caribou close beside the road, emerging like wraiths out of the thick mist. When I stop they trot quickly away, but they are the first animals we have seen and so I follow them for a time, coasting down the incline of the pass with the engine switched off. They seem to grow accepting of our reticent presence and after a time stop to watch us. When I stop they wait, then the lead bull walks off on a parallel path, about 150 metres off the road. We hang back, because each time I draw level with them they start to run.

They are six in all – three cows and three bulls – all with rich, new winter coats of charcoal-grey. Their coat runs to almost black on their bellies, and from their throat to their brisket hangs a mane of pure-white. The males have huge antlers that rise in a forward-facing arc to form wide, twin Cs above their heads. The antlers are clearly heavy, for the bulls carry their heads slung down and forward, moving them with a slow deliberation, as if the great branch of horn requires balancing and is a weight that their necks can only partially hold. Their gait reminds me of the rural women at home in Africa walking on narrow paths with pails of water steady on their heads, their hands at their sides. Only when they stop to watch us do they raise their heads above shoulder height.

The presence of the caribou charges the country: it makes the land crackle with the electricity of life. The calm is gone. They slow to cross a dike of loose, tumbled boulders covered in pale lichen. Patches of scarlet-red avens mark the ground, and far below, as the heights of the mountains ease to a yellow plain, a round lake reflects the silver sky. I find that their presence defines the land, bringing unswerving clarity to a borderless wilderness. They move fast across it, pausing to graze on a patch of lichen and then trotting on. We drift downhill. A camper pick-up stops and two German travellers climb down to photograph the small band, which moves off with the short white stub of their tails held erect. They are the first animals the Germans have seen on the whole road, and their eyes are bright with happiness.

A pick-up making its way uphill stops below us, but as I roll gently toward it the driver starts and drives away. Some 200 metres on, the caribou slow to contemplate a high ridge of loose stone before them, and I hang well back in the hope that it will turn them in the direction of the road. The pick-up reappears behind me, drawing up close to where I wait, but the driver does not switch off. After a time, he pulls out and, driving ahead, turns off the highway onto a track across the top of the ridge, directly into the path of the caribou. I follow, for the ridge is loose stone covered in pale cream lichen, with patches of low willow bushes turning orange and red in between, and would make a pleasing background against which to photograph the caribou. I am not yet where the pick-up stopped when the small group change their minds and head directly up the mountainside towards the road.

It is a setting fit for a caribou: the mountaintop draped in mist that hangs in tendrils like the trailings of a veil flowing down the cleft grooves of its gold-yellow sides. The valley below is bright with a silver evanescence, beneath the grey roof of the low cloud whose ceiling is below our height. On the road, I drive slowly closer and then stop as the caribou pause at the steep embankment of the highway. They are close and through the lens I can see the roll of the lead bull's eye as he lifts his head to look.

As soon as the caribou are across the road, the pick-up behind us pulls out to overtake us and drives up close. I pull up behind him and through my open window am filling the lens with the impressive head of the lead bull. The passenger door of the pick-up opens and a man kneels into a quick squat right beside the vehicle, no more than 10 metres from where I am photographing. He lifts a rifle to his shoulder. The explosion of its report crashes across the stillness. The caribou are running. At the second shot the lead bull stumbles, but he lifts himself and runs on. It is at the third shattering of sound that he falls, his legs kicking at the air. Quickly he stands again, but he cannot run. His head is held low, struggling for balance, his mouth wide and his legs splayed. One of the females tucks protectively close to his rump, turning frantically from one side to the other when the bull does not move. The man shoots again, and although I can hear the slap of the bullet, nothing changes. Again, the report of the rifle booms like a thunderbolt off the mountainside and the second bull in the group goes down hard, rolls to its feet, and stumbles on. As the sixth shot batters our senses, the bull falls. The man is laughing. I hear the bolt of the rifle slide back and metal crunch as another round is pushed into the breech. As he fires yet again, the last bull in the group falters in his run. It is not yet 80 metres away from the road. The rifle crashes once more, and as the last bull falls something in me finally gives way and my senses shut down. I register in some distant echo the rifle-man's laughter as I start the Land Rover and drive away. Beside me, Beverly is racked by silent gulping sobs. In less than one minute, half of our small group of caribou are dead or bleeding to death. Fifty per cent of all the life we have seen for hundreds of kilometres – obliterated.

For perhaps half an hour I drive in second gear slowly along the side of the highway. It is only when an overtaking articulated truck rocks the Land Rover that I realise I am dangerously unaware and pull off into a wide cutting above a river. After a long time, Beverly takes my hand, pulling me back from the unfo-cused whirl in which I eddy.

'What are you thinking?' she asks.

I am wondering at the way of the universe. How something seemingly so ugly can take something so blameless, so perfect, but I am too numb to speak. After a time, she shakes my hand that she is still holding tight.

'Say something,' she implores.

I turn to look at her red-rimmed eyes, tears running down her cheeks. 'There is no God,' I say. 'There is only us. And we are going to destroy it.'

29 AUGUST

We camp far to the west of where we were yesterday, and as we retrace our steps, climbing back up Wright Pass, mist lies low on a land with its palette of yellow, gold and red made warm by the sun's first light. At the saddle of the pass the cloud is close to the ground, but it clears shortly below. The sun is bright but the air remains bitterly cold. There is fresh snow on the distant peaks. As we pass the place of yesterday's slaughter, three small groups of ravens fussing about the tundra are the only sign of the killing ground. We chase the slow drifts of sunlight back and forth across the hills, as it spotlights the autumn colours of the leaves beneath the backdrop of the heights, but there is no joy in our work, and when the mountains recede behind us we are relieved.

Beyond the mountains the road approaches a short escarpment, the land ahead flat and endless. The Peel River runs wide in the foreground and Fort McPherson is a tiny closet of white between the unbroken trees. The road runs more east than north between the close spruce to the Mackenzie River. The ferry here makes a triangular circuit: from the road south it crosses the mouth of the Arctic Red River where it flows into the Mackenzie to the tiny settlement of Tsiigehtchic (pronounced 'shick-chick'), with its distinctive church perched high on the peninsula, before crossing the Mackenzie to link with the road to Inuvik.

The ferry is not fast and I chat with the engineer as we make our way across the broad current. I have noticed a condominium of swallows' mud nests under the superstructure of the bridge that straddles the ferry and I am curious to know if the birds nest while the ferry is stationary or if they follow it as it plies the river.

He tells me that they come every year in the spring and they build their nests and raise their chicks as the ferry goes about its business, back and forth across the river. The river here is perhaps 400 metres wide and I am fascinated that the birds have learnt to accommodate such a wide shift in their nest site. There is nothing like it that I have known before. I ask him, too, what happens with traffic in the winter months when ice must hamper the ferry's operation. He smiles at my lack of understanding. The river freezes. It is often colder than -40°c. By the second or third week of October the ferry can no longer operate, and as the ice becomes solid, snow is scraped into walls to make a dam the width of a road, and a few centimetres of water is pumped into it and allowed to freeze. By this increment an ice road is made across the river, which by early November can support the weight of cars and by early or mid-December will be thick enough to carry a fully laden articulated truck.

The road to Inuvik on the north bank passes through land as flat as a tabletop. Only the occasional small pond or lake breaks the mantle of the trees. Towards Inuvik, where the country finally rises to rocky hills, we camp in the Gwich'in Territorial Campground beside Campbell Lake.

As we walk beside the lake at dusk we are visited by an ermine, the first we have seen. It is a tiny creature of furious energy and is incorrigibly inquisitive. It approaches us with its head bobbing up and down, skipping from one rock to the next between fractional pauses. It comes close and I can see its small ears tight against it skull. At our movement it dashes away, peeps out from a rock and returns. It looks like a weasel but is no more than 20 centimetres long from the top of its brown head to the dark tip of its smooth tail. I could hold it in my hand. I know I should warn it against trusting men, that I should clap my hands at it and stomp after it into the bush, but I cannot, because in its quaint scampering antics, its bright curious eyes and its cocked head, it is making me feel whole again.

30 AUGUST

Inuvik is a town in the traditional sense. There is a school and shops and people in conversation on the street, and although the houses are all raised on stilts above the snow and unstable subsoil, they are homes, imbued with the taste and character of their owners and it makes the town quite characterful. There are no gardens to speak of. We are quickly through it and turn around.

On the ferry back across the Mackenzie River we meet Darius Elias and Joe Linklater, who are Gwich'in leaders from the town of Old Crow, far to the north and not accessible by road. They are some of the rare First Nations people of North America to show an interest in us, our vehicle or our work. We talk of caribou and wilderness and cold. Joe tells me of lying stretched out on his back on a frozen lake at home at 1 am watching the spectacle of the Northern Lights, his dog beside him. A shadow passes across his eyes when he talks about his golden retriever; he tells me that it has passed on. Darius talks of caribou seen on the plains of the Richardson Mountains. For a time, thousands of them, like cattle, spread out across the land. He uses the word 'breathtaking', and then purses his lips, 'awe-inspiring, actually'. Beverly touches on our experience with the hunter the day before yesterday and Darius reacts with, 'There should be no shooting from the road, or from a vehicle, we need to lobby to change that. At Old Crow there are no roads, we hunt properly.'

I have a thousand questions I would like to ask them, a hundred quandaries on which I would value their opinion, but the crossing is short and too soon we are shaking hands and saying goodbye. In the evening, we drive into the foot-hills of the Richardson Mountains. I look at them wistfully and try to imagine the caribou herds there. Some 200 metres before we are to pull off to camp for the night, a vehicle travelling fast in the opposite direction throws up a stone that smashes a hole in our windscreen. Beverly is driving and I am fortunate to be reading at the time, my head down to the book on freshwater fishes of the Yukon. My hair is filled with splinters of glass.

31 AUGUST

I wake to find the full moon setting over the mountains. I have seen the moon so seldom in the past months that I stand outside drinking my coffee, despite the chill of the early-morning twilight. Half an hour before dawn, however, the moon slips behind a high crust of cloud that grows across the sky like a veil, making the dawn flat grey and bitter cold. We head upward into the mountains made brooding by the sombre morning.

It is only towards mid-morning, when the cloud changes to patchy cumulus and sunlight runs across the land, that life seems to return. We find a bear feeding on berries in one of the creeks, swinging its head in a mowing action in thickets bright with the red and yellow of turning leaves. It moves between the widely separated berry patches at a fast amble, almost a run, as if it has an urgency to attend that the berries keep interrupting. Emerging from the almost-vertical climb up the bank of another creek, it pauses at the top with its nose to the wind and then takes off across the country at a run. It runs for more than two kilometres, and I am wondering at the source of the agitation that would cause it to expend so much precious energy, when ahead we see another bear. The second bear is feeding in one particular place and only notices the approach of the other when it is 300 metres away. It takes off at a gallop, running 200 metres before glancing behind. The first bear stops where the second had been feeding and I see it tear several mouthfuls from an animal skin before giving chase again. The bears circle away from the road towards the mountains and are disappearing from view when a third bear comes over the ridge, its head held high, searching the ground ahead. When it crosses the track of the first two it falls in on their path. We lose sight of the chase behind a distant low ridge.

We see no further life in the Richardson Range and, crossing the Arctic Circle, we drive south into the Ogilvie Mountains through squalls of rain. Shortly after we cross the Ogilvie River a fuel tanker, coming around a blind corner on the fringe of control, forces me to swing wide onto the outward-sloping verge where, when we are almost at a standstill, its second trailer throws up a fist-sized rock that smashes into our windscreen. For the second time in twenty-four hours we dust glass shards from ourselves and the front of the vehicle, washing out Beverly's eyes with water because she has a scratchy pain beneath her lids. We are hundreds of kilometres from any medical assistance, and although I examine her eyes as best I can, the small splinters of glass are difficult enough to see against the black of the dashboard. In her eyes, I can find nothing. Whitehorse, the closest place to repair the windscreen, is about 600 kilometres away, so we tape it up as best we can. Fortunately, the driver has an almost unobscured view.

As we emerge from the moonscape-valleys section of the road, and drive on towards Blackstone River, the sun comes out and we see a creature cross the road ahead. As we draw slowly level with the spot, a lynx is sitting in the sunlight watching us. It drops to a belly-low stalk into the shadow as we stop, and although we wait it does not return. I do my best rabbit-dying call and wait again, but nothing moves. Finally, I decide to try on foot to find it, but I have only taken five paces from the Land Rover when I see its movement 30 metres to my right. It has returned to close beside the road. I am arrested in mid-stride, the cat poised, head turned, watching. It is pale-grey, a shadow among shadows. It hesitates a moment further and then it is gone with its considered, tentative step into the brush. It does not come again.

Right beside the Blackstone, its cold surface black and gold, the colour of the sky, we make our camp. Late in the evening we photograph the moon's rising over the mountains reflected in the river's surface. We have seen more of the moon in the last twenty-four hours than we have in the past months. The air is the bitter cold of clear, snow-clad mountains at night.

1 SEPTEMBER

We stop at Blackstone Lodge, where we ask permission to photograph the river up and down from their camp, which is bounded by high banks bright with the colours of the changing leaves. The woman who comes out asks whether we have seen any caribou, and when I say we have not she tells me the hunters staying at the lodge have just witnessed the slaughter of an entire herd of forty, females and calves included, by two Gwich'in hunters. When we are done I walk into the dining cabin to say my thanks and goodbye. Its close warmth is sudden after the hard cold of the outdoors, and I find a room of hunters gathered at their ease around long ranch tables. One of the hunters is a pretty young blonde woman with a rifle on the table in front of her.

Dawson City is a town of two parts. Approaching from the east, one drives through a valley so reworked by massive earthworks that I am for a time speechless at the magnitude of the unrepaired devastation. The entire valley floor, with the exception of the roadway, is a turmoil of bulldozed heaps and deeply gouged gullies, the remnants of placer mining. Placer mining gives an almost attractive name to a horrendous practice in which the entire floor of a valley is essentially bulldozed aside and sifted for the alluvial gold that has been brought to the surface by tectonic activity. The mounds of boulder-littered earth stand as high as quarry workings, the channels between deep and wide enough for small craft to manoeuvre through. Indeed, it is only the still, crystal-clear ponds that lie in the depressions, and the sparse growth of trees, that soften the abuse that humankind has inflicted here. What disappoints me most is that none of the wealth extracted here was used to rehabilitate the land. The land was simply plundered and left.

Beside the Yukon River, Dawson City is at pains to present another face. The gold-rush town, with its grand public architecture and picturesque-fronted buildings of lapped wood, is maintained to historic tradition even down to the hand-painted signage. The effect, although it smacks of tourist trap, is quaint and attractive. On the far side of town, I feel like I have passed through a movie set rather than a place that people call home.

We catch the ferry across the Yukon River. On the far bank the road winds steeply into the hills. When we finally issue out on top of the road, known locally as the 'Top of the World Highway', it is almost dark and a thin rain foreshortens a view that falls without interruption to a deeply creased land far below.

2 SEPTEMBER

I wake to a damp world encased in mist, the visibility only a few metres in any direction. It is a complete revelation, while we are making coffee, when the mist parts and suddenly our close world expands to encompass a view in which I can see the Tombstone monolith more than a hundred kilometres away. The mist soon closes in again, but as the day warms, the breaks in between become more frequent. We work furiously, making images of the vast reach of the land, its quilt of changing colour accentuated by the white of the mist that lies far below in the twisting valley bottoms. From the high vantage of this singular road, it is possible finally to grasp the scope of the land, its unfettered reach, its wildness.

Hill after hill after hill recedes finally to a blue opacity, and steep-sided valleys fissure every facet, carving even the distant Ogilvie Mountains to the northeast into a high, serrated barrier. It feels as if one could walk out from the road and be alone until the end of time.

But man has been here too. On the climb up into the mountains I noticed a sign stating, 'No Caribou Hunting'. It was a little faded, flaking in places, with the vegetation growing up to obscure it. It seemed to me more a statement of hope, for there are no caribou here. They were killed, shot out. The Fortymile Caribou Herd used to roam here. In the 1920s, the herd was estimated at more than half a million animals strong, roaming between the south-central regions of the Yukon and Alaska. When gold had been discovered in the Klondike River Valley, Dawson City boomed to a population of thirty thousand miners, hungry for meat. The caribou died in their thousands, hunted without pause, until the herd was but a trace of its original number. But even when the gold boom proved short-lived, the caribou did not enjoy much respite. With the construction of roads through the range of the Fortymile Herd, both in Alaska and the Yukon during the 1960s, hunting pressure resumed and what little was left of the herd was reduced by 1973 to a mere six thousand five hundred animals. It is staggering to me that it was not that six hundred thousand caribou had been killed, but that there were only six thousand five hundred left to kill that finally brought men to their senses. In the forty years since 1973, the herd has recovered only slightly, to twenty-three thousand animals, a fraction of the original half a million, and their range has been reduced to such an extent that those that remain seldom if ever visit this part of the Yukon. Perhaps their numbers are just too few, perhaps without their lead animals they have forgotten their ancient routes, but it seems more reasonable to assume otherwise.

All animals learn by experience. A herd of caribou will begin to avoid a place where it is known to suffer slaughter, and if this slaughter persists for thirty years or more it does not seem impossible that the changes the caribou make to the course of their annual movement through the land to avoid it will become permanent. But even if the 'No Caribou Hunting' sign is just a hope, it holds no teeth, for there is in truth not one square centimetre of the Yukon that is designated as sanctuary for the animals. Nowhere. Not one single safe and certain refuge. From the Top of the World Highway the wilderness is vast, but in fewer than eighty years humankind has changed it, stolen the light out of its face. To what end? Mansions in a faraway town, a bright glint in someone's teeth, an unseen foil that conducts electricity? On what skewed balance do we weigh the worth of wild? There is but one answer: by and for our own esteem. This makes us proprietor, judge and jury, and we remain therefore innocent until history proves us guilty. But then it is too late – our crimes are done. Why do we not grasp that one cannot replace a wild eye with glass?

There are still great herds of wild caribou in the Yukon and they are fine barometers of the wildness of the land, of the magnitude of natural, untrammelled country. For now, it is like that, but the time is not far away when people will challenge that, and then the advocates of wilderness must put their backs against the wall of reason. They must enter a court where man's right is the bias perceived as truth, where a mountain may no longer exist because it does, but only because of its worth to man. To my mind there is no greater sacrilege than this delusion by which we presume the right of authority over all creation.

It is to the scientist, to the researcher, that we must turn for our defence, and I challenge not only science's authority to prove worth here, but also the basis of its measure. Several times during my time in the Yukon and Alaska I have heard scientists say that the numbers of animals in the caribou herds have always fluctuated. 'Always' is a big word; it encompasses all time. The accurate records of the numbers of animals of each caribou herd can only be done from the air and are even then still approximate. Any count done from the ground is a guesstimate at best. We are left, then, with an accurate record of perhaps the last seventy years. To prefix this knowledge with 'always' is dangerous. We have no real idea of historical caribou herd numbers, just speculation based on the best facts we can find.

Looking out on this mist-shrouded morning, over the red- and yellow-hued quilt of land that stretches unmarred by road or power line as far as the eye can see, I feel that the caribou did themselves a great disservice in their annual aggregations in the spring to calve. I know that these great gatherings of caribou serve an evolutionary purpose by simply overwhelming predators with the sheer number of calves available. There are just too many for all to be taken. But how are they to learn when there is a negative aspect, only recently arrived, that may prove more deadly to them still?

Research counts caribou when they come together. And because the numbers counted run into the hundreds of thousands, science hides behind the apparent security of this figure and pronounces them healthy and fit to hunt. But the truth is that no one really knows what is healthy and what is not. By counting caribou when they come together we are, too, ignoring the other portion of the equation, the land that supports them. When we divide the land by the number of animals, we are left with an entirely different picture. Not one of mass, of sure tenure, but rather a plaintively fragile presence. One animal per how many thousand hectares of wilderness? But we will not make our scientific judgements based on this fraction, because it runs contrary to our greed and so we render science pliable, a farce at best, a deception at worst. But, deception or not, we want to believe science because in doing so we excuse ourselves and do not have to forsake our appetites. The great auk is gone, as is the passenger pigeon, both wiped out by humankind encroaching on their country. Each of these two species still occurred in their millions little more than one hundred years ago. Millions, one hundred years – there is no security in numbers when we do not understand the tipping point.

There are still caribou in the wide land of the northern American continent. Let us watch them closely, for they are like trout in a river: they will disappear when its purity sours. Already the roads and tracks course through their country like poison flowing into the veins. It is just a question of how much the wilderness can take before it sickens and dies, allowing us to step into the void without guilt and render yet more ground to our design. I have no doubt that history will prove the caribou alive to be more valuable than the meat it renders. However, the question is – can we stay our appetite? It would be laudable indeed if we could let them be and it has always amazed me how close it lies within our grasp to be extraordinary.

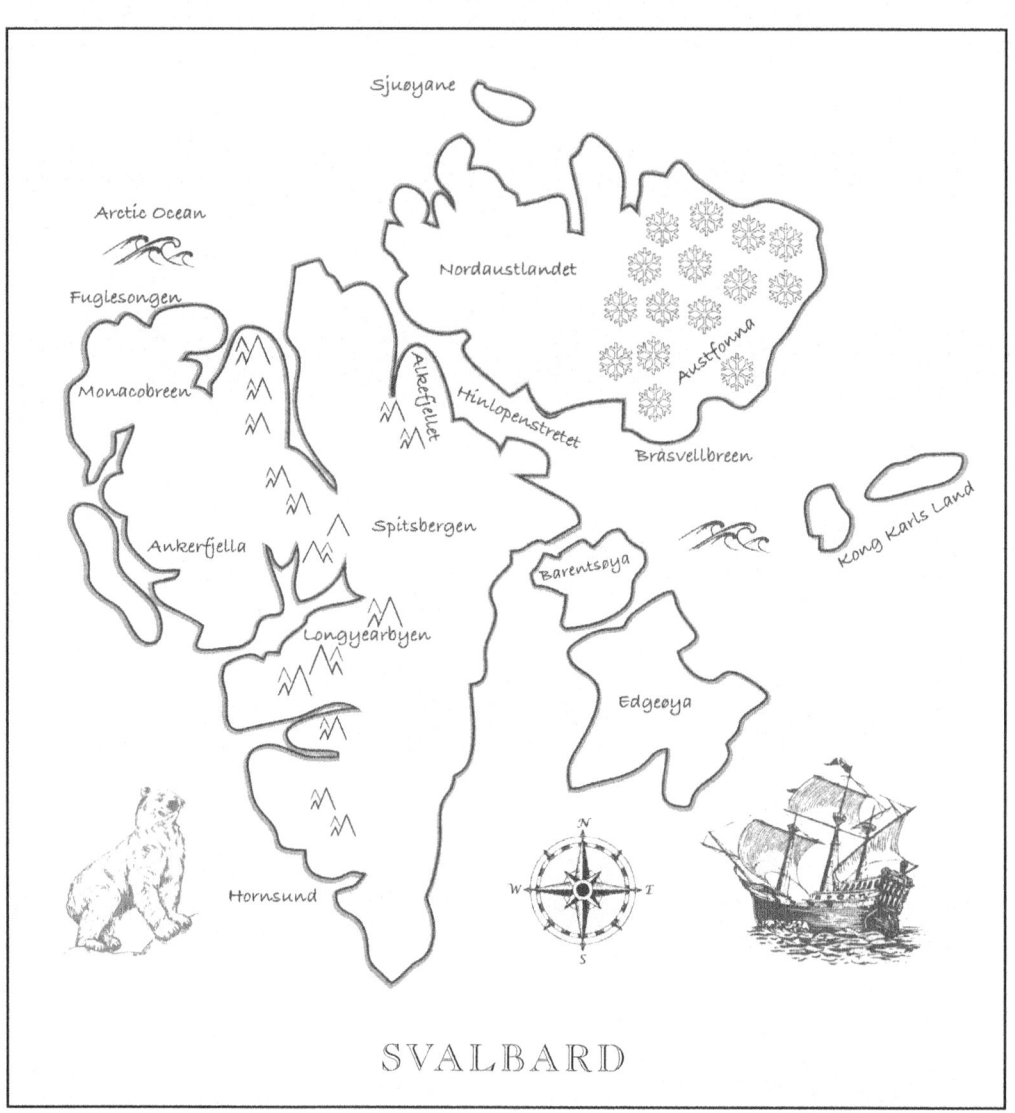

SVALBARD

IV

EUROPE
The Arctic

SUMMER 2013

*Over the past two and a half days, during which we have remained beside the
bear and his kill on the small patch of ice, we have drifted 27 nautical miles, or
50 kilometres. There is no sun, no point of reference, but as the bear swims away
he is heading only two points off due north. He is headed back to the land that is
no land at all, just ice.*

Departing the Yukon in September of 2012, we drove across continental North
America to Philadelphia, where we loaded our Land Rover and ourselves onto
a ship bound for Germany. After a curious fourteen-day passage, where our
ship was constantly speeding up to avoid storms approaching from behind, then
slowing down to avoid running into the aftermath of storms ahead, we landed
in freezing Hamburg in early October and put our vehicle into storage for the
northern hemisphere winter.

We returned with the last ice of winter still dripping from the leafless trees
and headed north into the coming spring of 2013. In the following months,
however, we discovered that what we sought – vast contiguous stretches of
land in as wild and natural a state as possible – was not to be found in Western
Europe, with wilderness areas fragmented into a scattered mosaic of small
enclaves. Summer-house cabins mushroomed as colourful as the first spring
flowers in every recess or final redoubt of the countryside, and while World

Heritage Sites straddled ranges of singular mountains, the valleys, rivers and lakes had been excluded to make way for hydroelectric schemes, fishing or other enterprises. Power-lines, roads and infrastructure blemished the land and marred their vistas. In Finland, vast tracts of wild forest were being cut down.

Leaving Lemmenjoki National Park one day, I stopped and walked through the 25-metre fringe of trees left standing to screen the devastation of the clear cut from the road and, walking to the closest pile of logs, chose one at random and counted the annual rings of the tree. The distance from the centre to the outside of the wood, excluding the bark, was 19.2 centimetres. For the first 2.5 centimetres from the centre the annual rings were too dense for me to count. In the remaining 16.7 centimetres I counted approximately three hundred and fifty annual rings, making this tree at least three hundred and fifty, possibly four hundred, years old. This meant that this tree was already two hundred years old when Helsinki was declared the capital of Finland in 1812, that it was three hundred years old when Finland declared its independence from Russia in 1917. This tree began its life before Finland was an idea and now it is dead for the sake of some small economic gain.

Our disappointment was further compounded by a reality that firmly dispelled our romantic notion of the Sámi people as practising a tradition in which wild and humankind were joined. The idea of an ethnically strong people following the annual migration of vast herds of wild animals in a relationship of evolving symbiosis does not exist. Instead, we found the whole of Finlandia – as the combined lands of northwestern Europe spanning Norway, Sweden and Finland are known – to be fenced with thousands of kilometres of Bonnox (game) fencing. This division of wilderness has its cause in the evolution of reindeer-herding from survivalist tradition into ownership. Ownership initially divided the herds, but soon it divided the land as well. Today, as a result, even national parks are divided by fences demarcating proprietorship rights, and what was once a free-moving migration has devolved into farming.

It can also be seen in the reindeer themselves. Reindeer are of the same family as caribou, but those that we encountered in Europe lack the tension and caution of their North American kin: the mistrust, hesitation and anxiety for every stimulus that holds the truly wild eternally bound to the edge of flight. Instead, reindeer have about them the air of domesticity, that wide-staring bovine ignorance of predation, the mute acceptance of the security of care and the mental lethargy that this seems to nurture.

The Bonnox fence now used to keep yours separate from mine is a matrix of 20-centimetre squares, raised to about two metres in height. In Norway, I once spooked a snow hare from its concealment in close proximity to a reindeer fence. The hare zigzagged away from me in a burst of speed, but was brought up short by the fence and, unable to find a way through, raced off down its line. If the fence is sufficient to prevent the crossing of a large hare, then its impact on wildlife must be extreme.

In all our months of exploration of the sparsely populated, non-agricultural-ised lands of northwestern Europe we saw a single moose and two foxes. I did not note a single scat or track of the other predator species purported to exist in the few national parks and wild places.

We were reluctant to accept the sum of the truths that were being laid on the table of our understanding. But as the weeks morphed into months, it percolated slowly into our minds that what land remained wild and pristine was separated and isolated by the sea of human endeavour, and cohesion as one identity was impossible. The land lacked the spirit of wild: its essential ingredient had been driven out by the attentions of humankind. What was once association had morphed into administration, fragmentation and finally utilisation. The land was God's no longer.

And, so it was, finally, on an evening when we camped in the lee of a ramshackle cluster of unoccupied boathouses with low cloud drifting curtains of rain across a high lake, where mergansers – sawbilled fishing ducks – hunted the bouldered shore, that we made the decision to turn away, turn north and pin our hopes on one last place. A place of extreme cold that lies out in the sea, the redoubt of the polar bear, the Svalbard Archipelago.

25 JUNE 2013

Through the downy blanket of cloud that carpets the ocean the Svalbard Archipelago rises like an abstract graphic, the sharp outline of the peaks piercing the deep mantle of snow. At the island's edge the white wall of the snow and ice is as clear in its outline against the blue sea as if it were etched with a knife. Nothing could seem more hostile yet more compelling, and from this view it lies serene. In fjords clear of snow, rivers issue a silt-laden stain into the sea. Little grows on the slopes of bare, dark brown earth.

Longyearbyen, the main settlement, was founded as a coal-mining town and that character remains. There is no greenery or vegetation to soften the wood and rusted steel, that is in use or discarded on the grey stony ground; as unlovely as any industrial wasteland. Mary-Ann's Polarrigg Hotel, where we are staying for the night, has an old, square red bus as a smoking lounge and a kebab of small, cosy rooms skewered along a passage.

26–27 JUNE

We board the Russian research vessel, *Akademik Sergey Vavilov*, on a One Ocean expedition and are thrilled that our compact but comfortable cabin has a wonderful large window. As we sail away from Longyearbyen the sun breaks through the clouds in patches and northern fulmars – the pretty blue-grey petrels of the Arctic – cruise beside the upper decks in a brisk, freezing wind.

In the morning, we anchor a distance off a cluster of bird-nesting cliffs, where Brünnich's guillemots line the narrow edges and kittiwakes circle against a sky drawn closed after the morning drifts of sunshine. An Arctic fox patrols the foot of the slopes along the shore. It is early in the season for chicks, but the fox's time will come.

The guillemots look like rows of tiny, dress-suited mannequins, with their stark white bellies and pure-black wings and upper bodies. The guillemot chicks will leave the nest before they can fly. Hopping off their high ledges, they will glide as best they can, but without feathered wings it comes close to free fall. Here the cliffs are quite set back from the sea and many will thud into the spongy slopes and then hop and tumble their way to the sea. Those that escape the foxes will be joined by their fathers and begin a remarkable epic. They will swim to Iceland, a journey of approximately 2 000 kilometres. By the time they arrive, they will finally be getting their full plumage and will soon be able to fly. On the way, the adult males moult their plumage, and on arrival in Iceland they too will again be able to fly. The females abandon this rigour to the males and fly away as soon as the chick has leapt from the ledge.

28 JUNE

The ship pitches to a slow pendulum as we turn east around the northwestern point of Spitsbergen, the largest island of Svalbard. The wind is gusting 32 knots. We land at Smeerenburg, on Amsterdamøya, where 17th-Century Dutch whalers came to hunt beluga whales. On the beach, square mounds of black, coal-like soil mark the remains of their blubber vats. From this desolate, driftwood-strewn strand the whalers rowed out a short distance in wooden longboats and harpooned the whales that fed in their hundreds in the bay. Today the whales are gone and there is just the wind. On the top of a hill beside a large rock is the communal grave of a group of whalers who died of scurvy while wintering here. It is marked with commemorative stones and wooden plinths. Of the thousands of whales that died as horribly in this place, there is just the earth blackened by their burned flesh.

In the afternoon, we make a landing on a beach of slippery, white, rounded boulders and hike through a hillside of fallen rock, softened with patches of bright-green moss, to a little auk nesting colony. These tiny, compact birds nest deep in the crevices of a slope of steep scree, a treacherous tumble of rock fallen from the mountain behind. The place is called Fuglesongen, and as the flocks pass overhead calling, they are answered from the hollow labyrinth beneath our feet, the birdsong echoing forth out of the earth. Across the sea, steeply below, the land is an amphitheatre of mountains white with snow, and through a thin cloud the water is silvered and gold. I feel I could dream up fairy tales here and the setting would all be true.

29 JUNE

The ship slides as if on rails through a mirror-calm sea, moving so slowly that we make no more wake than a rowboat. Drifts of ice, white and pale translucent blue, are the only reference to our gliding passage. We silently pass pairs of Arctic terns that rest on these cold islands, and a single walrus raises its head once, then sleeps again. Out in the Zodiac inflatable, we see two other swimming walrus on our way to land at Worsleyneset, but they are moving deceptively quickly and, knowing their reputation for aggression, we keep our distance. Soon all I know of them is the sound of their exhalations as they surface somewhere far out across the stillness. On the low rise of the island, which seems a flatland in the arena of mountains that surround us, four reindeer are grazing. Their legs are short and the whole animal seems, as a result, a miniature of their mainland kin.

Frank Worsley, explorer and Ernest Shackleton's right-hand man, named these places as recently as 1925, and there are men still alive who made the crude fox traps that we find along the shore. There is a reconstruction of a hunter's hut on the brown earth beach where we land, besides which stands a polar bear trap. Essentially, it is little more than a long box in which a rifle lies protected, its barrel tip standing free. A piece of meat was placed directly in front of the rifle and a string attached to the trigger. When a polar bear came for a meal, it shot itself in the face. Although basic, this method was so effective that greedy hunters often shot more bears than they could skin before the bodies froze. Once frozen, the coat – the only reason the bear was killed – could not be removed and the carcass was wasted.

It is as we are leaving this place, where red-throated loons make their raised nests in small, tranquil ponds, that a polar bear is spotted on a far-off island. The ship takes us closer and anticipation is high as we depart in the Zodiacs, rifles lying in their cases in the bows. We cruise among a litter of small ice floes in a brown sea as silt-laden as any river mouth. Three walrus are resting on a flat-topped drift only just big enough to accommodate their imposing girth. They are gargantuan and, viewed from our perspective, low on the water, they look like three rhinoceros reclining on their sides. They lie pushed up against each other, and when one moves, the others grouse, raising their heads in objection and revealing their tusks. The largest has ivory that protrudes beneath its brush-bristle whiskers at least twice the expanse of my thumb and pinkie finger wide spread, possibly three. Their eyes are small and blinking. One appears nervous and apprehensive, the other belligerent. The third dozes on, secure in the close contact. I am told that they are as tactile underwater, where they hunt clams, puncturing the shells and sucking the bivalve out with a slurp.

When we move on and continue our scouting of the low island, it becomes obvious that even in this place, where nothing grows taller than a few centimetres, there are plenty of nooks, crannies and depressions in which a resting bear could disappear. It is Beverly who spots it at the head of a small cove, asleep on a drift of snow in a lee of the shore. Our approach is measured, and for a time we drift slowly before a gentle breeze with the motor switched off. When we are 30 metres offshore, the bear opens its eyes and then closes them again, tucking its nose a little into the snow. Only the eyes and the tip of the nose are black; all the rest is a loose-skinned coat of thick hair that is not the pure-white I had imagined but stained more with a faint yet distinct yellow, the colour of butter. It is thin. Its lethargy, on what for it is a warm day, emphasises a weakness that I have seen in other animals in the first stages of a hunger that without luck will kill them. I recall a lion, too thin for the massive spirit of energy and violence required to make a kill, panting in the shade beside a waterhole, where in a desperate attempt to catch anything it was hunting guinea fowl and careless young warthog in broad daylight without success, weakening itself further with every successive dash. This bear seems the same, and now faces its leanest time. The ice, its required environment, where it hunts and kills seals that come to the surface to rest, is gone out to sea, retreating towards the North Pole. Now, each year, this summer break-up comes earlier and lasts longer. Marooned on land, the bear must either swim after the retreating ice or scavenge bird chicks and eggs, which are not enough to sustain so large a predator. If it is fortunate a reindeer will die or a whale will drift ashore. If not, the odds of its survival are stacked against it. Once in a while the bear stretches its leg. The paws are enormous, the size of dinner plates, and through the lens its dark claws stand out, long and distinct against the white of the snow. When we finally motor gradually away, the bear has still not moved at all.

I know that I am in part responsible for the earth's changing climate, both by what I have done and what I have not done, for the rapid changes that are happening here and whose legacy is a thin and listless bear. I dip my fingertips into the sea and touch them to my lips, asking the gods of luck a favour. And I make a promise to myself that 29 June will be a day that I do not drive. A token only, perhaps, but I have come to understand that there is a remarkable power in the action of one, and without it my gesture is not even a courtesy.

After dinner, we set out in the boats once more before the expansive, towering glacial face of Monacobreen. The inflatable hulls scrape and bump through a sea that is in wide swathes as much ice as water. The Monacobreen is in fast retreat and, nudging our bow into a small island of smooth, green-black rock, we look back to the tall craggy cliffs of ice that stand more than one kilometre behind us. Until eight years ago, the ice had kept this island secret. In a 90° corner in the middle of the face of the glacier where two valleys of ice collide, we find thousands of kittiwakes feeding on an upwelling created by a huge unseen river of fresh water. The kittiwakes paddle busily in circles on the surface, dipping after organisms I cannot see. The glacier booms and cracks like rifle shots, and to those shudders close enough to send vibrations into the water, the entire flock rises, mewing, into the sky. A thousand birds on white wings turn with a dancer's grace before a backdrop of striated, pastel blues, and I would not have the world any other way. The air is still and very cold, and to the north at the mouth of the fjord the sun is turning the low filter of cloud to gold.

30 JUNE

We cross 80° North and are now less than 600 nautical miles from the North Pole. We are chasing the edge of the sea ice that has broken free of the land and is retreating northward before a 20-knot wind. At exactly the spot where we come to its sharply defined edge, there is a polar bear and another smaller ship. We turn along the perimeter in the direction the bear is travelling.

It is several hundred metres distant, and although too far to see intimately, the bear's purpose and intent are clear. It tests the wind frequently, raising its head, and peers and sniffs around blocks of tumbled ice. It is hunting. I scan the ice and the sea edge for seals, but can see none. The ice is a vast sheet where there is no constant. There are small icebergs marooned in a crush of thin, flat pallets, areas of jumbled ice that look like Lego lying in a heap, areas where discs of ice islands lie in tight formation in oval lakes surrounded by water that is half ice, half liquid. Out in the distance of its expanse I can see leads and ponds of open water.

I look back to the polar bear and realise how difficult its hunting must be in so vast, variable and unstable a place. Chance would be useless: it would die. It must rely on its nose and ears. With the wind howling, as it is today, its nose would be the only faculty to lead it to success. It turns and ambles back past the smaller ship, now fast in the ice. Its pace is unhurried and yet it moves astoundingly fast. Despite there being no sole sheet of ice, more a variable cluster of chunks pressed together by the wind, the bear moves as if it is on solid land and not once do I see it enter the water. This is a bear in its prime and there is no doubt as to its capacity, its power. Even though it is reduced by distance to a small speck in the vast, bleached sheet of the landscape, its movement describes a competent, dangerous predator. We turn west along the ice edge to search for another bear, as protocol dictates that the first ship has rights to the sighting.

We pass out of the sunshine into a fog through which little can be seen. Together with the wind, it makes any place on deck bitterly cold, and within the hour only a few hardy stalwarts remain, scanning what can be seen of the ice in the halo of our fog-shrouded world. Finally, we emerge from the bank of mist, but the sky is overcast and dull and the ice a maze of shapes and forms. We see no bears and turn around to retrace our steps. By the time we return to the point opposite where we had seen the bear in the morning, the wind has driven the ice edge 3.5 nautical miles further north. Where the bear was is now open ocean.

1 JULY

Boarding a Zodiac, we cruise along the remarkable Alkefjellet – black dolerite cliffs where over sixty thousand pairs of Brünnich's guillemots have come to nest. On the sea, the birds are tolerant of our presence, and only move off in a heavy, flapping skitter across the surface, or simply dive away when we are closer than a metre or two. Above us they line the ledges, packed densely together. Those high enough take flight in a stylised leap from the edge, their slender wings held rigid in a deep V, flapping only when their speed is enough to give their wings purchase. Those too low plop into the water in a comic belly flop, making a great splash that seems somehow to please them, as if something special has been achieved. It is when the guillemot dives that its wings function best, and the bird is agile and fast beneath the surface. For all intents and purposes, the guillemot is a flying penguin. The only penguin or truly flightless bird of the Arctic, the great auk, is long extinct, killed in their millions by humankind. The guillemot survived because it could fly. If land predators had not persisted, it might well have also been on its way to becoming more a penguin than a flying bird. Its body is tightly compact, and seen in flight or swimming is the classic oval shape of penguins. Its feathers hug the body and the wing feathers are foreshortened, so that beneath the water the wings and their action are almost identical to those of a penguin, except that they remain elbowed, a necessity not for water but for the air. The bill is pointed, ending with a sharp hook, and the plumage, in its striking contrast of black upper body and white belly, has a familiar penguin quality. The cliffs echo loudly with their calling and, looking up, the monochrome grey sky swarms with their graphic black forms.

A fox hunts the heights where the towers give way to steep scree. Its half-grey, half-white coat makes it seem scruffy, as if it struggles to survive, but through the lens it is quick and agile, probing the fringes for any fallen morsel. Beyond the guillemot colony, a glacier flows like a frozen waterfall over the height of the lower cliffs to the sea. The rock exposed here runs in dramatic layers of white, black and white again. The white is limestone, between which molten lava forced itself before hardening. In the process, it transformed the limestone to marble.

Across the strait on Nordaustlandet we land at a known walrus haul-out, Torellneset, but the beach is deserted except for a carcass that has been picked of everything except the larger bones and the skin. Around the remains are bear prints, deep depressions in the gravel. The beach is of small pebbles rather than sand, and ice boulders protrude or crunch underfoot. Two walrus, a mother and calf, approach the beach cautiously, coming to within 10 metres of shore and eyeing us with circumspection. They cruise slowly back and forth several times before deciding to move on. In the shallows two phalarope, eye-catching in their russet-orange plumage, wade beside a slowly dripping iceberg stranded by the tide. The wind is bitter, and off the point of the beach it drives a stream of ice twisting out into the sea. The land behind is as bare as any desert, its pale, tan-coloured soil a place of bones: a long-dead whale and the small skull of a polar bear cub. A mist shrouds the island's rise, the tendrils flowing downwards over a smooth mantle of snow.

After we depart a polar bear materialises out of the void, walking alone into the cold silence that we have left behind. We are far from shore when we see it, a single movement in the stillness. I am filled with an ache to return, to be quiet and close, to know.

3 JULY

Bourbonhamna, in Bellsund, was the first hunting lodge in Svalbard. The business collapsed long ago, but the building still stands, if only just. It is small by today's standards, but in its day this double-storey lodge, with a brick chimney and a view out over the fjord, where four glaciers issue from the high valleys, must have been quite grand. Its foundations have sagged downhill into the permafrost and it leans now at a crazy angle, shored up by bracing beams. It is the stuff of trolls and fairy tales. Looking at this bleak, cold place, I wonder at the dreams that must have filled the head of its builder. I struggle to conjure in my mind the conviction he must have had that he could lure clients here to hunt polar bears in sufficient numbers to make it profitable. Even by today's standards, it is remote.

There are many for whom Svalbard was to transform grand dreams into folly. One of the most startling tales is that of an English businessman who came to these shores in the early 1900s to mine marble. He arrived early in the summer, which is still bitterly cold, and set up his works on the shore. He extracted blocks of marble, crated them carefully and shipped them home to England. When the crates were opened on arrival all they contained was a mess of white pebbles. He had mined frozen gravel, and the ice had slowly melted on the journey south. Not to be outdone by this unfortunate turn, our entrepreneur built up a fabulous yarn about his marble mine on these hostile and remote shores and sold the entire enterprise to some unfortunate who took him at his word.

We walk around the leaning lodge, staying clear of its downhill side, and head up towards the heights, where a few reindeer are grazing. The ground is sodden and boggy. The reindeer are inquisitive but shy, and they repeatedly flee a short way, stop to graze, then trot directly back towards us with their heads stretched forward and their nostrils flared to test our scent. We climb a tall, steep slope of jumbled glacial scree. From the top, we can look down on the glacier that brought it here.

Quite suddenly, when I have been standing for a minute or more looking down at the brown-mud delta below, the term 'glacial retreat' loses its scientific meaning and becomes a shocking understanding. The glacier that brought the scree to this place now stands kilometres distant to my left. I need binoculars to study its surface for fox or bear. Below me, the valley falls at least 200 metres to the delta that now issues silt into the sea. The rocks where I stand, strikingly patterned in wavy strata, are clearly formed by earth and pebbles under extreme compression. In its heyday this glacier, which is now not even an ice sheet but more a frozen moraine that is slowly melting, stood more than two kilometres to my right, at least five kilometres beyond its current foot. But it is the height of where I stand that brings home the staggering reality. To deposit rock made by compression in the quantity to build the steep hill of moraine that is my footing, the ice must have towered above my head. This now-unremarkable river of ice must once have stood in the region of 250 to 300 metres thick. I shake my head involuntarily, struggling to come to terms with the magnitude of this: a moving body of ice taller than the average height of the buildings in downtown Manhattan and two-thirds as wide, now gone. I do not even know its name.

4 JULY

At the western mouth of Isfjorden the mountains of Alkhornet end abruptly, as though cleaved by an axe. The high face of the final cliff is a mass of nesting sea birds, and as we cruise steadily in to anchor, they fly past us low over the sea, issuing out of the haze as if they had been made in great batches and set free to race to this certain destination.

We land on a small, stony beach and clamber up the shaggy rock to a place where, more than a century ago, Pomor hunters from Russia built a rudimentary stone shelter from which to hunt walrus. The Pomors came across the open sea in small crude boats, but of all those who came here to take the wealth and wildlife, they alone seemed to have understood the idea of a sustainable harvest; the Pomors hunted only walrus, changing sites from year to year and apparently killing only the best of what they found. All others killed without regard. The Pomors hunted here for decades before others intruded and discovered the wealth of life here. After the Pomors came sophisticated ships and their smart captains keen for riches, and between 1869 and 1878 the Svalbard walrus was brought to the brink of extinction.

The Pomors' camp is set against a small cliff with a fine view across the wide fjord. There is vegetation here, a spongy, lime-green moss and other small plants growing close to the ground, that lends the mountains scope and the steppe a vivid green, which is unusual in this place of bare, brown earth. At the base of the cliffs the green is darker still, where the guano enriches the impoverished soil, and we see a fox patrolling there for eggs and chicks fallen from the nesting ledges high above. Its coat is a patchwork of grey, black and white. It is nimble over the high, steep terrain, and when it pauses it is hard to distinguish it from the rock. Reindeer graze the slopes: cows with curious youngsters in coats of long hair that are a clean pale-grey, fringed with white, and bulls with nearly fully formed antlers who are more reticent about our presence. Without ever looking at us directly, they manage to keep moving away as they graze, keeping us at a constant distance. In a wide snowdrift, they paw the snow and then lie down. They nibble at the icy crust where they lie, holding their lips away from their teeth as they eat.

I look up to the cliffs with the silhouette of thousands of birds circling against the low sky. I can hear the babble of their calling: the croaking, guttural moan of the guillemots, the plaintive mewing of the kittiwakes. I am glad they are here, that we have left some facet of what was so magnificent in this landscape.

I am uncertain about the fate of the great white bear. With the exception of Native North American communities, we have stopped shooting them, but are killing them now with something more insidious. We are changing their country, their place, which is not land at all but rather the ice that surrounds it, and our pollutants are contaminating the seas on which they depend. Perhaps it is too late, but that does not excuse apathy, because we do not know what can be done until we try. If I have learnt anything on this voyage around these cold, snowbound islands, which less than two hundred years ago teemed with life, it is that by shrugging off what is obviously wrong, as either not our concern or beyond our capacity, or too late, what we harvest is silence. By our apathy we are making the world empty of life that is wild. Change does not require greatness or miracles; it is the responsibility and possibility in each of us.

6 JULY

In the afternoon, we board the *Ocean Nova* for our next circumnavigation of Svalbard with the PolarQuest expedition company. As we sail, the sun breaks through in patches, and Beverly and I sneak out of the orientation lecture to photograph the cliffs of Isfjorden shrouded in mist. We return in time to be introduced, and I give a brief overview of our project to our fellow passengers.

As we round the mouth of Isfjorden and turn north the ship begins to roll and pitch to the small swell. I fall asleep as my body seesaws up and down between my head and my feet.

7 JULY

The ship is rolling heavily to the swell of the open sea and the dining room is half empty. As a chronic sufferer of seasickness, I again bless my friend, Alex Taylor, who years ago in the Florida Keys made me an old-fashioned, Chinese pressure-point bracelet to help me during a particularly bad crossing of the Caribbean. I have never looked back.

In the afternoon, we sail into Krossfjorden, then on into Lillihöökfjorden. In the Zodiacs, we divert to an orange-faced bearded seal sleeping on drift ice, before landing at Signehamna. It is ten days since we were here last and the ice on Lake Hajeren has melted in a ring of clear water, about 20 metres wide around the entire edge. The frozen centre is no longer white but pale-green and wet. The kittiwakes still roost on its chill platform and bathe enthusiastically in the clear water at the ice edge. It is a pleasant, easy walk around the lake perimeter and I fall back so that I can hear the country. Rain is falling and a mist shrouds the mountains all around.

The sea is green and calm as we motor to the face of the Lillihöökbreen. Fulmars fly low over the mirrored surface, racing their reflections on fixed wings. Beside a cave in the glacier face, a huge block of ice falls into the sea and lifts a flock of kittiwakes off the surface like a towel lifted to the wind. The ice is white and pastel-blue and booms like cannon fire across the fjord as the glacier inches forward.

8 JULY

We are woken at 4 am at our anchorage off Danskøya by the news that a polar bear is swimming past the bow of the ship and that Zodiacs are being put into the water. Half an hour later, the entire passenger complement has turned out and we are navigating through a sparse drift of small ice floes towards a string of islands. No one talks; all eyes are on the shore. The water is oily calm. A fog is lifting off the peaks. We know a bear is near and the air is potent with its presence. We flank the islands with two groups of Zodiacs. The bear is not on the first island. Nor is it on the next. The mainland lies beyond. It is a rocky shore, with stone outcrops emerging from a slope, strewn with sharp-edged boulders. I scan it with our long lenses, but it is only patches of snow and yellowed rocks that arrest my eye.

Suddenly the bear is in view, stepping out from behind huge boulders on the crest of a distant ridge. It is carrying something in its mouth and moving fast. It disappears for a time and then reappears on the ridge behind. It pauses to look back before starting up the steep slope of loose rocks the size of ovens. And then a second bear is spotted, closer than the first. This one is more leisurely in its pace, investigating the ground every now and then before it too vanishes into one of the many folds in the land. The higher bear is now right at the foot of the scree slopes of the peaks, moving with speed and agility over the tortured terrain. We watch it for some time, and through the magnification of the lens it is clearly agitated, aware of pursuit. Whatever it is that it is carrying in its mouth must be food. It vanishes once more after a while and does not reappear.

The decision has just been made to return to the ship when the second bear reappears, closer still. It is heading toward the shore. I can see its back just behind the shoreline ridge, the shoulders rolling powerfully beneath its cream coat. Its gait is slightly swaying, just as lions do in their prime. It climbs a gentle rise, issuing slowly into our view, until it is complete, the mythical icon of the northern ice not 50 metres from our boat. I hear my own breathing in my ears, short and excited. As it pauses at the tideline and sits, raising its nose to test the air, I realise I am holding my breath. It is large, a healthy male, its buttocks rounded by the fat of a successful winter's hunting. The long coat hides the muscle and makes the animal seem softer, an adjective that we associate with gentle, but the black pits of the eyes are inquisitive and entirely without fear. When it begins to walk along the water's edge I purposefully take my eye from the camera and watch. It is alarmingly close, its huge, black, padded paws moving without a sound upon the rocks.

I find myself glad that the era of hunting in Svalbard has closed. The hotels, museums and public places of Longyearbyen are full of images and paintings of dead polar bears. Sepia-toned men pose with rifles beside animals sprawled in the snow. A dark, bearded man pares the last of the skin from a bear lying on its back, its legs pointing rigidly into the air, the body naked without its luxurious coat, leaving only streaks of dark muscle and white fat and sinew. A mural in white and blues of a bear slouched unnaturally into its shoulders, lying in a heap upon the white ground, the blood pooling beside its face, a violent scarlet in the pastel reconstruction.

I prefer the animal like this, walking unafraid and vital along the shore. Sticking its nose after a scent between the rocks, pawing a beach-ball-sized boulder out of the way and licking at what it finds, raising its head and its chin up high, its tongue curled at the tip and somewhat purple underneath. It alters the land, makes it potent. It is a beast in the real sense of the word and its presence requires a land unconstrained, primal.

The bear changes the place metaphorically, just as dragons changed the fairy-tale lands of my childhood. It makes what we came here for real: wild land. In all that has gone from this place at the hands of humankind, the bear is our one concession. If what I am watching, a wild, unconstrained animal, can bring such a dynamic charge to the chill silence of this towering landscape, then what we have done in the past is unforgivable. There can be no excuse for those who took away the magnificence here. How can one lifetime of wealth for one man, or even a shipload of men, be considered reasonable exchange for a whale-filled ocean? It is to me utterly remarkable that we still pursue this course, still find this argument not only reasonable but also, more incredibly, persuasive. Greed blinds us to history, and it is making our place in the world all the more lonely.

I am here because I believe in my gut that we will not stop, that what I am seeing will soon be the last. For now, the bear is on the shore, a short distance across the water, and every eye is fixed on it. Our voices are stilled because it is a myth come to life, a living dream. The bear does not stay long, but before it leaves it rubs its chin and chest luxuriously along a shoreline snowdrift, pushing itself along with its hindquarters. It then turns onto its back, twisting its spine on the cold, frozen shore with its paws paddling at the sky.

We land in Virgohamna to visit the sites where Walter Wellman and Salomon August Andrée commenced their attempts to reach the North Pole. Wellman led two polar expeditions, from Svalbard and Franz Josef Land and, in 1906, attempted to reach the North Pole in an airship. On his third attempt, he was forced to turn back due to mechanical failure after 65 miles.

Earlier, in 1897, Andrée and his two companions, photographer Nils Strindberg and engineer Knut Frænkel, had attempted to reach the North Pole by balloon. They failed, and historians have written both ridicule and praise of his efforts, but the aspect I like most is an aside. In their final attempt, after the balloon crashed, they made a remarkable journey back over the ice, but were marooned by the onset of winter on Kvitøya, an island off northeastern Svalbard. It is believed that Strindberg died shortly after reaching land, and it is not known whether Andrée and Frænkel died of carbon monoxide poisoning from a faulty Primus stove or from trichinosis, from eating polar bear meat, or just plain exposure. They were found side by side in their small tent thirty-three years later. Strindberg's fiancée, who did not know the fate of her intrepid, waited thirteen years for him to return before finally counting him as lost and marrying an Englishman. When the remains of the men were found, they were taken to Stockholm for burial. On her deathbed, Strindberg's fiancée gave the instruction that her body was to be buried in England, but that her heart should be taken to Sweden and buried beside her first love, Strindberg.

9 JULY

Beyond our porthole window the fog lies low over Monacobreen. Kittiwakes stream past over a silty sea. The fog lifts slightly as we motor out in the Zodiacs, the air growing colder as we near the face. Birds are everywhere. In the corner, where the two glaciers collide, the face of the ice has changed dramatically, but the kittiwakes still feed in their thousands on the unseen upwelling, turning this way and that, picking minute invertebrates, plankton and fish from the water. From time to time they take to the air, the mass of wings rising synchronised from the water and then fanning out like the fabric of a balloon being inflated. I cannot touch the sense that forewarns the feeding, for they take off intuitively a moment before the glacier booms, or a car-sized block of ice falls from the mouth of the blue cavern behind them. Twice, to our right, the main face calves giant blocks that raise an explosive splash to three-quarters the height of the ice cliffs, sending a huge wave surging into the calm bay. The birds return immediately to the site of the crash, and I see two pluck stunned fish from the surface. Gulls and a skua squabble over the prize.

Ice is everywhere. At times, it is a dense slush of small pieces but mostly it is a tight scattering of bergs the size of our boat. The larger icebergs, which after a time turn turtle under their own weight, are the perfect blue, the colour one would dream a lover's eyes. The fog lifts slowly and drapes the surrounding sharp peaks with layers of gossamer trailings, as light and translucent as fine silk.

Our afternoon stop at Vulkanhamna in Bockfjorden is raw by comparison. A red sandstone mountain leaches a series of frothing, brown cascades into the bay under the low drizzle. The sea is rust-orange and as cloudy as milk. A small geyser issues mineral-laden water down the lower slopes of the distant extinct volcano after which the bay is named.

10 JULY

Our ship is anchored close off the shore of Phippsøya in the Sjuøyane, a small cluster of seven islands north of Nordaustlandet. The swift tidal current carries the ice sheets scraping along the ship's hull. Looking out the window gives the impression that we are moving, and it is only the stationary island that proves otherwise. It is snowing and 2°c, with a wind that drives the chill through my parka. A group of walrus swim along the shore, raising their heads to gaze at the ship.

Later, in the Zodiacs, we find them again and when they dive as they swim towards us I peer into the swift-running sea in the hope of making them out as they pass beneath us. The sea is crystal clear, but the bottom is dark except for the occasional paleness of a rock. My fingertips, as I wait with the camera ready, ache from the bite of the cold. I watch two black guillemots swimming tamely ahead of the bow. I can see them for a metre or more as they dive swiftly down, flying underwater with beats of their wings. There are more walrus and seals both in the water and on the ice, but no bear. Every bay and shore is fast with ice sheets and rafts of ice. The wind is blowing, driving the falling snow pellets at a slant.

Returned to the ship, we raise anchor and head east. Our progress is gradual and the ship heels as the captain weaves through the larger floes, ice dragging against the hull. I am busy writing when the ship shudders from a contact, the bow dips alarmingly, and we stop within the space of a few metres. I rush to the bridge to find that we have rammed the ice edge. Ice stretches ahead and there is no way through. We back out and must return on our course to a point where we can head south and then east, before turning north once more. Looking out over the still, white sea, with long icicles hanging from the railing and the ship unable to move ahead, I feel, finally, as if I have come to the place of Nansen, Amundsen and Scott. A frontier. And alone on the open bridge I share their yearning to be upon it, for the trial and adventure that are certain to be there.

Our speed is reduced by the ice, and we slow down even more to circle around groups of walrus hauled out on the floes. After being slaughtered to the brink of extinction, the walrus was finally protected here in 1952. They are making a slow recovery, which is not surprising given their fifteen-month gestation period and the fact that the pups remain tightly bonded to the mother for two years. It is estimated that there are now two thousand walrus resident around the Svalbard Archipelago.

We thread our way to the east of Martensøya, but as we move deeper into the ice the fog settles lower and lower until visibility is not much more than a hundred metres on either side of the ship. Beverly and I stay on the bridge, scanning our close world for polar bears. The hours drag on. After four hours, we retreat to the cabin for a break. Then, as I remove my boots, our expedition leader comes on the public-address system to announce that a bear has been spotted.

It is a female, collared for research, sleeping with her feet tucked beneath her. She raises her head to regard the ship 200 metres distant on the open water, then snoozes on. Her haunches are rounded and fat. After a time, she rises with a yawn and a stretch, and, glancing sideways at the ship from time to time, leisurely makes her way through the patchwork of water and ice that is the drifting ice pack. She crosses through a ridge of tumbled blocks of ice and then starts walking in earnest. For a time, the ship keeps pace, but when the bear lies down abruptly to sleep again the ship moves on.

An hour later we find another bear. It is far out on the huge drift of ice, but it is walking directly towards where the ship is nosed into the close ice of the drift to hold it still against the rising wind. It is a distance of more than two kilometres, but the bear advances on us without deviation and the gap closes quickly. The cold wind is biting, and as the bear draws closer I notice that it frequently tests the air. I am intrigued, because I realise that it is hunting. It is walking on the downwind edge of an ice drift at least five kilometres across. To rely purely on chance to find a seal hauled out on the ice would be a fruitless enterprise, as the place is vast and the seals we have seen are separated by distances measured in hours of sailing. By walking on the downwind edge the bear is testing the air coming the entire distance over the ice floe, cutting the need for its patrol to a fraction of the energy that random searching would require. It also gives it the advantage, when finally smelling a seal, of approaching from downwind where it can be neither smelt nor heard. The sniff is an almost incidental action. It has me riveted.

The bear approaches without hesitation until it reaches the open water 50 metres downwind of the ship. It hunches down in its shoulders to sniff the sea and only then turns to the ship, raising its nose high into the air. Its neck is surprisingly long and, standing with its forepaws on a low ridge of ice, it hints at a size that one does not fully grasp when it is on all fours. It comes forward. Again, I am awed by the fact that curiosity is its overriding demeanour. There is circumspection in its now more measured advance, but the fact that it approaches an icebound ship without any show of fear is to me a measure of the certainty of the bear in its place as a predator. It comes towards us until someone accidentally kicks something metal that clatters across the deck, and it flinches and turns away. Once past the ship, it continues on without a backward glance. It is hunting again, our presence having warranted only a cursory investigation. I watch it until it is no more than a speck on the ice, squinting against the cold wind and white of the sky and the ice-covered sea. Each time I come into contact with this ephemeral life I find myself wondering which bear will be the last one I ever see.

Polar bears need ice, and it is only when one comes here that one fully comprehends what is happening. It is too late. What has been set in motion cannot be reversed in time. The ice is melting. Polar bears have been around for at least 200 000 years and in that time have survived warming periods, but because humans have now occupied all the temperate lands that surround the Arctic, polar bears have nowhere to go. Over the past fifty years, temperatures have risen in the Arctic more than twice as fast – and in Antarctica five times as fast – as the rest of the world. Like most people, I have heard these facts, read the grim predictions, and then moved on with my life without any concerted effort to change. It was only when I sat at the foot of a glacial face 60 metres high, eight kilometres across and 100 metres above where I stood on the mountainside that dwarfed me, and could see a delineation as sharp as black on white of where the glacier stood previously, that the reality of change hit me in the chest.

There has not even been enough time during the change in the Monacobreen for moss or lichen to take hold on the ice-scraped rock. Even if you do not like or want to believe the science, there is the testimony of the animals themselves. Ocean birds of the more temperate regions are being seen further and further north. Gannets are nesting on Bjørnøya for the first time ever recorded in our history; whales are expanding their ranges north and south as the waters they can tolerate move closer to the poles. Here on Svalbard, between forty and seventy female polar bears used to den on Kong Karls Land. Last year there were two. There was just not enough snow. Also, if a female polar bear is not in prime condition, if she has not caught enough seals before denning, she will abort. The seals she feeds on need ice to rest on, and the ice is melting.

It is not this alone, however, that effects this trend of polar bear decline. There are other complications. Svalbard is administered by Norway under the terms of the 1920 Treaty of Versailles, and all of Svalbard and Nordaustlandet on the eastern and western shores of the Hinlopenstretet are designated national park, yet it is only the land on either side of the Hinlopenstretet that is national park. The sea is not, and Norwegian trawlers fish here. If polar bears require seals, and seals require fish, then Norway stands responsible for placing commercial interest ahead of conservation.

It is bitterly cold on deck, but as I turn away from the vanishing speck of the bear and my reverie on its survival, a fellow passenger approaches me. He is a tall Swede with silver hair and a near-silver moustache. His green-khaki outdoor clothing is matching and immaculate. He has the appearance of a man who likes order. We fall into conversation about wildlife. He asks if I had seen wolverine while I was in Sweden, and when I say no he tells me he has pictures of them taken with camera traps in the forest near his home. He seems like a man who likes to walk, and when I mention this he says he is a hunter. He hunts moose. He sets the camera traps to find the places the moose frequent and happens to have accidentally captured images of wolverine. He also hunts the grouse-like capercaillie, and he laments that their numbers seem to be declining. He is convinced that the few images he has of wolverine represent an increase in the population of this shy, small, bear-like predator, which he claims are responsible for eating the eggs and chicks of the capercaillie. This year he plans to shoot wolverine in the hope that the numbers of capercaillie will increase. Too incredulous to continue the conversation, I blame the cold for my need to retreat.

11 JULY

We retreat into Hinlopenstretet to avoid a blustery 35-knot wind that is blowing cold out of the northeast, off the ice cap. The temperature is 2°c and it is snowing. We take to the Zodiacs in a lumpy chop and cruise downwind off the castle-like basalt cliffs of Alkefjellet. Fog surrounds the high turrets. The guillemots fly upwind, coming at us low over the water in tight squadrons. Their numbers seem without end and there is no patch of sky that is not crisscrossed by flying birds. Close below the cliffs the air is loud with their raucous clamour. I scrutinise each tightly packed ledge but see no chicks yet. I am glad that the ship meets us on the downwind side of the cliffs. To motor back would leave us drenched, and the wind is sharp and freezing cold.

We sail back up Hinlopenstretet into Lomfjorden, seeking protection from the weather, and drop anchor in the shelter of Faksevågen. We climb the hill overlooking the fjord. Small groups of reindeer males circle us, their interest apparent but their bravery hesitant. When I walk out to the side of the group and sit still and quiet on the damp moss, they are emboldened and approach to within two metres, pretending to graze but watching me constantly. From time to time, when my head is down, they raise theirs in a characteristic gesture, head held low and neck outstretched, to look at me directly. Their bulging eyes have that bovine quality of seeming near-sightedness. Moulting fur hangs from their flanks in long threads that flutter in the wind like prayer flags. The moss is bright-green, thick and soft, and wet.

We walk up the spongy hillside to the crest and from the saddle are able to look down two valleys. It is obvious that both have until recently been covered by glaciers. Dirty-grey moraine stands like peaked dunes across the valley floor and shallow braided rivers run silver in the sunlight through wide deltas of silt. The remnants of the glaciers are now little more than ice caps that sit as rounded domes in the upper valleys.

We have three scientists qualified to comment in the field of geology among the passengers. Jacek Torbicz, the geographer, has raised my ire during mealtime conversations by claiming that climate change as a result of human activity is merely a propaganda ploy by governments to levy more taxes and put a tradable value on carbon dioxide emissions and carbon. He claims that the earth has always been subject to climatic changes and not all of them have been gradual. He cites the most recent cooling period, during the 11th and 12th Centuries, when the earth's climate changed during a period of fifty or one hundred years. Glaciers, he argues, are not affected by warming or cooling but by precipitation. It is, however, his statement that carbon dioxide emissions produced by humans are an inconsequential factor in climate change that brings me into direct conflict with him. He tells me that I am suffering, like many others, from indoctrination and that a single volcanic eruption far outweighs any human output.

I turn to Dr Hinrich Bäsemann, a considered German geologist, with a silver beard, who looks over his glasses as he talks, and Liz Nesbitt, Associate Professor in Earth Sciences at the University of Washington, for clarification. Hinrich, who works with the Norwegian Polar Institute, speaks deliberately when he explains that climate change has always been a gradual process throughout the earth's history. Those changes that have happened more rapidly are as the result of some marked change in the planet's system of naturally occurring cycles, such as the ocean currents, or volcanic activity blocking out the sun. We are, he says, currently in a period of rapid change and its cause is man-made pollutants, not only in the atmosphere but also in water. The oceans then, too, are changing temperature as a result, with the consequence that ocean currents are in jeopardy.

12 JULY

At Torellneset, there are once again no walrus on the beach and we sail on, passing a single fin whale on our route to Wahlbergøya, where there are three walrus hauled out on the sandy peninsula of the beach. We approach them as a group. I am surprised when the walrus tolerate our presence without flinching and allow us to approach to within 30 metres. I lie on my belly on the shore and have not been photographing long when another walrus approaches from the sea. From my low perspective, the animal's bulk is impressive. It lifts itself from the water, resting every few metres, as if the effort of moving its one-and-a-half-tonne body overland has exhausted it. By the time it has negotiated the five-metre strip of beach and climbed the small tidal step to join the others, another has arrived at the water's edge. By the end of our sojourn, there are seven. They lie in a huddle, close against each other, those in the middle squashed tightly between their outer companions. From time to time they raise their heads or shift their bodies, some flat on their backs with tusks pointed skywards. They frequently rub the wrist of their fore flippers along their muzzles, the back of their heads and their chests, where the hide is pink with small protrusions. With their hind flippers, which open in a great fan-like spread of pink webbing between brown digits, they scratch their bellies and their penis sheaths. These are all males, the females with their calves forming separate harems on their own where the males don't intrude. By the time we leave, Beverly and I have been lying at the water's edge for more than two hours and our biceps ache with the spasm of holding the heavy lenses unmoving for so long.

Our passage to Vibebukta passes through a calm sea adrift with large icebergs. The ice here is more large chunks than sheets, and in the sun take on myriad forms. Some are finely sculpted, others a tumble of geometric and rounded shapes in shades from white to deep turquoise-blue. The sky is cloudless and the sun reflects bright off the sea. The spouts of two whales divert our course and we come upon two humpbacks feeding. They rise to the surface head up, their throats bulging as wide and distended as that of a bullfrog calling. Kittiwakes fill the air above their heads, diving into the krill-laden wake of their turbulence, to feast. The whales feed towards us, coming to within a few metres of the ship. We are looking directly down upon them from the height of the top deck and can see them coming up from the depths, jaws agape. They rise up between icebergs, at times hemmed in on all sides by blue. And on the blue sea, beneath a blue sky, I find myself caught in what seems a figment of my imagination. One comes so far, searches so long, to find a time like this where the earth is magnificently primal. Only three hundred and fifty years ago this place teemed with life. It is my hope that someday it will again. Today, I have a sense of the remarkable character of its return.

It was in 1986 that the International Whaling Commission (IWC), in response to the devastating state of whale populations worldwide, issued an international ban on whaling. Norway, one of the world's principal whaling nations, along with Japan, registered an objection to the ban, but complied nonetheless. In 1993, however, Norway broke the ban, citing its objection, and began whaling again. Japan continued some very controversial whaling in the Southern Ocean under the guise of scientific research. With time, the quota of whales Norwegians allow themselves to kill each year is increasing. The ban remains in place. Their argument, that populations of minke whales can support a limited harvest and that they do not intend to return to commercial whaling but are merely allowing long-standing family traditions to continue, is, however, fraught with paradox and irony.

Between the IWC ban of 1986 and 1993, traditional Norwegian whaling families were unable to practise whaling. This meant that for seven years they were without work. This is too long a period to endure without seeking an alternative livelihood, unless they were extraordinarily wilful about not taking other work; this means that the whalers now killing whales are not doing so because they have to, but because they want to. If, on this premise, they were then killing whales because the hunt and the resultant whale meat were part of a community tradition

– as they are among Native North Americans – then the few whales that would be killed to satisfy this need would in likelihood be seen as an acceptable harvest by many more. However, when hundreds of whales are killed every year, the practice loses any pretence of community-based tradition and takes on only a singular purpose: commercial exploitation. Bearing this out is the fact that Norway exports up to 80% of the whale meat it harvests to Japan. This means that the whalers of Svolvær, a bustling port in its own right, with a lot more industry than just its whaling entrepreneurs, are in the business of killing whales, not in the tradition.

I am compelled then, to contemplate their action in terms of financial necessity, beyond their own family survival, if one is to seek a rational justification for killing whales. Norway is per capita one of the wealthiest nations on earth, thanks to oil and a population of only five million people. Norway's Oil Fund for future generations, which can be seen as surplus funds after social welfare, education, defence, infrastructure and free healthcare are attended to, now stands at over $450 billion. The Norwegian government therefore cannot possibly be encouraging whaling for funds vital to national or personal wellbeing. It is not money that drives whaling to continue. It is just more money.

Beyond its commercial conundrum, whaling is also a highly charged emotional issue. Whales are rare, benign, long-lived mammals that present almost no conflict with human life. Add to this their great size, and suddenly it reduces most people who enjoy an encounter with them to a profound sense of privilege that reaches inside to touch our spiritual selves. To kill a whale takes a hardened heart, and to advocate the killing of whales is a completely heartless decision. Norway, however, practises one of the most active refugee-assistance programmes in the world and presents itself as one of the most human rights-conscious nations in the world. Humanitarian aid is the conscience of compassion, but I am confused as to where this compassion retreats when it comes to life other than human?

The whales are becoming satiated. They lie on the surface and arch their backs as they defecate a rust-coloured stain into the water. One can see the whole length of their bodies, the tail and the long, slender flukes showing white beneath, as they scull at the water. As we sail away I keep my eye on the bustle of the kittiwakes that hover in a milling flock above the whales feeding. From a distance their white column looks like the spout of a whale lingering in the air.

At Vibebukta we come ashore onto a stony polar desert. What plant life there is hugs the ground closely. The land is flat, the colour of suede, and ice forms sheets on the distant slopes as the ground rises in distinct steps towards its low horizon. The stones underfoot are filled with the fossilised remains of sea life, molluscs and corals, from times when Svalbard lay in the tropics. High on the land, a kilometre from the shore, a massive whale jawbone is decaying into the sterile ground. Tiny flowers hug tight against its shelter, feeding on the few nutrients that the bone leaches into the ground.

As we motor back to the ship we come across a polar bear swimming between the ice floes. When it sees us, it turns directly towards us, so that we have to back away. It swims easily, without haste, its chin raising a small bow wave, nostrils just clear of the water. It pauses for a better view, treading water and balancing with its forepaws as it raises its neck and shoulders, until its head is nearly 60 centimetres above the surface. It circles our small flotilla until it finds a pancake iceberg. It lifts its forepaws onto the ice, craning its neck and head forward. The forepaws stretch out as far as they can, dig deep into the coarse surface, and the bear hauls its body up and onto the floating platform with water streaming from the long belly hairs. It walks a few paces, shakes itself cursorily, and then steps to the edge of the ice pancake to survey the strange presence that remains tantalisingly close but retreating. It must feel all our eyes upon it for it yawns widely, its left forepaw against the shin of the right front leg.

I could not have dreamt a more fitting animal for this place. The shore behind it is bleak and pale tan, and I conjure an image of woolly mammoths walking there in single file. Although extinct, they too seem to me to be a creature that was made with this singular country in mind. The bear returns to the water, backing in gently off the iceberg, hindquarters first. It swims directly at the closest Zodiac and we move further away, and it leaves its inquisitive pursuit of our presence and continues on towards the shore, making a cursory check of each iceberg it passes. Out of the water, its body is more yellow than white, the colour of fresh cream, but in the water only the head shows and it is as white as the ice. As it moves away it becomes difficult to distinguish between it and the litter of ice chunks that drift in the sea. Only its movement betrays its presence. We are moving away, but my mind is lingering behind. There is a vigour coursing through me, and I feel like I do when I stand alone on the top of a mountain and stretch out my arms in the wind.

The feeling is still with me when, hours later, we cruise along the face of the Bråsvellbreen, the 30-metre-high ice wall of the Austfonna Ice Cap, the seventh-largest mass of ice in the world. I lean back against the ship's super-structure and watch the waterfalls dropping rivers of ice-melt into the sea. And when fog lingers in patches off the ice, and the sun to the north casts it in gold, and fulmars and kittiwakes materialise slowly through its curtain, flying directly towards us, there is something in me that yearns to hold this still. A painting I could visit when I needed to.

14 JULY

The ship anchors in Hornsundfjorden just as the sun breaks through a thin mist that hangs around the mountain peaks. The mountains are yellow, black and red, their lower slopes ochre-yellow. Drifts of snow lie in the hollows of their sides, and where the two valleys of the heights issue to the sea, a glacier grinds and booms, sending out shockwaves that reverberate through me as I sit writing in our cabin.

We motor through to the face of Hornbreen, the ice on the southeastern shore now standing on rock. Hinrich has been coming here for twenty years, and on his first visit the ice stood where our ship is anchored, far out in open water. Perhaps it is the sun and relatively warm atmosphere, but the glacier face is los-ing more ice than any other I have sat and watched. During the two or so hours that we cruise slowly down the length, we see this ancient ice collapse into the sea seven times. On four of these occasions the calving is large enough to raise a splash almost as high as the face of the glacier, and creates a breaking wave whose swell reaches us a half a kilometre out into the bay.

In the afternoon, we land on the rock-strewn ground that the glacier's retreat has left exposed. We climb to the ice and walk out onto a frozen landscape, where rock, warmed by the sun, drills holes into the surface as it melts it, mak-ing the surface deeply porous and running with tiny rivulets of water. Higher up, the snow has frozen into hail-size pellets, and near the top it is deep and soft underfoot. I fall into a small crevasse, and my foot disappears into a void. It is only my elbow hitting the top edge that arrests my fall. Reaching the heights, I turn to the view of Samarinvågen: jagged, towering peaks reflected in the mirror calm of the deep blue of the bay.

15 JULY

In St Jonsfjorden, we go ashore at Ankerfjella and climb to the plateau, beneath orange cliffs where a colony of kittiwakes is nesting. Some of the party climb the steep slope to the base of the cliffs, but I have noticed a fresh bird carcass in a jumble of rocks. Hoping that it is a fox den, Beverly and I content ourselves with staying put and photographing two impressive-antlered reindeer bulls that graze the rich vegetation below the cliffs.

When the people around us spread out, and become ultimately quiet, as they look out over the views from our high vantage, a kit fox appears in a crevice in the rocks, just below the bird carcass. It is cautious at first, peeping forward then drawing back. Soon, though, it grows bolder, and finally comes out into the sun to sit at the mouth of the den. Its coat is pale charcoal-grey over its back and cream on its belly. The fur is a dense fluff, except on the head and down the legs, and the eyes are quick and bright. We wait for others to emerge, but none do, and after a time the kit tires of our attention and retreats into the den to sleep. It sleeps a long time before its head and shoulders appear once more in the narrow opening to the den. It yawns expansively twice, showing rows of fine pointed teeth, then lies down to doze with its head on its forepaws.

Out in Forlandsundet the wind is raising white caps on the sea. We land in a persistent rain at Alkhornet and walk along the flat land below the triangular cliffs. A pair of long-tailed skuas, with a nest or chick we cannot see, object to our procession and dive-bomb us from behind. We follow the female that is feigning injury until she has led us far away from her brood and takes off. The rain falls harder, and no one is feigning enjoyment, so we return to the Zodiacs.

Beverly and I have arranged for a transfer at sea back to the *Vavilov*, on an Exodus expedition, as their voyage has begun this same evening that ours is ending. Both ships pause in the middle of Isfjorden. Henrik Løvendahl, the expedition leader, interrupts his final presentation to hug us goodbye. The whole staff line up to do the same and the passengers send us off with loud applause. Beverly and I are too moved to speak. We board the Zodiac, with our luggage, in a choppy sea and make our way across to the familiar lines of the *Vavilov*. Viktor, one of the ship's crew waiting on the gangway, is as taciturn as ever. 'Viktor!' I announce happily as I see him. 'Welcome back,' he says.

16 JULY

I find Captain Beluga on the bridge and thank him for collecting us the previous evening. I am flattered when he remembers me, for he is the epitome of an old sea dog, with little thought for anything but his ship. He asks me a lot about the other ship and listens attentively to my impressions. Finally, he asks how long we have been working around Svalbard, and when I answer just over a month, he nods and says, almost under his breath, 'Fantastic!' He turns his attention immediately back to anchoring and barks a command down to the foredeck, while I hover in the shadowed recess of the bridge and watch. His presence charges the room with authority and I search for what it is that makes it so. Most noticeable is his formality. I have never seen him out of uniform, and his posture is stiff and upright. Perhaps it is his economy of attention, for he never allows distractions to reduce his vigilance.

17 JULY

The day begins with cruising Smeerenburgfjorden, looking for bears. We find one right at the head where the fjord joins Sørgattet, but the bear walks over a rise of smooth glacial ice and behind a hill and does not reappear. As we sail over to Raudfjorden, Ian Stirling, retired Professor of Biology at the University of Alberta and a world-renowned polar bear biologist, gives a lecture on polar bears. He speaks off the cuff, and once he switches from bear biology to its dependence on ice, the room becomes sombre. Bears hunt on what is called annual ice, ice that melts and reforms each year. This is because their principal prey, the ringed seal, uses this ice as cover. Ringed seals have claws with which they open breathing holes in the annual ice. Over time, these holes become covered in snow, and the seal creates an igloo-like cave beneath the snow where it comes to breathe and as a place to haul out with its pup. Permanent ice, which is ice that has not melted and to which layers are added each year, is too thick for the seals to claw through, and so is, in terms of the polar bear, a virtual desert. It is the formation of this annual sea ice that is being affected by climate change. Annual sea ice figures for the Arctic, measured at the end of summer, are available from the early 1900s, when it stood at about eight million square kilometres. This figure remained essentially stable until 1979, after which there has been a steady reduction. By 2012 the average sea ice cover at the end of summer has shrunk by 50% to only four million square kilometres.

Polar bears do not hibernate. Winter is their prime hunting time, when the ice lies fast against the land and the ringed seals are confined to breathing holes for air and nursing their pups. Summer, when the ice melts and retreats towards the pole, is the polar bear's time of privation. Seals can surface anywhere and the polar bears are unable to find food. The seals, when they do haul out, do so on pancakes of ice where they can see an approaching bear from a long way off. The lucky bears may find a carcass, but the vast majority spend summer becoming hungrier and hungrier, living off the fat reserves they have built up over winter. It is a starvation.

Historically, the polar bear has always endured this, but it is the duration of this starvation that the decline in annual sea ice is so dramatically affecting. A decline in the amount of sea ice measured at the end of summer has two important impacts on the life of the polar bear. Ice forms late and melts sooner, and this extended period of iceless sea means an extended time during which bears must go hungry. Clearly, it also affects the seals upon which the bears depend.

Research has shown that for a period of up to 120 days without food, the majority of healthy bears can survive, with a mortality rate that applies only to the old and sickly. As this period expands, however, one must remember that it cuts two ways. First, the bear does not enjoy as long a winter hunting season in which to build up its fat reserves, and, second, it must survive a longer period of starvation on less fat reserves. It is clear that this spirals in on itself. At 150 to 160 days, the number of bears that will die the slow death of starvation rises dramatically to between 28% and 48%. Ian's photographs are not pretty: a bear as lean as a city-dump mongrel, a male with the huge head of a bear in its prime, its body too wasted and emaciated to stand. At this point the bears come into conflict with humans and yet more die. I ask him why polar bears are so threatened today if they have existed for between 100 000 and 150 000 years, as is popularly believed, and have survived warmer periods in the past. His answer is simple. There were no humans then to inhibit their movement to suitable habitats and wildlife was abundant, both in the sea and on land. He shows us images of a whale carcass that sustained a population of bears on Svalbard for three years before it was all gone.

I try to imagine the 1560s, when this place was discovered and whales filled the whole fjord, where today I witness silence. When thousands upon thousands of walrus lay on the beaches, where I walk on ground too thick with the bones of their slaughter to touch the soil. The polar bears have survived the changes of the past because the earth was abundant. When I walk out onto the deck to look at this quondam land, it is with the conviction that the polar bear is our most iconic nemesis, for behind its future lies every other magnificent life form with which we have been blessed to share our place in time.

In Raudfjorden we go for a long walk among the stony hills in warm sunshine.

It is windless and from across the water I can clearly hear the explosive boom of a glacier that issues blue between the peaks. In the sand of a small tidal lagoon the tracks of a polar bear, an Arctic fox and a Svalbard reindeer run side by side.

18 JULY

At 5 am we reach the ice sheet west of the Sjuøyane. The morning is calm. A light breeze runs in riffles across the smooth, dark surface of the sea and in the high cloud the sun shines silver in patches as it thins. We see our first bear within twenty minutes, and the *Vavilov* turns and pushes its nose into the ice. It is not solid. There is about 35% to 40% open water between the ice pancakes that drift clustered together. Some are as large as a tennis court, others the size of a dining table for two, and over the distance, blocks of permanent sea ice have broken off to form icebergs that puncture the expanse with the height of their jumbled forms. Captain Beluga exercises the caution of experience and we motor at not more than a knot or two, so that our contact with even the largest plates does not jar against the ship. As we cleave our way into the pack ice, the ice drags and scrapes loud against the hull, and seals surface in the opening created by our passing. The bear has lain down to rest, and only when we are 300 metres away does it lift its head from its sleep and what we thought was one bear becomes two. They are circumspect about the ship, and although they raise their noses to test the wind, they circle around us at a distance of a few hundred metres from where we have stopped, before lying down to sleep again. Scanning the landscape of the distant islands, I spot another polar bear about 1.5 kilometres distant that is feeding on the ice. It is the flutter of gulls in attendance that first catch the eye. I am standing beside Ian Stirling, listening to his explanation of its behaviour, when another bear appears from behind the obscuring ice ridges, walking towards the kill.

We will see an incredible thirteen bears, about 0.25% of the world's population, in the space of a few square kilometres of sea ice today. Bears appear behind us on the icebergs, where they have slept unseen in deep crevices, while we passed no more than a kilometre away. Others walk into view, coming steadily across the ice. Some remain distant, crossing our far horizon to disappear in the jumbled white landscape. The feeding bear finishes its meal and swims across an open lead towards us. It hauls itself out onto the ice with a fluidity that belies its bulk and pads determinedly over the unstable pancakes in our direction. Ian confirms that it is a male in his prime, and his prediction that the bear will come right up to the ship to investigate proves true. At only 40 metres from the bow, the bear stops his rolling-shoulder advance to test the air. He comes a little closer and then turns around the bow, pausing briefly once or twice more to ensure our presence does not signify any food source, before ambling on into his drifting white world. In his wake, the tension of silence gives way to a hush of electric whispers, like a single wave running up a sandy beach.

The second bear that follows in his trail is a sub-adult male and far more circumspect in his investigation. At 300 metres out, he begins to circle wide, stopping frequently to raise his nose high to the air. On our downwind side, he comes closer and closer, and finally certain that we hold neither threat nor food, he sits on his haunches with a disappointed air and looks about himself. His indecision appears quite comical, but eventually he opts just to lie down and sleep. He is no more than 150 metres from the stern of the ship and remains there for nearly three hours.

Another bear, approaching from a distance, greets the prone bear with a brief nuzzling, which Ian feels indicates a past mating association. The newly arrived bear walks around the ship and finds a comfortable sleeping hollow on a raft of jumbled ice, 200 metres off our port bow. There are bears asleep on both sides of the ship, and we can still make out the male walking steadily away, as well as yet another swimming down a long lead in the ice. At a distance, the swimming bear's head looks like little more than an out-of-place duck cruising the ice edge, but the seals are not fooled. I can see them lift themselves high in the water, craning their heads to look after the retreating threat, slipping back beneath the surface with a caution that hardly raises a ripple.

In the early afternoon, I am alone on the bow. Everyone else is at lunch. Beverly and I are taking shifts, watching a bear directly ahead as we push slowly towards it. The bear is about 250 metres away, standing square on, facing our advance, when the first of my fellow passengers comes on deck. I turn to greet him, exchange a few brief words, and then turn back to the bear. It is gone. With binoculars, I search every cluster of ice where it stood and then look to the maze of water passages between the ice, but can find no sign of it. More passengers appear on deck, and with everyone watching, the bear continues to avoid detection. I am teased openly about my ghost bear, and it takes ten-or-so minutes before I am vindicated and the bear climbs out of the water onto the ice, 400 metres ahead. It is walking away faster than we are pushing through the ice. We find it again a few hours later, resting on the ice. This time the foredeck is full of passengers, when at 250 metres distant the bear eases itself into the water and, swimming behind an ice floe, vanishes. The entire ship's complement is watching, many with binoculars, and the bear evades us. I cannot help smiling, not only for its kind vindication of my spotting ability, but also for the humbling lesson of what we are actually dealing with. The bear is a predator, and this animal's example of its skill in the stalk, of making its 400- to 500-kilogram body vanish in a white landscape so thoroughly that so many pairs of eyes, with the advantage of seven decks of height, cannot see it, chills me with the thought of being the seal that it had in its sights. It is fluid, understated and devastatingly efficient. Some 200 metres to our left, off the port side, it climbs onto the ice again, shakes the water from its yellow-white coat and heads towards the open sea. Watching it, we can see huge icebergs, like lone ships on the open water of the horizon, separated from the sky by a rope of blue-grey fog. The fog lies in a band all the way to the seven islands of the Sjuøyane in the northeast, severing their connection with the sea so that they sail disjointed in the sky, like an armada appearing over the earth's curve. The light is warm, both in colour and on my back.

19 JULY

We have spent the night stationary in the ice pack, and now push our way slowly through to its edge, then turn north, the open ocean on our port side, the ice to starboard. We are cruising at three knots, but the still weather has left the ice spread out and we must look far over a maze of ice and open water in the hope of spotting the buttery dot that, I have learnt, is a distant polar bear. Towards midday the fog closes in, and what was difficult becomes impossible. We press on, going a little slower, for navigating through ice in fog is a precarious endeavour, even with the convenience of radar. At these times, the bridge is hushed to silence, except for the commands of the officer on duty, repeated by the helmsman.

A few hours later we emerge once more into the sunshine and a sea burnished gunmetal-blue. We find nothing except pods of harp seals that break the surface in union, crane their necks to look about, and then dive once more beneath the ice to hunt cod. We turn around, and about halfway back along our morning's exploration, turn east into the ice pack. I stand on the bow, peering over the edge with boy-like fascination, to watch the huge pancakes of ice split at the nudge of the pointed bow, the water appearing dark-blue in the lightning-jagged cracks as they open. The ice drags against the hull, its edges crumpling up in flat painter-like palettes where it collides with other ice. From time to time we see bear prints in the snow on the surface of the ice, but no bears.

It is late in the afternoon when Noah Strycker, the expedition's birding specialist, spots a bear in the far distance. It takes us more than an hour of gentle progress through the ice to reach it, and on the way it becomes apparent that the bear is feeding on something. I chafe at our slow progress, especially so when Ian Stirling tells me that typically a bear will finish a kill within an hour. Captain Beluga remains inscrutable behind his wraparound sunglasses. On this occasion, I am happy that the bear we are approaching is the exception to the rule. It continues feeding. Consulting his notes later, Ian tells me that it fed for an incredible four hours before resting.

By this time, we are stationary again, held in the ice at a vantage that looks down on the bear. It is a huge male, fat, every contour rounded, and his gait is a slow-rolling, wide-legged swagger, like John Wayne wearing chaps. To be in close proximity to this animal feels like arriving in some place normally reserved for others beyond our station. It reduces conversation to whispers. When he raises his head and looks at the towering apparition of the ship, his eyes are small black pits that register no expression. His head runs straight into his neck and his body is squat, as wide as it is tall, and his posture brooks no retreat. Looking through the lens, I feel like I am squaring off with Muhammad Ali and am glad for the distance that keeps us separate. The bear has stripped the skin off the seal. It is peeled clean away, attached only over the skull and marginally at the tail. The muscle is exposed in an arch, dark red-black, like meat hung out to dry. The bear is fastidiously paring the fat from the skin with his front teeth, cleaning it with a skinner's care so that the pelt is left shaved of any residue. Ian has told me that this is their habit. It is the fat they consume; the rest of the seal's body is discarded. Ivory and glaucous gulls wander about on the ice, stepping closer to steal cautious pecks at the exposed meat when the bear has his nose buried deep in the far side of the carcass.

In the early evening another bear is spotted, approaching from the northeast. It must have picked up the scent of the kill, for it is coming fast, almost at a run. As it closes, its agitation is clear, its steps a rapid shuffle where the ice thins between floes, and twice its hindquarters slip into the water where the ice is too thin to hold its weight. It is only when it circles to a point directly downwind of the other bear and the kill that it pauses in its headlong rush and stands with its neck fully extended to test the air. It works its way closer, sniffs at the ice, swings its nose in an arc, and comes forward, its agitated anticipation tangible in its quick short steps. It is a female, fat but still not much more than half the size of the male that made the kill.

The male finally succumbs to his bellyful of seal. He is panting heavily, both from the effort and the physical discomfort of eating so enormous a meal. In what Ian describes as extremely unusual behaviour, this bear lies down only a few metres from the carcass. 'Lies down' is, however, not entirely correct. At first, he sits on his hindquarters and remains still, while panting heavily over his distended belly. Then he lifts each of his forepaws and licks them clean. His tongue is exceptionally long and black-purple in colour. He grooms carefully between his long claws, the base of his pads and the long fur on his forelegs. Then, with his forepaws forward and hindquarters raised, he rubs his chin and neck on the snow to clean them. Finally, he lies flat on his belly, his chin hard on the ice, his forepaws folded back beside his body, and his back legs splayed out and stretched backwards, pads uppermost. It reveals this seemingly awkward, bulky animal as remarkably supple. Later I watch in fascination as he twists his neck with apparent ease to points where my own would have snapped, and grooms the more difficult-to-reach parts of his coat.

He rolls onto his side and gives a luxurious stretch, rubbing his cheek up and down on the ice. When the female arrives, he is quick to his feet, and the female, 25 metres downwind, stops dead in her hurried approach and lowers her head. Neither bear moves. Circumspectly, the female raises her head. The male walks a few metres forward. I am fascinated that his defence has nothing of the strut, the commanding arrogance, I have seen employed by other male predators to intimidate approaching rivals. This bear just walks, and it emphasises, more than any strut could, his absolute confidence in his ability to defend his kill. The female retreats across the ice to another floe, about 50 metres away, and lies down. The male waddles back past the gulls that fuss and squabble over the unattended carcass and once again lies down to sleep.

The female, however, is hungry and after a time rises onto her haunches and looks towards the prize. She closes slowly, sniffing at the snow. She is at the edge of the intervening floe, one foot on the ice drift with the kill, when the male raises his head. This time he charges: a few running strides and then a double planting of his forelegs hard on the ice. He skids to a stop in a low spray of snow.

The female turns and runs, her hindquarters tucked in and her head down, like a scolded street mongrel. But she is not cowed, and when she is certain that the male is not following through, she turns at the edge of the floe and looks back. The male has come to where she put her foot on his ice floe and is scenting her track. She retreats a little further and once again lies down. Over the next few hours she makes several attempts to approach, but is rebuffed each time. When the male comes across onto the floe one short of where she had been resting, she moves a safe distance away before lying down once more to wait.

It is after midnight and the sun has softened, drawing longer shadows, when the male finishes his second bout of feeding and moves off 20-or-so metres from the carcass to sleep. The female's approach is almost a stalk. She watches the male more than she looks at the food, and she follows an exaggeratedly zigzagging path as she works her way forward. She steps onto the floe with the kill on the side closest to the ship. Her footfall is tentative, like a child touching a stranger's beard. The ship, for all its bulk, is remarkably silent, with the watchers holding their breath. The squabbling of the gulls cuts loudly through the tension. The female steps forward following the edge: the kill 12 metres upwind, the male beyond. I am amazed that she continues to follow the sea edge of the ice floe, for if the male wakes and charges she leaves herself no alternative but to swim. She is at the narrowest point, between the edge and the kill, and as she steps forward toward the prize the male wakes. He charges, and to my complete astonishment she stands her ground. Ian is shaking his head. The male, at nearly twice her size, is no contest. Ian has photographs in his book of adult bears that have been killed and cannibalised by other bears. She would make a substantial meal. However, the male stops just short of contact. They both hold their jaws half agape and circle the other's head, separated by only a few centimetres. The female turns her neck to the male and lowers her head in what I recognise as a submissive gesture. He lifts his neck, head cocked downwards, lips drawn a little back. Her head drops lower and she takes a step back. There has been no actual contact.

In his thirty-odd years as a polar bear scientist, Ian has never seen a female take such a risk. As with all the patient explanations of this quiet-spoken man, his assumption of her desperate bravado is pragmatically simple: hunger. Both of us are perplexed by this, for the female is in good condition and relatively fat for this time of year. I can sense Ian searching for some more removed reason whose con-sequence would be this dangerously desperate act. The male follows her retreat, sniffing down at her paw prints and lifting his head high, as if testing the air. The female retreats to a floe, from where she has a high vantage of the world around, climbing on top of a jumble of large blocks. Throughout the night she probes at the male's vigilance, but never as closely again. Without fail, the male wakes to her intrusion, even when the weather closes and the wind begins to blow.

In another anomaly of his experience of polar bear behaviour, Ian is surprised that this male has remained with the carcass. Almost without exception, bears making kills will eat all the fat from the seal, then abandon the remaining meat to walk a kilometre or so away from the carcass to sleep. To sleep in the vicinity of a carcass invites interaction with other bears and runs the risk of injury and death. The meat on the seal carcass is rarely eaten by the polar bears and is most often left for the foxes and gulls. This male, however, is eating as much of its kill as it can. Towards morning the female manages to reach the floe and not be chased off by the male, who is resting on a jumble of ice, the carcass safe right below his feet. She finds a string of intestine that the gulls have been picking over and manages to make a few mouthfuls of the viscera that surround the bowel. She stands looking longingly towards the carcass, but advances no further.

By early morning the clouds have closed to within a hundred metres of the ground. The world is once more an opaque ghostliness. The wind raises white caps in the short stretch of open water beside the ship. Rain is falling, and the wind chill on the bow is measured at -18°c. The female bear huddles, curled up, in the lee of a distant block of ice. After twenty-six and a half hours of being awake, I leave Beverly on deck and stumble, numb and ecstatic, to bed.

20 JULY

Two and a half hours later, the weather is as I left it. Beverly and I begin a long day of alternating shifts on deck, trying to catch up on our sleep deprivation. However, it seems that each time we undress and climb beneath the covers there is a staff member at our door to say the bears are moving. We take to catnapping with all our outdoor gear on, only taking off our boots. The male continues with his cycle of feeding and resting. The female circles the prize, like a cat around a fishbowl on a pedestal. At one point her curiosity drives her to investigate the ship, walking to within 15 metres of the hull, directly beneath the bow. As I lean over the edge where the wind accelerates, the exposed tips of my fingers pass through the painful burning stage of being too cold to the point where they become so numbed by the chill that I cannot feel the contours of the camera. I must take my eye away from the viewfinder to place my fingertip on the trigger button. I can feel nothing, and only know whether the shutter is firing by the sound and the brief moment of blackness in the viewfinder.

We are all in a mild sleep-deprived daze. Ian places his field notebook, containing fourteen years of accumulated periodic notes, on a hawser port, failing to note the drainage hole at the top of the guide plate. His notebook and years of fieldwork tumble into the sea. He is fatalistic about it, but his devastation is clear. As the ship drifts slowly backwards in the strong wind it reveals something remarkable: Ian's notebook is floating. I have from the outset been impressed by the way Captain Beluga and his team of officers have handled the ship, but when, in a wind of 30 to 50 knots, they manoeuvre the 6 718-ton vessel to bring the port-side bow between the bear and the book, their true finesse is revealed. Standing on the point of the bow, keeping watch on the male bear close below, I can hear John Rodsted, the expedition leader, close behind, giving instructions to the bridge: 'Two metres forward. Okay, a little more. One metre to port. Perfect!' With the wind gusting and ice pushing hard against the hull, the ship is manoeuvred gently enough to pinpoint something as small as a notebook in the sea. John drops a net on a line over the side and retrieves the red-bound notebook. I turn and make a photograph of Second Officer Vitaly looking down from the bridge. It is the closest I can come to a salute.

Late in the afternoon the female makes a final attempt at what remains of the carcass. Rebuffed once more by the male, she wanders off slowly into the ice, passes her resting place, and keeps going. Her form grows smaller, a yellow movement in the patchwork of white, grey and blue. Then the rain draws a curtain across the scene and our world is reduced once more to the male and his entourage of gulls on the red-stained ice floe beside the ship. In the evening, it begins to rain in earnest and I find myself content to accept this excuse to abandon the world outside and sleep. Through my porthole, as I close the curtains, I can see the male lying asleep on his side in the ice, 50 metres away. The rain paints rivulets across my view. I am cold. The bear looks at peace.

21 JULY

Opening the curtains to the morning, the bear remains just 50 metres away. The sky is still close and bleak, and the bear is asleep. The seal carcass is now little more than a red-tinged skeleton, the ribs starkly raised against the ice. Vociferous glaucous gulls and delicate, pure-white ivory gulls are picking them clean. Except for its eye, yellow bill and black legs, the ivory gull is at times almost invisible against the shadowless white background. It is small in comparison to the glaucous gull, perhaps half its size, and it is amusing that although it gives way to the glaucous gulls, it is far braver than the larger gull in its advances to the bear. It is also most interesting to see that squabbling over the prize is almost entirely constrained by species. Glaucous gulls are vociferous when another approaches, calling with their necks arched and their wings spread and attacking any bold enough to land too close, but ignoring ivory gulls that do the same. The ivory gulls are all wary of the glaucous gulls, but with other ivory gulls their defence of this bloody patch of ice is vehement and persistent. The bear is unconcerned with either. When an iceberg, close behind the bear, loses a chunk with a splash into the sea, he picks up the skin that is clamped beneath his giant feet and tears at the cape, scraping away the very last residues of fat and flakes of meat. On the far end of the skin, an ivory gull pecks and pulls in the opposite direction from the bear.

With the skin clearly exposed and untwisted on the ice, we can see that it was a bearded seal that the polar bear caught. The skin of the bearded seal is almost twice as thick as that of ringed seals. At about one centimetre, it is also markedly tougher, and perhaps explains why polar bears do not often eat these skins, while the thinner skin of ringed seals is often eaten along with the fat. At the end of the skin I can see the skull of the seal, with the stiff long whiskers that give it its name sticking out to the side. The bear worries the skin for another hour then drops it, stands up, walks to the edge of the floe on which he has become marooned and steps into the sea. The fog is low, with an icy, light precipitation, and our whole world is 400 metres across. Around us, the sea is an ominous grey-green, on which a few tabletop-sized pieces of ice bob in the light chop.

Over the past two and a half days, during which we have remained beside the bear and his kill on the small patch of ice, we have drifted 27 nautical miles, or 50 kilometres. There is no sun, no point of reference, but as the bear swims away he is heading only two points off due north. He is headed back to the land that is no land at all, just ice.

With visibility reduced to zero by the fog, there is little point in searching for other bears, and we turn south towards Spitsbergen, once more nosing our way cautiously through a white world, on a white sea, beneath a sky that clings low to our advance and closes swiftly behind. On the bridge, conversation is terse and the duty officer stands before the big windows with his hands behind his back, watching. His gaze switches from the luminous glow of the radar to the sea.

22 JULY

Sorgfjorden is as serene as a glass tabletop. The mountains that rise around its western perimeter are layered with mist trails. The air is 4°c and still. About three hundred years ago this place rang with cannon fire when the French tried to oust the Dutch from control of the whaling here. Fifteen ships were lost. Today, it is laid with a quiet that even the occasional bird, flying above its own reflection, does not disturb. On shore near the point is a cross, held upright by a stone cairn, and below it walrus lie hauled out on the smooth-pebbled beach. Their stench wafts down to where we land, the pungent, gagging odour of seal and old fish. The walrus lie so close upon each other that it is impossible without looking down on them from above to count how many there are. A fair estimate seems to be around fifty. They are males. Their tusks vary with their ages, from toothpicks to impressively curved ivory of 60 centimetres or more. Interestingly, the tusks have little functional purpose except as a visual attribute when males are courting the harems of females during the cold dark months of December, January and February. I have seen tusks raised in a threatening posture during jousting in the sleeping huddles and once saw them used to prod a neighbour into moving, but walrus are not known to use their tusks to fight to gain mating rights. Instead, they sing. On land, they utter seal-like barks and grunts, but beneath the water they have a repertoire of utterances that they string into song. They have a tap, a knock, a sound rather like a strummed string instrument along the lines of a zither and, most intriguing of all, a church-bell-like gong, complete with long-fading resonance. Males sit in the shallows beside colonies of females and sing their own unique song. The song is repeated precisely over and over, and research shows that with each new breath a different verse is sung, but that it is always the same verse after the same breath. Breath one, verse one, breath two, verse two, and so on. One male was recorded as singing for sixty-two straight hours before taking a break. Right now, they do little but sleep, and the heat rising from their tightly crammed bodies makes the view of the distant mountains shimmer.

Across Hinlopenstretet, we do a Zodiac cruise around the barren islands of Murchisonfjorden. The islands are small and without any sign of vegetation, the milky-tea-coloured stone bleached to a grey-white at the crust. The land here was once buried beneath huge plates of ice, thick enough to push down on it so hard that it actually sank the land into the earth's crust. As the ice melted, the land, freed from the pressure exerted by the ice sheet, rose once more to its natural height. Science has termed this resurfacing of the land isostatic rebound. Many of the beaches and coastal plains of Svalbard are areas of isostatic rebound, and the whalebones lying kilometres from the coast and the fossilised corals in the yellow stone are proof of their aeons beneath the ocean. We find a small group of walrus on a beachhead, and in every cove Arctic terns dip and dive into the icy sea. Beneath the water, rectangular jellyfish with long tentacles light up in a flashing run of neon lines that pass from their dome to their foot.

23 JULY

We wake to high cloud that, breaking open to the south, draws a tempting layer of golden light along the horizon of Hinlopenstretet. We are stationary off the bird cliffs of Alkefjellet, and the cloud is high enough that the dark peaks and columns stand clearly visible. We take to the water for an early pre-breakfast cruise, but the mist is not easily fooled and we have only just begun when it materialises to hang its pennants from the heights and basalt pillars.

The day is calm and birds mill in a raucous melee over every inch of sky. Those descending into flight from the cliffs skim close above our heads as they gather speed. They mill about the sea below the cliffs in meandering flotillas. And when they drift too close, they dive beneath the turbid, snowmelt-blue water or flap laboriously away, beating at the surface with their narrow wings like nesting birds feigning injury. The sheer dramatic rise of the black cliffs out of the sea, until they touch the clouds above, the graphic form of the dark rock, with the staggered, horizontal symmetry of the guillemots huddled tightly side by side on the narrow ledges, and the noisy cacophony of life, give the feel of sitting for a time in some other world. A place of the imagination, vivid in its detail and fantasy.

After two hours in the cold we return to the ship for breakfast, then, like Romans gorging at a feast, return once more to the Zodiacs to cruise one last time past this fairy-tale place. We have just arrived at its northern end when a fox is spotted high above, searching the base of the cliffs for fallen birds or eggs. Its coat is charcoal, grey and white, and when it pauses in the jumble of the steep, bouldered scree it becomes, in its stillness, quite invisible. It searches animatedly, agile across the strewn boulders, pausing to sniff in crevices or to turn alert to some sound from above that I cannot hear. It comes right down to the sea's edge, where the wavelets shatter in a burst of white against the smooth rocks. It acknowledges our presence 50 metres offshore with only the briefest glance, then crossing a patch of old snow climbs nimbly once more up the steep slope to where the cliffs end. The narrow steps at their base prove no hindrance to its search, and I watch, holding my breath, as it jumps upwards from ledge to ledge across a low face to the scree above. It is disappearing into the heights when a polar bear is spotted on the shore.

It is a young bear, more white than yellow, and much smaller than those we have recently watched out on the ice. It is eating a guillemot on the rocks just above the sea, holding the small carcass between its forepaws and tearing away the flesh. The bird does not provide more than a few quick mouthfuls. The bear patrols the boulders along the tide line, and then, encountering two columns rising sheer out of the sea, climbs up the steep green slope beside one of them. At the top of the columns glaucous gulls are nesting, their fluffy chicks sitting vulnerable and exposed on the outer ledges. The bear appears directly over their pinnacle and the adult gulls dive-bomb and worry the bear, swooping to within less than a metre of its head. It pays no mind at all and peers down onto the chicks, coming cautiously right to the edge, its head craning forward. The drop is sheer to the sea. The bear hesitates, waving its head slowly from side to side. The earth beneath a front paw gives a little beneath its steep purchase and a few stones clatter onto the nest ledge. The chicks sit still, while the adults hovering just above call loudly, their bills wide open. The bear is stalling; the drop is its body length plus a little. Finally, it decides against it and turns away, sending a small shower of stones onto the ledge.

Between the rock pillars, the bear descends to the sea and, probing into a crevice with its nose, ignoring the wash of the waves, reaches in with its paw to where it had been trying to find a scent. The paw does not fit so it changes feet and tries the other. When this too does not work, it shifts its position, and with the paw held at an awkward twist it reaches once more into the crevice, pulling its shoulders into an uncomfortable skew. I am reminded of the first time I heard my father curse, standing beside him, holding a spanner, as he contorted himself in the narrow cubbyhole beneath the bathroom basin trying to repair a leak. The bear withdraws its paw slowly from the crevice and with it comes the extended end of a wing. It grasps this in its jaws without pulling and then reaches down once more with its paw. Gradually, it works the bird free.

On the next column, it finds a steep access down to the first ledge, but there is no nest there and it crosses this narrow place to the slope beyond, like some circus animal of the high trapeze. It makes its way across the slope, the steepness apparent by the small slides that run from where it plants its broad feet. When it climbs straight upwards it is almost standing vertical, and I can see the apprehension in its shoulders when it is pulling hard upward and the earth begins to give. It finds a place of long grass and rubs its chin and chest on the soft green carpet, as I have seen others do on patches of snow. It rubs its side, pushing hard, arching its chin and its neck up and away like my dog does when I tickle her waist and flanks. It finds another dead bird of which I can only see the feathers on its muzzle. This snack, too, takes less than a minute to eat. The bear works its way across the scree slopes towards the nesting cliffs, but finds its path blocked by a cliff ledge cut by a small waterfall. The distance across is not far, but the slope is extremely steep on both sides, and after probing the hindrance from several angles, it opts not to try the jump and heads back downhill to the sea. Just above the water, it must come down a low rock face before the relative flatness of the tidal boulders. It does this head first, dislodging stones with its splayed, creeping advance. From the last ledge, it must make a jump to the rocks below. It does not hesitate. It walks fast across the last open ground separating it from the nesting cliffs and, without pausing at their base, begins to climb. There is none of the bulky, waddling gait of the bears we saw on the ice in this recently weaned female, which Ian estimates to be two and a half years old. She heads determinedly straight upward, probing to the side wherever there is a ledge wide enough to fit her bulk. I watch amazed when she rises to her hind feet in a place too narrow for her to turn around, and sniffs at a ledge where a cluster of guillemots have just given way. She climbs steadily higher, and coming out on the top of the cliffs, clambers over their narrow crowns to their face and out onto the ledges, one above the narrow nesting places. It is a dizzying precipice to the sea below. Guillemots fall from the cliffs like rain, blackening the sky like clouds that sweep over our heads and out to the sea. The bear squeezes herself down onto a lower ledge, but the eggs are still too far below and she climbs back and out behind the cliff's dome. A shower of guillemots to our left warns of her advance in a higher place. She climbs on a narrow swathe of earth, cliffs above and cliffs below. At 180 metres above the sea, the ground is so steep that she walks with her shoulder pressed against the side, her outer feet placed in half steps lower than those closest to the cliff. Again, she comes to the edge, causing another fall of guillemot rain. One particular ledge of birds, however, remains steadfast. Separated by no more than two metres, birds and bear eye each other, both with their necks extended and heads cocked to the side. The bear retreats on all fours, in reverse. There is no other way. We finally lose sight of her, diminished by height, behind one of the castle-like basalt turrets of the cliff. The guillemots return fluttering to the ramparts as we drift away, too animated to talk coherently.

24 JULY

Our pre-breakfast excursion to Monacobreen is a slow meandering through brash ice in a cold wind. The Zodiacs have just reached the main face when a colossal glacial column topples into the sea. The explosion ricochets off the wide valley and the spray blossoms into a bomb-like blast, arcing about and away from the face. Thousands of calling kittiwakes take to the air, lifting like a gossamer fabric from the sea and spreading the fine veil of their myriad wings against the powder-blue backdrop of the ice. The wave it creates breaks white and frothing through the debris of floating ice, its motion causing several icebergs, the size of buses, to turn turtle. By the time we have motored slowly along most of the length of the face my fingertips are numb with the cold. Although I have not measured it, the air is decidedly colder closer to the glacial face. The sea temperature is 3.8°c and the brash ice bumps hard through the fabric keel of the Zodiac, thudding against the floorboard beneath my feet.

After a warming breakfast, we return to the glacier to photograph the swarms of kittiwakes feeding on the upwelling at the foot of the ice face. We remain a healthy distance off, and during the pauses between the rifle-shot crack and thunderous boom that sends the kittiwakes skyward, I look at just how much the face of this gigantic ice river has changed since we first came here, almost a month ago. The deep ice cave in the western face, where we had first found the giant flocks of kittiwakes feeding, is entirely gone. In the southern face a shallow, bowl-shaped U, a kilometre across, has developed in the centre of the main face, the edges advancing now as jagged pincers. Ice falls constantly. From the centre of the U a silt-laden underground river is flowing into the sea at a current of several knots, over an area perhaps half a kilometre wide. We must keep the motors running in gear to hold steady against its flow. The kittiwakes circle in a ceaseless revolution: landing at the foot of the face, feeding on the current's upwelling, and then taking to the air to return to the face when it has pushed them too far out. It is impossible to follow a single bird; they mill in their thousands, both on the water and in the air.

In the afternoon, we land near Texas Bar, a little way back down Liefdefjorden from the glacier. There is a dead polar bear on a domed island close to the shore. It lies as I have seen the other bears lie on the ice: its hind legs splayed flat out sideways, the paws facing upward, its belly pressed to the ground. The forepaws are the same, so that its chest and chin lie hard against the stony ground. A few small clusters of purple saxifrage bloom beside its jaw and nose. Its spine, hip-bones and ribs stand raised in relief against its skin. It has starved to death. Ian Stirling has removed the small, white tag from its rounded ear, and he relates the information of its life history as I stand downwind of its sad, still carcass, wind caressing the long, yellowed hair.

It was a sixteen-year-old male, a bear that should be in the prime of its life. For the past four years it had been resident in Hornsundfjorden, to the south of Liefdefjorden, on the southwestern shores of the main Spitsbergen Island. It was last tranquilised and its measurements taken on 13 April 2013, when it was still in Hornsund and in average-to-good condition. This year, however, there was little or no fast ice in the fjords of western Spitsbergen and the bear had no suitable habitat for hunting seals. Without the ice cover, the seals were not reliant on breathing holes and could surface anywhere, making the bear's task impossible. It is only reasonable to assume that it was hunger that finally drove this accomplished adult male to abandon its familiar territory and march north towards the ice. The fact that an adult male bear found healthy in April then starved to death three months later, 290 kilometres (as the crow flies) to the north, is testimony to the severity of the landscape it crossed on its journey and the singular lack of food to be found in the absence of ice. The name Spitsbergen means 'sharp mountains'. This bear would have had to cross numerous high ranges, ice caps and glaciers as it moved north, and would have had either to walk hundreds of kilometres of detours to circumvent fjords or to swim across stretches of ocean, where from its departure point it could well not have seen the opposite shore. The swim across Isfjorden would be approximately 20 kilometres and Bellsund

25. Ian has records of a polar bear that swam 680 kilometres in a single journey. It lost 20% of its body weight in the eleven days that it took to make the swim. The bear at our feet used every ounce of its reserves to survive. Not only is the fat entirely gone, the muscle is so atrophied that there is no covering of the bones at all beneath the skin.

I have only truly considered hunger once in my life, when I found an elderly man making a meal of rats he had caught in his field. Hunger is one of life's most powerful forces, at the base of Maslow's hierarchy of needs. And it is hunger that drove this healthy, experienced predator to a desperate march, the demands of which sucked the very life from its body. On a hillside, with a group of people looking on, when Ian finishes his chronology there is no sound save a skua calling from the sky. There is nothing one can say when one finally stands before so literal an epitaph to the disappearance of ice.

25 JULY

On the beach at Alkhornet, I pick up one of the smooth, rounded pebbles. It is black, the size of an egg, broad whorls of white quartz running crosswise. Wet, it shines with a dark lustre. I weigh it on my palm, turn it slowly held in the tips of my fingers. It would make a fine gift for my mother, who loves such things. But as soon as I have this thought, it alters the stone, and I drop it back in the small surf and watch as it rolls in the wavelets. I will tell her about it, describe it as if I held it then in my hand, and that will be enough.

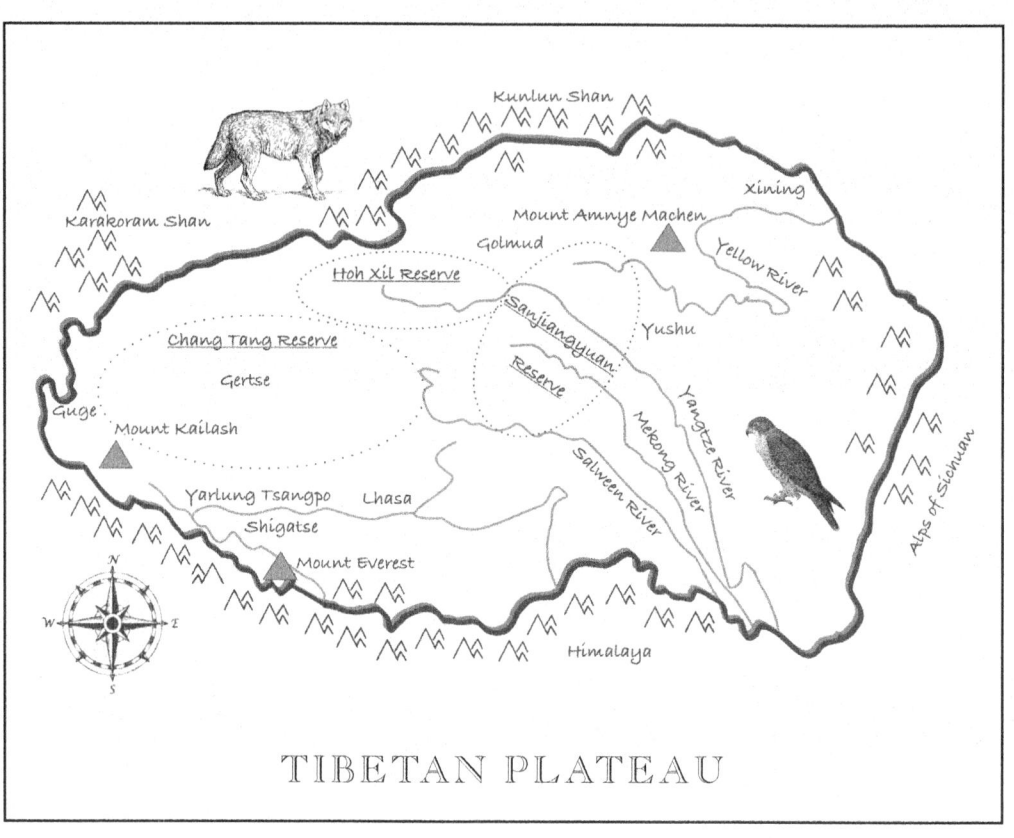

Kunlun Shan

Karakoram Shan

Xining

Mount Amnye Machen

Golmud

Yellow River

Hoh Xil Reserve

Sanjiangyuan

Yushu

Chang Tang Reserve

Reserve

Gertse

Guge

Mount Kailash

Yangtze River

Mekong River

Salween River

Alps of Sichuan

Yarlung Tsangpo

Lhasa

Shigatse

Mount Everest

N

W   E

S

Himalaya

TIBETAN PLATEAU

V

ASIA
The Tibetan Plateau

AUTUMN 2013

My own voice is quiet this morning, stilled by the cold and the sound of the kiang across the valley. I have heard them close in the night and in the black of the land. Now they call from all around. The cold is a vice on every exposed part of me.

17 SEPTEMBER 2013

The small yellow sign, with red type and red trim, above the bath in our hotel reads:

TIPS FOR SAFETY.
DO NOT TAKE A BATH IF YOU ARE:

(1) With Diseases in Heart. Blood Pressure. Or Under No Good Health Conditions.
(2) Under Age 12 or Pregnant.
(3) Drunk.

We are in Xining, China, a country of steep and orange-brown hills, where the dust blown constantly onto the city sidewalks leaves one feeling the need for a bath. Xining lies in what was historically Tibet, and the present-day Chinese province of Qinghai has cut through traditional boundaries with the same regard with which the colonial powers divided up the lands of Africa and South America. Less than 12% Tibetan, the city feels wholly Chinese, its austere high-rise centre bedecked in gaudy neon signage. With no more than 2.3 million people, Xining is small by Chinese standards, but at 11 pm, with the window of our smart downtown hotel open due to the failed air conditioning, the traffic runs a solid river of light and sound twenty-two floors below. We have eaten dinner in a near-empty Tibetan restaurant, where I was glad to discover that I liked yak-butter tea. In the weeks to come, I am certain that politeness will demand that I drink it frequently. We are to spend the next thirty days exploring as deep into the backcountry of the Tibetan Plateau as roads, our legs and Chinese authorities will allow. It is with images of vast steppe grasslands fringed by rugged mountains, of nomads with yaks, of earth's last great open country, that I turn out the light, my head aching slightly from the effects of high altitude.

18 SEPTEMBER

In a hazy dawn, we are met by Tenzin, our guide, and Gongpo, our driver, in the Land Cruiser that is to be our vehicle for this part of the journey, as we are not permitted to travel in our own vehicle on the Tibetan Plateau.

At the height of our climb over the Nyida (Sun and Moon) range, the altimeter reads 3 400 metres. On the far side, the road descends to a wide steppe, with mountains curtailing its far reach in every direction. The land here is abused by overgrazing, and sand forms small dune fields in the distance. I count four sets of power lines that run parallel with the road across the flat country. I notice that the land seems recently fenced, a fact I find perplexing given that this is nomad country. Already the presence of nomads is evident, their distinctive white tents dotted randomly across the sweep of the steppe. I learn later that the fencing is a measure brought about by the government, as far back as the 1990s, seeking to parcel out land to the Tibetans. Perhaps the intention was to induce through ownership a sense of stewardship, or to encourage Tibetans to become small-time ranchers, but whoever the architect of this plan was, it could not have escaped their notice that the fencing of open land would ruin any continuance of a nomadic tradition.

At the small town of Pelka (as this is a journey across the Tibetan Plateau, I have chosen to use Tibetan names or those commonly used by the Tibetans), we encounter the remarkable phenomenon of an almost entirely new town. The road is blocked by slow-moving construction vehicles and is running with mud after a recent downpour. Set 25 metres back from the highway, kilometre after kilometre of identical two-storey buildings face the road across a boggy quagmire of construction debris. The buildings gleam with fresh paint, new roofs and glass still marked with large white Xs. They are all empty.

The old part of town that runs off at a 90° T-intersection is not half the size of the new, unoccupied extensions. It is a mud street crowded by a jumble of small-fronted shops where the distinctive features of its inhabitants are clearly Tibetan. Tenzin points at the red flush on his own cheekbones as the mark by which one can distinguish a Tibetan from a Chinese. But it is more than that: the Tibetan skin is sallower, the eyes more rounded and tapering longer, above pronounced high cheekbones. They are a striking people. I step gingerly across the muddy street to photograph a truckload of sheep and a woman loading coal, while a teenager with dark mascara and black plastic high heels sucks on a pink ice lolly. Many of the women have their heads almost entirely covered by a headscarf, with only the eyes and nose showing. Unlike the severity of the burka, however, the coverings are vibrant greens and pinks or pure shiny white. The women smile shyly whenever I catch their eye. Some wear heavy ornate gold and silver jewellery.

As we travel, we see more and more development. Some settlements are only a few hundred metres on either side of the highway, others whole towns radiating out like the spokes of a wheel from what are established villages and towns. All of it is new, most of it vacant. I am told that the government is building these expansions to encourage nomads to settle, citing education and healthcare among its motives. Another programme is to resettle nomads in towns and give them limited subsidies in a purported effort to protect the land. But there is resistance. The nomads are attached to their lifestyle. The crux of its principle is that they be free to move. I understand their confusion at the government's lack of understanding that if each family of nomads, with their herds of several hundred sheep and yaks, were to gather all in one place for anything more than a short period, there would soon be no grazing. And yet the government persists, building rapid expansions to traditional settlements and denying healthcare and education or imposing fines on the nomads who do not use the settlements.

In the late afternoon, we descend into the spectacular Chuchen Gorge.

Towering earth cliffs, eroded into spires and arches, stand like warriors along the edges, with herds of yak grazing the flush of green at the cliff bases. Through the next gorge, tucked behind a tall, freestanding mountain of weathered grey granite, we find the Dakar Tredzong (White Monkey Fortress or Monastery). The monastery, too, is under construction, with three new terraces of monks' apartments being built. Overhead, Himalayan griffon vultures and a lone lammergeier are circling against a bright grey sky. Those lower down twist and weave through the fissured contours of the stone towers of the hill, turning above the monastery to disappear once more between the granite monoliths. The vultures are here to feed on the dead. In accordance with the Buddhist belief in the Circle of Life, the dead are carried by the monks to a special site on the mountain where the vultures descend to consume them, thus completing the circle. It is an auspicious burial, and the monastery receives much of its support through bequests calling for this ceremony and the attendant prayers.

In a courtyard beside where we park our Land Cruiser, two bharal (Himalayan blue sheep) are feeding. At my approach, they jump nimbly onto a low structure, and from there onto a wall, onto the roof and down the other side. I follow them through the muddy maze of temporary roads and litter-strewn rubble onto the hillside beside the monastery. I must stop to catch my breath; the monastery is situated at an altitude of 3 500 metres. Looking up, I see what I have not noticed before: a herd of bharal grazing on the close cliff, two silhouetted against the sky. Their colouring is so cryptic that even at these close quarters I can only discern some of them when they move. The herd is mostly females with young lambs. I see two males, their forelegs, belly and rump marked with distinct black stripes. The older has widespread horns that are laterally flattened with two tight spirals. They are incredibly agile on the narrow footholds of the cliff and, secure in this unattainable refuge, they allow me to approach quite close. I make images of them between draped prayer flags, the vultures turning high overhead. Later, we climb the monastery steps to an empty courtyard where carved dragons and lions stand silent sentinels. A young monk comes bounding up the steps with more energy than my panting lungs could hope to muster. He is clutching his cellphone and the keys to let us in.

The huge, brass-studded, red wooden doors swing open and we are greeted by a dimly lit pillared vault ablaze with vibrant colour. The pillars are wrapped in a bright orange, carpet-like tapestry down which snarling black tigers descend with claws extended. On the floor, mats have been laid in long lines beside the pillars. They run the length of the monastery from the altar to the entrance. All are covered for their entire length by intricate tapestries that are now worn with use. From the ceiling hang a multitude of bright pennants, some plain, others embroidered with saint-like images; they have the texture of silk. The young monk lights candles along the altar end of the hall. They illuminate framed portraits of the Dalai Lama, the Panchen Lama and the Lama of the Tredzong.

The hotel at Tsigortang is remarkably smart inside, given the dusty, unkempt streets onto which it faces. Our room has fake-gilt mirrors, an orange ceiling and luminous purple curtains. The restaurant we eat in is traditional Tibetan, with an iron stove by the entrance and a brass kettle simmering. The proprietoress wears large, gold pendant earrings and sports two gold teeth. She shows us to one of the carved, low-bench alcoves while the few other patrons of the restaurant stare open-mouthed at us. We are seated at the street window and within minutes several passersby have stopped and are gawking openly. By the time our meal arrives the restaurant is full and every patron takes their turn to sit with us and have their picture taken. They are quite unabashed about the intrusion, and five or six cellphones choose their angles about two metres from our faces. I have yak dumplings and the food is hot and good.

19 SEPTEMBER

Changlulin Pass, in its long, gradual, rather straight ascent, hardly seems like a pass at all, except that its highest saddle stands at an impressive 4 400 metres. We descend quickly, and on the flat plain beyond find a place to pull entirely off the road to photograph the mirror-calm Duktso (Poison Lake), with the reflections of the mountains and clouds that stand at its far shore. For the first time we notice the small, large-eared ice rats that we recognise as plateau pika scurrying for cover on the open grassland every time one of us moves. I get down on my belly and work my way closer to one that sits motionless at its burrow entrance. It has a head reminiscent of a rodent, but with large, round, unfurred ears and the body a little bigger than a hamster. Their burrows are everywhere, and once one realises the scale of their occupation of the land it becomes apparent why they gained a reputation as a pest, adding to the pressure of overgrazing by yaks and sheep. Through the lens I watch one nibble grass tufts with the panicky urgency common to all rodents out in the open.

We turn southeast off the highway onto a dirt road, where we stop to photograph yaks grazing on a hillside covered in prayer flags. Raptors are suddenly everywhere, and so are the pika. Walking out to photograph two upland buzzards squabbling over a pika that the first has caught, Beverly sees a third make a successful stoop on yet another pika that has wandered too far from the sanctuary of the burrow. Underfoot the ground often collapses into their tunnels, and the yaks I am photographing stumble as frequently as I do.

Stopping at the tiny new red-brick settlement of Tawo Zhulma for lunch, I fall in love with the long-sleeved embroidered chupas, or woollen coats, worn by the locals. They are sheepskin, dyed dark-brown, edged with striped, neon-bright brocade. They are mostly handmade and each is embroidered with a distinctive pattern, some lavish, some sparse. They are bound at the middle with a belt as fanciful as the owner's whim, often with shiny metals sewn to the cloth. The men often wear their chupa with one shoulder and arm free, the sleeve hanging loose down to their knees or tied to the belt at their waist.

Beyond the town we drive on a narrow road into a valley flanked by snow-capped peaks. The sky is broken cloud and shafts of sunlight strike the treeless landscape, where herds of yak graze almost to the snowline. The valley floor is cut deeply by a small stream, and the eroded grey and pale-yellow earth cliffs are cut into such narrow peninsulas by the slow erosion that they resemble a line of sculptures made by a more purposeful hand. We pass the Guru Dzong, its squat four storeys all reds and white with gilt-golden roofs, where the monks are hurrying to prayer. High on the extremely steep hillside directly behind the monastery, a triangle of red prayer flags, 80 metres long on every edge, flutters like a field of birds' wings in the stiff wind.

The peak of Dirgdela Pass is just a few metres over the 4 800-metre mark. I have a mild headache from the lack of oxygen. The pass flutters on both sides with vast lines of prayer flags tied between wooden masts and cairns of rock. Lifting and waving in the occasional breeze, they have a festive, happy atmosphere. There is a brief window of sunlight between the clouds and Beverly and I try to hurry to a good photographic vantage point, but are breathless after even a brief exertion, and the best we can manage is a steady walk. A small lorry arrives from the opposite direction, wheezing smoke. It is carrying three horses wrapped against the cold in bright, floral blankets. The owners stop and climb down, and, walking to the incense cairn, shout as they throw tiny squares of

I look up from my work above Salisbury Plain, on South Georgia Island, where hundreds of thousands of king penguins gather to breed and raise their chicks.

The final reaches of the Sierra de los Andes, on Tierra del Fuego, are notorious for vicious katabatic winds that accelerate down from the heights to pummel the lakes and hills of the Karukinka wilderness at its feet. I, like John Muir, am invigorated by storms; they draw me out to stand in the face of the power of the elements.

Beverly stands braced as our ship heels more than twelve degrees in a storm that batters the vessel with winds in excess of 80 knots off the Antarctic Peninsula.

Barefoot, I walk the tidal flats of the Estero los Patos, Chile, where we found dolphins hunting shoals of fish in the shallows, and a sense of quiet serenity that lured us to stay.

On the northern boundary of Parque Nacional Perito Moreno, Argentina, where the 3 706-metre peak of Monte San Lorenzo stands obscured by cloud, I return defeated after tracking a puma that outsmarted me.

Beverly walks beside the Helio Courier aircraft that dropped us and our gear in one of the remotest places we were to visit, the eastern Brooks Range of Alaska. Note the outsized balloon tyres, which enable it to land where no landing strip exists.

Beverly and I stand beside the Jago River in the Arctic National Wildlife Refuge. The head nets are used to keep away the swarms of mosquitoes that are at times so thick one cannot help but breathe them in.

Having hiked far into the hills away from the road to Deadhorse on the North Slope of Alaska, Beverly waits with our camera set up for the herds of wild caribou to pass by. The caribou came in waves that were hundreds strong, and crossed in minutes what it would take us almost a day to walk.

At the Upper Sagavanirktok tributary in Alaska, a musk ox watches me intently as I back cautiously away. I had spent hours getting the animal to accept my presence until it became so curious that it kept approaching me to investigate.

On an exceptionally warm day for the Arctic, I cross the top of a small glacier in Svalbard. A few moments after Beverly took this photograph, I fell through the thin surface layer of snow and ice into a narrow crevasse. It was only my elbow hitting the edge that arrested my plunge into the pale blue chasm.

Two Buddhist monks share the images Beverly has made of the lavish interior of the Dakar Tredzong (White Monkey Monastery) in northern Tibet. The monastery is known for its esteemed Buddhist 'air' burial practice, in which the dead are placed on platforms high in the surrounding tall granite domes to be consumed by vultures.

I stand with a group of yaks on the high Tibetan Plateau where the average altitude is over 4 000 metres. These yaks, with their traditional load saddles, were returning after delivering loads of cheese to a nearby settlement. They were some of the very few we saw still carrying loads, a task now mostly done by motorbike in the remote lands without roads.

In the central Kaokoland Desert of Namibia, Beverly is encouraged into a dance by the women of a traditional Himba onganda (homestead). Himba society is strongly matriarchal and, with the power this wields, Himba women often choose to remain steadfast to cultural traditions.

In the trackless central Namib Desert, part of the aptly named Skeleton Coast, we crest a 150-metre-high sand dune to slide cautiously down the forty-five degree slip-face of the far side.

paper prayers high into the air. Their shouts echo back off the close mountains. The peak of Amnye Machen, Snow Mountain, at 6 282 metres, was considered for a time to be the highest mountain on earth, and today the clouds hide its heights behind an opaque grey veil that intermittently drops light rain on us and snow on the heights. As we descend, both of us develop more severe altitude headaches and for a time Beverly, usually stoic, is clearly in severe pain.

On the last stretches of the steppe before we reach the highway we find a lone goa (Tibetan gazelle), and I leave the vehicle to gradually try to work my way closer to where it grazes on a low hill. After half an hour, I am 30 metres from it, and although it grazes steadily away from me, it does not run as I make images of it silhouetted against the stormy, late-afternoon sky. An empty articulated truck clanging down the road below finally puts the gazelle to flight, but as I walk down off the hill my mood is buoyant despite my headache.

The country is open all the way to the mountains to the south and there is wildlife roaming freely here. The hotel in Mato is fully booked with highway construction officials. We find a second-rate alternative where the rooms are clean. The shared bathroom, however, with its stepped-level floor of tired and cracked blue mosaic, with two squat stalls perched on the highest level above the basins, is far from inviting. Mops, a shaving mirror and an overflowing ashtray occupy one corner, while the odour of ammonia and past ablutions make the air hard to breathe. There are washing bowls in our room, though, and we fill these with water and retreat to our small sanctuary.

20 SEPTEMBER

A light snow has fallen, its white cloak softening the raw, utilitarian ugliness of the town. A group of drunks, whose revelry kept us awake most of the night, lie sprawled fully clothed across their hotel bedroom, with the lights on and the door open. We leave town early as the clouds become tinged with pink, following a narrow tar road out to a set of twin lakes, known as Ngoring and Gyaring, or Eling and Zhaling in Chinese, away to the northwest. These lakes lie within a wetland protection buffer zone, in the Yellow River catchment, part of the vast Sanjiangyuan National Nature Reserve.

Recognising the value of the Tibetan Plateau to the ecology of China, the government has established three sprawling, contiguous nature reserves: Chang Tang (334 000 square kilometres), which means 'Northern Plain', in the west;

Hoh Xil (100 000 square kilometres), also known as Kekexili, which means 'Blue Ridge', in the centre; and Sanjiangyuan (94 500 square kilometres) in the east. Sanjiangyuan means 'Three River Headwaters', as the three main rivers of China – the Yangtze, the Yellow and the Mekong – rise in this reserve, thereby protecting China's principal watershed.

Although these reserves will protect the range of the endangered chiru (Tibetan antelope) and other associated wildlife species, there is a downside. With half the Tibetan Plateau being set aside as nature reserves, and with traditional pastoralists being moved off the land and resettled, the sustainable coexistence of wild animals and human pastoralists that has persisted for thousands of years is ending. The proposal to set up a UNESCO World Heritage Site on the Plateau has drawn attention to the debate around the zone system proposed by the government for these reserves. There will be a Core Zone (all residents resettled, no grazing or people allowed, wildlife protected), a Buffer Zone (partial resettlement with limited and rotational grazing) and an Experimental Zone (partial resettlement, eco-tourism and green industries). The Tibetans are counter-proposing community control and management of land, with some set aside for the protection of wildlife, maintaining healthy rangeland, taking fences down, and patrolling the land against intrusion by poachers.

The road is peaceful and the country open and it is not long before we start to see wildlife: a group of goa far out on the plain, and later two kiang (wild ass), but the light beneath a brooding, low sky is still too dark for photographs. Thin tendrils of smoke issue from the occasional, distant nomad tents, the travellers' flocks still sleeping in close huddles around their centre. There are many small ponds and lakes beside the road and the black silhouettes of ducks and geese stand out against their silver surface. As the day lightens, we find two rare black-necked cranes, only 4 200 of which are left in the world, feeding towards the tailings of one of the shallow lakes. We walk out onto a muddy plain and work our way slowly towards them. The air is chill, snow thin on the ground. On the lake fringe a low succulent grows; it turns red in the autumn and, against its bright background, the crane's graphic black-and-white plumage and yellow eye make a striking image. The geese on the lake take flight as we approach and the cranes raise their heads in alarm. For a time, they stand erect, honking softly to each other before they run a few steps and follow the geese.

The clouds thin and the sun warms the ring of mountains that surround the vast plain in which the lakes lie, the fresh snow lying right to their foot. We walk out onto the steppe towards a large herd of kiang. They are tentative about our approach, trotting away, to circle back and stand snorting. When we walk away at a tangent, they follow. It is a band of mares and some foals with a single stallion. Another stallion circles the group, keeping a hundred metres or so distant. Neck, chest and belly are white, the mane and the tail almost black, and the rest is a dark russet-brown. They stand against the yellowed grass of the steppe, with the rise of snow-dusted mountains capped by cloud behind.

An upland buzzard calls from a nesting post and then flies off low over the land, sending the few plateau pika, emerged to enjoy the first sun, diving for the cover of their burrows. Nesting poles dot the entire steppe. I noticed them yesterday for the first time, and now when I glass the far reach of any plain I can see that they have been erected everywhere. Beginning in 1962, the Chinese government launched a massive campaign to poison the burrowing pika in an attempt to control their devastation of the grassland, but bears, foxes and raptors that ate the poisoned pika died too, and pika proliferated. The nesting posts are the alternative solution. The poisoning has stopped in this area and raptors are everywhere.

Present-day scientists argue that the pika is a keystone species and that its destruction will be detrimental to the food chain. In fact, recent studies show that the pika chooses already degraded land, as opposed to long grass, to burrow, as it can view its predators from a distance. The burrows also aerate the soil and provide channels for the heavy rains to penetrate the earth rather than creating dangerous runoff and floods.

As if in affirmation, we have driven only a short distance after leaving the kiang when we find a Tibetan sand fox hunting in an area of taller grass. It is stouter and stockier than other foxes I know, its fur long and dense. The face is almost owl-like: a roundish plate with the cheeks flared wide and a shortish muzzle, with two pointed ears protruding above the disc. The muzzle, ears and back are a darkening yellow-orange, the rest of the body a grizzled grey-black. It trots away at our approach with the typical light-footed, slightly bounding gait of foxes, pausing occasionally to look back towards us. Its colouring is a perfect camouflage, and soon it is lost in the grass.

At the lakes and in the streams emptying into them we find a variety of birds. Prettily marked bar-headed geese, with graphic grey-and-white plumage on their heads and necks. Ruddy shelduck with tan bodies and white heads, and roosts of cormorants on small exposed sand-spit islands, pure-black in silhouette against the bright sparkle of the water. There are nomads here with huge herds of livestock, thousands of animals that in the distance drift as dots across the rolling of the hills, the yaks black and the sheep white. There are small herds of kiang and goa too, whose white rumps show distinctively as they trot away.

Towards noon the sky draws closed again and rain falls in squalls. We head south to Yushu, the 400-kilometre journey taking seven hours, with the highway interrupted by detours, construction and degraded surfaces. Beverly and I take turns to catch up on our sleep.

The country is all tall hills and distant mountains, and nomad tents and healthy herds of yak and sheep occur between villages. Construction is going on almost everywhere, and on most roads we have travelled a new and better one is being built alongside. Every town and village has new building under way, with bulldozers on the street and towering cranes perched like storks above the skyline. There are new power lines and crews high on craggy hillsides cutting footings for more. Everything related to construction in China seems to be to the power of ten to anywhere else in the world. Gravel works have heavy tip trucks and bulldozers line up in rows of twenty, too deep to see how many stand behind. Every 50 kilometres or so, there are temporary, specialist concrete factories making precast bridge sections. The factories are as big as any permanent works I have seen anywhere and the finished bridge sections stand stacked in yards half a kilometre long, awaiting transport. The scale is overwhelming. The workers are housed in towns and, beside the highway, in rectangular compounds complete with entrance flags, huge solar power panels and temporary mobile phone transmitter towers. At 8:30 pm they are still throwing concrete beneath spotlights, or planting grass sods on the banks beside finished sections of the road. In my own tiredness, I wonder what time they get to sleep.

Towards evening we cross the Yangtze River. Already impressive, its muddy brown surface roils with the power of water moving at speed. Upstream, prayer flags hang in drooping loops from the arches of an old bridge; downstream, the construction of a massive new high-speed highway bridge is almost complete.

Yushu (Jyekundo in Tibetan) is a town of some eighty thousand inhabitants, and in the restaurant where we eat dinner there is an eclectic mix of Tibetan nomads and monks, Muslims and Chinese crowded around the tiny tables where an excellent, spicy fried chicken is served. It is hard to describe Chinese cities without familiarity with Third World urban environments. Beverly describes it as 'very Nairobi', which is a perfect description. It is somehow an uncomfortable and yet fluid juxtaposition of First World technology with Third World chaos. The result is untidy, often dirty, or ugly, but surprisingly efficient. On an unswept, extremely dusty street where traffic weaves to a set of constantly evolving rules, one can stop mid-lane, climb over a mound of building rubble, avoid the slop being thrown out of a dingy, grimy-paned restaurant next door and walk beneath a bright flashing neon sign in human-sized Chinese lettering into an immaculate, marble-floored, glass-countered shop selling mobile phones and offering high-speed internet service. Our hotel is one of these institutions, overlooking a busy street, beyond which yaks graze on a rubble-strewn riverbank where a huge caterpillar-track crane is scraping the rubble flat.

21 SEPTEMBER

We spend the morning shopping for a chupa for me and provisions for a night's catering for ourselves. It is fun to see the effect that Beverly and I have on the other pedestrians as we walk down the street. They literally stop dead in their tracks and stare in slack-jawed amazement. Some follow us, walking close by our shoulders, staring at our faces.

Headed south, we follow a narrow country road into the steep valleys of the Kunlun Shan range. The fences disappear. Waist-high bushes grow on the steep rocky slopes, their leaves starting to turn a rust-orange. The hillsides are mostly too steep for yaks, and the heights of the peaks shut out the sky. The road climbs higher. Nearing the crest, we spot a herd of bharal resting on the grassy knoll of one of the pinnacles. There are about thirty animals, but I cannot distinguish sex, although they seem to be principally females, as there are quite a number of juveniles. Stepping out of the vehicle we see what we could not from inside: the sky is full of raptors. Vultures are thermalling high over the close ridge, and out to the side a single eagle is coursing the updraft on broad wings, the digits in sharp silhouette against the cloud. A lammergeier cruises fast across the hillside, sweeps wide to the other side and then, dropping low, descends into the valley's depths.

Around the 4 890-metre summit of the Kulha Pass, which means 'Image of Buddha', isolated rocky peaks rise from the high country, their orange, black and grey granite faces holding a delicate lacework of snow on the narrow ledges. At the apex, there are prayer flags and a pile of red-and-white painted prayer stones. Shortly after we begin our descent yet another lammergeier appears, its languid, drifting flight carrying it low over our heads as it patrols the foot of the rock spires. Following a small clear stream, the road descends lower and lower into the most traditional Tibetan country we have encountered so far. Low stone houses with stone-walled compounds are spread widely throughout the valley. Most are single, others a tiny cluster of two or three houses, the ground outside the walls trampled where the stock sleeps. Large black mountain dogs are chained outside the entrances and motorbikes rest on their pedestals beside virtually every house.

The motorbike is the vehicle of choice for the nomads. On occasion, even some of the herding we have seen – where yaks or sheep are being driven a long distance – is done by motorbike. I have seen them used to transport beds, yak calves and quite frequently families of four. Returning from town they are invariably overloaded, and on the passes they shortcut the hairpin bends, going straight up and down on footpaths between the roads. Almost everyone wears a chupa and many of the men wear wide-brimmed hats at rakish angles.

Where the high valley we are following is still narrow, we stop to photograph two women milking their yaks. The women squat on their haunches beside the yaks, and I must assume that this is because the animals are too small and low to comfortably use a stool. They reach under the long, obscuring hair and milk into a beautifully patinated wooden pail made of tightly fitted planks bound with rawhide. We ask to see how they make butter, and inside their tent we are shown a modern, hand-cranked cream separator and a twenty-litre plastic container with a lid in which they shake the cream rhythmically until it turns to butter. On the stove a large wok of white, chunky cheese is curdling, while outside a chained dog, its head the size of a lion's, growls and barks as it charges against the restraint of its chain. I am distracted from their explanations, translated by Tenzin, by my hope that its peg holds.

We turn onto an even smaller road, following the now-growing river. It passes through a wide, flat-floored valley where a thousand or more yaks graze. We stop to photograph a particularly striking woman who is carrying two water buckets up from the river, hung from the split-bamboo pole she hefts on her shoulder. She wears a large turquoise brooch in her braided and plaited hair.

Around her neck is a necklace of red and yellow amber beads with silver and turquoise. She is shy of our attentions but claps with delight at the images I show her. The patriarch of the household, too self-conscious to smile for a photograph because he has missing teeth, tells us we are on the wrong road. We are headed into the Tibet Autonomous Region and four kilometres down the road we will require permits to pass through the government security checkpoint. We turn back and, emerging once more from this valley, climb yet another hairpin-bend pass. Right at the top we are struck by a violent hailstorm. Visibility is reduced to no more than a few metres, the clatter of the hail on the roof so deafening that we must shout to be heard. Within minutes the countryside is white and the narrow, steep-edged road treacherously slippery. We advance at a snail's pace until the hail grows intermittent and then turns to rain.

Once more we enter an enchantingly traditional valley, and arrive at the Surmang Monastery in the late evening. The monastery is under construction and most of the monks have taken the opportunity of the interruption of its function to travel to family or other monasteries. Those left behind are either unwilling or disinclined to take the authority or initiative to find us lodging and we are turned away. It is getting dark and raining hard without pause. We turn around and head back to some of the nomad houses we saw on our way in. Tenzin braves the dogs and emerges from the first house of our inquiry with an invitation to stay.

The kitchen is a long, oblong room with a metal stove at its centre. Wide wooden couch-benches line two of the walls, and a large, wooden display cabinet occupies virtually the entire length of the other. Water comes from a plastic drum, and on the stove an enormous, Buddha-shaped pot sits over the open flame of the yak-dung fire. Three other kettles hiss on the steel-plate top of the stove, and we are immediately offered tea. The only occupant is a tiny, wizened crone who wears her strong but greying hair in two untidy ponytails. She is dressed in a red chupa, the sleeves dark with grime and the skirts flecked and smeared with fresh mud.

After a dinner of tasty vegetable noodles prepared by Gongpo, Beverly and I are given the family sleeping room. It is dominated by a bright, tall shrine with photographs of lamas, the family, and religious texts. Brass candleholders and a brass teapot are arranged neatly on its mantle. Four couch-bench beds line two walls. They are ornately carved and painted in bright blues and reds.

Coarse-pile tapestries in the same colours and sporting intricate graphic designs cover the seats. This house is one of the homes constructed by the government for the settlement of nomads, and although the workmanship is rudimentary I am impressed by its spaciousness and the fact that it has 12-volt power provided by solar panels and battery packs. To my final question before turning in, our hostess informs us that there is no toilet.

22 SEPTEMBER

Our hostess fetches two twenty-litre buckets of water from the river, 250 metres away, and the only help the bent old lady asks is that we lift them to pour the water over the high rim of the tank in the kitchen. We share what food we have for breakfast and enjoy her yak-butter tea, then express our thanks with gifts of food and money as she waves us away.

A mist is trapped between the high sides of the valley and it drifts and wavers, pasting the valley with patches of light that run across the landscape like the pass of a paintbrush. Smoke is issuing from the chimneys, and young girls are moving between the tethered animals, collecting dung to be dried for fuel. I am photographing a small squat stone dwelling when a woman emerges through the curtained doorway and invites me in for tea. I decline the tea, explaining I have just had some, but saying I would love permission to see inside her house. I round the low, mud-brick compound wall, and pass through a neat arrangement of tethered calves.

Inside the modest hut the air is smoky from the mud stove, on which a large lidded wok and a kettle are simmering. The roof is low, my head almost touching it, and the smoke escapes from an open grille in the roof above the stove. The only illumination is the flicker of the flame in the open fire door and light coming through the roof grille and the doorway curtain. The woman is sitting on a low, mud-moulded seat-bed where a toddler, naked from the waist down, is playing with some long-expired piece of electronic gadgetry. The bed is piled with a crumpled duvet, as if someone had just woken up. The woman is churning butter in a plastic container similar to the one I saw yesterday. She waves me to sit down on the bed and when I do the toddler hands me his toy. The wall has a hand-painted motif of blue and red on a white background, about a metre wide, running around three of the inside walls. Another mud bed occupies most of the short side wall, and looking it over I see that there is a baby asleep there, a string of prayer beads with green tassels hanging close above its head. The woman's face is broad and when I show her the pictures I am taking she smiles, which makes her eyes narrow to slits, and dimples appear below the rosy blush on her cheekbones. Only when the toddler reaches for her does she pause in her slow rhythmic shake of her butter churn.

We return over the pass, where the ice from yesterday's hailstorm still lies white on the ground. Turning west, the road descends, only to start climbing again to the height of the 4 428-metre Shatrila Pass. We climb through the mist that is trapped by the heights and look down on a patchwork landscape. Sunlight plays on undulating hills, their slopes covered in a quilt of hail-ice and grass, while clouds lie like cotton wool in the valleys. The descent is steep and winding, but the road straightens along the narrow valley floor, and we find a commune of chiwa (Himalayan marmot) on a rise where a small stream is running fast and cold. The chiwa are sunning themselves at the entrance to their burrows, the more industrious a few metres away from safety, foraging in the grass and between the low, sharp-leafed bushes. The chiwa is a squat, compact animal, a little bigger than a rabbit, but looks more like a guinea pig. The coat is tan and dense, with the back grizzled to black. There are several in each hole and they come cautiously to their burrow entrance, where they huddle in close contact, often resting their chins on each other. During the time we are with them they remain entirely silent.

The same cannot be said for a covey of sikpa (Tibetan francolin) that we find on a slope between nomad houses where dung has been spread to dry. The small, finely marked bird with black barring and a chestnut back twitters constantly as it feeds and utters a harsh low croak when alarmed. They are remarkably tame, and for the umpteenth time since we arrived I find myself in a state of thanks for the Buddhist mantra of respect for all life.

We have been told that the Dalai Lama's speech of 2006 urging Tibetans to respect wildlife had a profound effect, ending any hunting that was happening, and was so taken to heart that people even destroyed the skins of animals they had hunted. It is quite wonderful and entirely unique to find such an abundance of wildlife in all its forms coexisting with a rural people whose livelihood comes from what they can derive from the land. Everything we have seen in Tibet is on a land shared with people, and in this place where no physical boundaries are imposed I imagine the land beyond the reach of the roads to still be rich in its diversity.

The road towards Nangchen leads down to a wider place, where the rivers are sweeping and the valleys extensive. We stop for lunch in what was a village and is now an almost completely unoccupied town, with new streets lined with square, plain-faced buildings. Our guide says, 'Tibet is under construction,' and there is more import to his statement than he intends.

Leaving this strange town on a new, single-lane concrete road, we climb once more. The hills rapidly become mountains with a Yosemite feel about them, and have us looking rapt out of the window, bending low to see their peaks high above. Layers of eroded granite and sandstone rise through the earth like walls. Some are only a few hand-spans wide, others 10 metres, but all run straight from top to bottom. They are covered in grey-and-orange lichen and have been eroded to a ruggedness that makes some seem as if they have been built. The streams and rivers have cut narrow, sheer-walled gorges through the stone, and the Tibetan mantras 'Om Mani Padme Hum' and 'Om A Hum Badza Guru Padma Sudu Hum' have been carved by hand into many of the cliff faces within reach of the road. On one face a two-storey image of Buddha has been carved in relief into the rock, then painted in the gaudy colours so favoured here. The face is glittering gold.

23 SEPTEMBER

Departing Nangchen to the east, we find a nearly full moon bright in the still dark sky. At Xiaolong Gorge, a tiny stream has cut a gorge 25 metres wide through the rock that now towers 200 metres above the road. It is dark and cold in the gorge, and the tiny stream runs swiftly beneath Buddhist mantras carved into the rock. The road climbs steadily, finally becoming a switchback pass before delivering us onto a high-altitude steppe of wide plains and gentle hills. There are few nomad camps here. I count only three in the more than 50-kilometre vistas before the rise of more mountains, yet yaks are grazing in groups scattered throughout the steppe. There are goa here too, in sparse herds. They are too cryptic to discern in the distance, but I count more than forty dispersed over the close hills and wet boggy lands of the various stream edges.

We are animated: this is the country we came to see. I feel I can breathe here. There is no restraint. Not one manmade impediment mars the integrity of this land. To the west the mountains rise again and clouds mushroom white against the sky. We cross the plain, stopping to photograph nomads in a traditional yak-hair tent. Tenzin tells us it takes ten women two weeks to make a yak-hair tent. The nomads are on the move and all their possessions lie in bundles outside, in preparation for dismantling the tent. Inside, the stove is still lit and cheese simmers in a large pot. The woman of the house offers us a thick yoghurt. Outside, another woman is chasing yaks with a traditional sling. She whirls the twin straps around her head and then lets the stone fly with a whip-like crack. The stone whirrs with speed and cracks off the ground to the side of the yaks, 40 metres away.

Beyond the nomads we enter a mountainous country where the land is steeply folded. By the time we stop for lunch I have lost count of the number of passes we have crossed, some of over 5 000 metres, with mountains reaching 6 000 metres into the sky. The road is narrow and endlessly twisting, and from time to time we must stop and back up to let ancient, gaudily decorated trucks crawl past. The drop-off at the edge is often sheer but none more so than our climb out of the valley of the Kyichu, which means 'Big River'. The river has cut through a whole range of mountains, leaving sheer cliffs hundreds of metres high. From our vantage at the neck of the pass, we look down on the river more than a kilometre below. The peaks are eroded granite and on the slopes a stunted pine grows. A kyaka (blue jay) hops on the periphery of where we sit. When it flies out over the gorge its wings flash white and dark electric blue. I can see more than 50 kilometres in all directions. Prayer flags flutter behind me in the breeze and vultures turn above the peaks. It seems a fitting place for spiritual affairs, for prayers cast to the wind.

The Kyichu emerges in the middle distance from a crumple of mountains that rise slowly behind it in a broad cup shape, their deep-creased surfaces mottled by drifting cloud to black, blue and emerald green. Where the river emerges, the valley floor has been terraced into fields. Two small villages, one on either side of the river, nestle in its sanctuary. The houses are all traditional, stone-walled and flat-roofed. It is the first place we have seen without construction work. No electricity lines divide the view. The small, rutted dirt road is the only outside convenience and the vehicles it has brought stand shiny silver beside the brown-earth adobe of the homes. In the foreground, the 70-metre-wide river makes a dogleg to the right, far beneath the road, and cuts then through a gorge of sheer rock perhaps 800 metres high. It is as if the mountain is swallowing the river whole through two gigantic gates; a tiny gompa (place of prayer) is perched precariously on the peak of the closest. Tenzin laughs, and when I look at him I see he is looking at me. I am standing with my camera forgotten in my hand, shaking my head with a small smile on my face. My privilege is sometimes beyond my belief.

We descend the torturous, narrow hairpin-bend pass to the village where the inhabitants are harvesting tall fields of barley with scythes. Everyone is involved. The women and men cut and lay the barley down while the teenagers and young children gather it into bushels. Some families carry it to their compounds where it hangs draped from high wooden frames to dry. Others gather their bushels into stands in the fields, which they tie at the top so that they resemble flared skirts. Even in this back-breaking work, some of the women wear jewellery that sparkles silver and red as it swings from their necks.

The road follows the river into the mountains. At times its narrow track cut into the sheer edge is so eroded that I cannot look down. The scenery is fairy-tale. The mountains tower above, with striated layers of weathered rock cutting through the mantle of the soil like the blades of spears. In the river, some of the rocky outcrops have over time been carved by the current into spires, and between other steep edges footbridges have been suspended: cable-supported wooden poles lashed side by side high above the rush of the current. From the hand-support cable, prayer flags flutter, white and trembling.

In the mid-afternoon we stop at a small village gompa, where a stuffed wolf hangs suspended in the middle of a prayer hall. A single monk sits in one corner beating a drum as he chants prayers. When they are found dead, wolves and other animals are sometimes stuffed by the monks so as to have the animal's spirit assist in their protection. This eyeless specimen has a collar of tinsel around its neck. The small monastic village is old: the buildings of mud and stone have been stained with a maroon-red powder, now faded to pale rouge. Steep wooden stepladders made of hand-cut planks lead to small living quarters, above dark storage rooms.

A storm catches us halfway to the pass around the Kyichu Gorge, cracking lightning between the bare peaks and drawing a black-grey curtain across the valley. Rain patters on the roof and the track leaches a flow of orange water into the opaque green of the river. By the time we reach the narrow gate of the pass the storm has moved on, hail lies thick on the ground and a herd of seventeen bharal graze on the upper side of the road.

The descent back into the Kyichu Valley is now slick with ice and wet mud, and several times I find myself clutching the vehicle handholds as we slip towards the edge. Goa have moved closer to the road, but it is overcast and almost dark and we drive on, into the night. I decide to count the number of hairpin bends we negotiate but lose count after ninety. We drive for three hours through the darkness and do not see a single night creature, only dogs.

Dogs are everywhere here in association with human settlement. In the walled central yard of a gompa under construction, I once saw fifty dogs or more sleeping, walking, and preening in the temporarily abandoned space. The dogs are almost identical: big-boned, long-haired mountain dogs with the head and droopy eyes of the St Bernard. The dogs are kept ostensibly to protect the stock from wolves. But on the several occasions I have seen packs of four or five running at a trot through the hills, I would suspect that hungry dogs are more a threat to stock and wildlife than the now-rare wolves. As we drive through the almost deserted streets of Nangchen at 10 pm dogs are the only creatures on the prowl.

24 SEPTEMBER

We return to Yushu on the main highway, and although the landscape is arresting, it is marred by up to four sets of power lines, construction and litter. Every principal road we have travelled has been fringed with litter. The development along the highway is on a staggering scale. At one point on the road there is a new nomad settlement town nearing completion. It runs for at least three kilometres beside the highway and is about 600 metres wide. Rows of new unoccupied houses, each home with its own well outside, stand facing the unfinished streets. There must be 8 000 houses here and not a single tenant is to be found in the wide-open country that surrounds it.

Driving out of Yushu we pass through the completed section of the new city centre, the old having been entirely demolished following the damage caused by the disastrous 2010 earthquake. On every city block cranes stand amid vistas of unfinished concrete and steel rods pointing skywards. The curious three-wheeler, heavy-load Chinese vehicles with their single-cylinder diesel engines and open, Vespa-style handlebar cabs, chug down the street, ferrying blank-faced labourers and loads of building rubble along the new, wide, virtually unused river promenade. There are ten or twelve blocks of buildings of three to eight storeys, complete with video-access security and patriotic flags flying in front. Not one has an occupant. It is astounding, and as we pass into the hills I wonder what the motivation is behind this incredible spending. If it were pure philanthropy, then the gesture would be excessive in the extreme. Perhaps the goal is to attract Chinese immigrants to these mostly minority-group towns. A cultural invasion is far more effective than military conquest, more permanent in its change than any rule or law could induce.

Ever since the occupation of Tibet by China in 1950, there has been resistance. The Tibet we are seeing is culturally intact. Customs, religion and language remain unchanged. Within this strongly individual identity of a steadfast culture lies the breeding ground for dissatisfaction with foreign rule. As South Africa learnt with apartheid, when resistance lies in the strong cultural identity of a person's heart, it is impossible to defeat. Tibet has been occupied for more than seven decades, and between 2011 and today there have been more than one hundred and twenty officially recorded self-immolations, perhaps the most shocking form of passive resistance. All have been in protest of the Chinese occupation. And perhaps the government, realising its continued failure to be regarded as a benefactor rather than an invader, is now seeking an alternative, more insidious method of changing the Tibetan mind. Cultural invasion is the only sense I can make of spending that must be measured in billions of dollars. There are only 6 million Tibetans, of whom 150 000 live outside of Tibet. The population of China is 1.38 billion and there are already 7.5 million Chinese settlers in Tibet.

The small Longbaotan Lake and Marshland, in the Longbaotan Nature Reserve, is bordered on one side by a highway under construction. The Longbaotan Marshland is a critical breeding habitat for the endangered black-necked crane, and I am staggered both by how small the place is and that there seems to be absolutely no protection in place. The marsh is ringed and divided by fences and there are nomad houses at intervals, completely surrounding the lake and marsh. The cranes, however, are also here, and breeding. Within the first half an hour we see four pairs, three of whom have a single, fully fledged chick with them. We drive to the far side of the lake, away from the highway, where we photograph the cranes and the sand foxes that patrol the marshland. At dusk, we pitch camp on a small rise by the water and eat a dinner of spicy noodles, with the treat of a chicken option, sitting on the icy ground while a cold wind blows up the valley. We have just turned in for the night when it starts to snow.

25 SEPTEMBER

In the night, we are battered by a wind-driven storm that started as snow, turned to rain, and finished with light snow. Our tent is crusted in ice when I climb out into the dawn made beautiful by the white-coated hills and clouds tinged with the warmth of the sun. The clouds, however, soon close out the sky and we huddle around the stove clasping mugs of hot coffee. The tents are too wet to pack, so we leave Tenzin to care for the camp while we set off to explore the roadless lakeshore.

We quickly come upon more pairs of cranes and, using the cover of a grazing herd of yaks, we are able to get close enough to photograph. This morning two pairs we encounter are chickless and we find only one pair with a chick. Tibetan sand fox and red fox are common here. Perhaps they are responsible for the crane's vital failure to raise chicks, for the foxes scour the marshland at a run, leaping high over the water runnels from one tuft to another. But the crane is a large bird and I am sure quite capable of keeping both species of fox from its nest. It seems more the human encroachment, with yaks trampling through every secluded reach of the marsh, and the unconstrained dogs, four times the weight of a fox, roaming the land that spoil the success of these graceful birds.

On our return from the head of the lake, the day has warmed and chiwa are now on the land. They sun themselves at the entrance to their tunnels, and the larger ones forage close to the burrow in their distinct creeping waddle-and-pause. With their bulky bodies and squat limbs, they look rather like tan-coloured koala bears. A large male stands on his hind legs and chirrs his alarm at our approach, the snow-powdered hills and dark sky behind. On our way out, we see two red foxes. One is particularly unconcerned with our presence, and when he disappears behind a low ridge as we drive parallel with him, I take the opportunity to run to a point ahead in the hope of having him come close to where I lie. But I fail to take into account that we are at an altitude of 4 200 metres, and by running in a crouch, I double my exertion. By the time I reach the spot where I lie down, I am panting so heavily that there cannot possibly be any element of surprise for the sharp-eared fox. No one is more delighted than I am when my huffing attracts rather than scares the fox and it approaches to within 15 metres, but at no point with more than its eyes above the rise of the land.

At Zhiduo we turn north and for a time follow the broad turquoise waters of the Nyichu downstream. There is a road inexplicably being built halfway up its steep hillsides. We cannot fathom the purpose of so expensive an exercise when the valley floor is broad with a road already constructed until Beverly suggests that a dam is intended here. Neither Tenzin nor Gongpo know of such plans, but when the Nyichu joins the Drichu (Yangtze River) and the new road crosses its flow on six columns towering more than 100 metres above the existing bridge we realise there can be no other reason.

In Chumarlab our hotel is the smartest in town, with a brass plaque designating it as registered for foreigners. The staff is eating dinner in the foyer, seated on faux leather sofas. The receptionist wipes the grease from his fingers on the back of his hands and reaches for my passport. After we check in we are told that there is no electricity until 8 pm. In our room, there is no water and the toilet bowl contains the unflushed remains of the previous occupant. We change rooms. In our new room, the toilet leaks badly onto the floor and one of the beds has been slept in. We use one bed and make our ablutions wearing waterproof hiking boots.

26 SEPTEMBER

The morning is cold and it is snowing. On the deserted main street out of town, as we leave in the first hint of dawn, the golden yak statue on a pedestal at the junction has a mane and ruff of pure white. We take the secondary road to the east of Chumarlab, where the Hoh Xil National Nature Reserve stretches for hundreds of kilometres, but it is snowing so hard that we can see little beyond the verge. We are at 4 300 metres and crawl slowly into the whiteout. When the snowstorm eases there is a steep, white landscape on either side of a broad valley and it is freezing cold.

The deeper we move into the country, the more frequently we see small herds of goa. In some places, they are dotted across an entire valley. It seems that the snowstorm has driven the raptors off the heights, and they perch beside the road. The most common are the large upland buzzard, which is dark-brown with an orange breast, but comes in a darker morph too, and the Tibetan saker falcon. It distresses me, when I watch them hunt the flocks of sparrow-like birds, to see them have to interrupt their stoop to swerve over the box-grid fencing that cuts through the countryside.

There are nomads here but they are decidedly fewer, and tethered yaks lie as still mounds in their coating of snow, blinking at our approach, like emerging mushrooms. The land is too covered by snow to discern the state of the grass, but the vegetation underfoot as I walk out to photograph is grazed as short as everywhere else.

After several hours on the road, we pull off into the compound of a single nomad winter home to request hot water for tea. Two dogs lie curled in the snow beside the door, their bodies mantled in white and only their eyes moving to our arrival. A bright-eyed woman invites us in, too busy tending her stove for the usual demure reception. Inside, her home is light and well kept, and on the central yak-dung-fired stove three kettles are boiling. In a large wok over the open flame curds are separating from whey as she makes cheese. One kettle has just water in it, the other two salty water, for yak-butter tea. In the corner, by the window, she pours lightly boiled milk through a cloth into a battery-powered separator. The cream that collects in the stainless-steel receptacle is dark-yellow and the consistency of cold honey. She offers me butter for the toast I make on the stovetop and it is rich and delicious. She has two twenty-litre containers of milk – two-thirds full – from which she is making butter and cheese. The third container, she says, shaking her head, one of the yaks kicked over this morning. I ask her how many animals she milked this morning. Nineteen, she says, without pausing to count. She sells the butter and cheese in town: 100 yuan, about US$1.47, per kilogram, for the butter and between 60 and 100 yuan per kilogram for the cheese. In the hour or so that we enjoy the warmth of her home it becomes apparent that she is one of the most quick-witted people we have come across in Tibet, grasping concepts and ideas despite the language barrier.

She is stirring the cheese when her cellphone rings and interrupts our conversation. She has a long and animated conversation without pausing in her constant bustle of boiling milk, separating butter, curdling cheese, and serving us. Towards the end of her telephone call I notice that there are tears running down her face. She is crying into her sleeve. We are uncomfortable and suggest we leave, but she waves us down, explaining it is her sister who is living in Zhiduo and she is missing her terribly. Zhiduo is less than 150 kilometres away and she has not seen her for two years.

Down the road, we turn north onto the rural road that leads to the source of the Machu (Yellow River). The snow falls steadily. Raptors abound, and we photograph a small owl that perches on the riverside cliffs in the snow, its feathers so fluffed out that it seems as round as a ball. Plateau pika punch holes through the white crust and surface cautiously, remaining motionless for minutes at their burrow entrance before darting out to forage. The weather is too inclement for the raptors to hunt, so they sit on their perches and bob their heads to the mice activity close around.

We stop at the tiny settlement of Tchudil, where the only restaurant is closed. We buy instant noodles from a sombre, slow-talking shopkeeper who offers us the use of his kitchen-cum-communal room to eat. While we are still in the shop, two nomads enter looking for all the world like Wild West cowboys. Both wear their trousers tucked into their calf-height, high-heeled, brass-studded boots. They sport wide-brimmed, Stetson-type felt hats with thin leather bands, onto which several gaudy, bright ornaments have been fixed. One has an ornate belt of realistic plastic-cast bullets sewn onto a broad, red cloth band that is edged with gilt brocade. He wears a striking, zebra-striped facemask. His companion's is plain, but both dark pairs of eyes – the only part of their person revealed apart from shoulder-length, jet-black hair – scan the room as they pause in the entrance. All that sets them aside as distinctly Tibetan is that they both wear thigh-length chupas of ornate gold embroidery with paisley whorls and graphic patterns on dark-red cloth. The sleeves are long and conceal their hands. When Tenzin breaks the silence, they loosen one side of their facemasks to reveal broad smiles and a natural curiosity about the foreigners in their midst.

Over our instant noodles we ask the storeowner about wildlife in the vicinity, about the chiru, describing its long, straight horns, and about wolves. For the antelope, he says, we will have to travel deeper into the Hoh Xil, but of the wolves there are plenty. I say to him that virtually all the valleys we have travelled are quite populous with people, stock and dogs – not the wolf's preferred place. He tells me that one must head into the back valleys, that there are many there. I ask if he has seen any and he harrumphs and waves his hand. 'Packs,' he says, adding that there are bears too. I know that bears require at least shrubs and berries, and would struggle for both food and cover in this high steppe, with only short, stiff-stemmed grass available. Bears, he confirms, then changes the subject to talk with Tenzin of news from elsewhere.

Outside, the snowfall has lightened and as we drive on the sun can be felt, tentative through the land-bound cloud. There are goa and stock but we see no sign of wolf or fox. At a switchback pass the snow lies too deep for us to get through, and we turn back. The snow abates and the sun starts to come through in patches, the land bright and sharp against the eyes, the mountains at the heart of the Kunlun Shan in the distance. Buzzards and falcons hang on the updrafts above the hills, stooping on the plateau pika, whose sorties to feed leave them in sharp silhouette against the cold white. Everywhere the land pulses with their darting.

Freed from the close mist of the snowstorm, our drive back to Chumarlab is quintessential Tibet. Low dwellings with smoke issuing from the chimneys, herds of yak and sheep in broad valleys, with tall hills covered in snow rising to the clouds behind.

27 SEPTEMBER

The snow lies brittle on the hills, the valley's yellows warmed by the sun as we depart into a perfectly clear morning. It is a rural idyll, sparsely populated, with the treeless lift of the countryside to the tall hills and the Kunlun Shan beyond. But there are the same fences and chronically overgrazed hills, incessant roadside construction and an astonishing flow of traffic. Whenever we stop to photograph, we must walk well clear of the road to avoid being sprayed with mud and the inevitable hooting that is the practice for passing vehicles in Tibet.

Because of the sun, the chiwa are out, and places that yesterday were barren of life today bustle with their crouched, nose-to-the-ground snuffling around their burrows. The road climbs higher without the dramatic passes of the past few days and the land changes slowly to a flatter place where the valleys are less distinct and the hills lower. The vegetation begins to resemble a mild alpine desert, the grasses forming tight tussocks between bare, gravel flats, and yet we see a herd of close to a thousand sheep right across the land, like an earthbound white cloud.

I tell Tenzin that overgrazing will soon be a major issue facing Tibet. Unless it is brought under control, Tibetans will exhaust the resource and the result will be catastrophic, both in terms of stock loss and permanent damage to the land. He tells me that controls are already in place in some areas to limit the nomads' stock holdings, and that even here in the Hoh Xil the government imposes rotational grazing of fenced areas to protect the fragile grasslands. To my eye it does not seem nearly enough.

We stop to photograph yaks grazing amid the ruins of an old mud-brick town. Through eroded arches, distant snowcapped hills are framed like paintings. When we return to the vehicle Tenzin tells me that this was the original site of Chumarlab until the government relocated the town in the early 1990s. He says that it was to establish the town closer to Zhiduo, but that seems a strange reason to forcibly relocate an entire town, yet no other reason is apparent.

At midday, we lunch beside the wide gravel bed of the Serwochu. The wind is blowing and we shelter behind the vehicle, looking upstream. There is a lone adobe house on a far bend of the river, hills opening beyond. No fences. Sitting looking at it, I know that what we are seeking lies in those hills. Behind us, a kilometre distant, where the road bridge crosses the river, there are the twin steel silos of a cement plant and a gravel-processing yard where water used to sluice the gravel runs a rich red-brown directly back into the clear blue river.

By the time we arrive at Chumar, a small town on the upper reaches of the Drichu, Beverly and I are irascible with disappointment that we have yet not found any place that feels truly wild. On our map, I note a designated track that leaves this town to the north and enters a vast realm devoid of any notation. Together with Tenzin, we inquire whether this road connects somewhere to the north. To our delight it does, but we are warned that it is bumpy and slow and that it is far faster and easier to use the main road. Within the first 10 kilometres we are rejoicing at our choice. This is little more than a two-wheel track. The fact that it is roughly gravelled is all that marks it as any more than a nomad country track. Everything falls behind – construction, traffic, settlement and, to our great relief, fences.

The country is far flatter here, a gently undulating blanket of low hills that are an arid pale-yellow. Far, far in the distance we can see mountains, but despite its low aspect the steppe is high and the altimeter does not drop below 4 300 metres. We pass a nomad encampment with a large herd of yak, more than a thousand animals, and I am troubled by the fact that this high steppe is already semi-desert, the grass just thin tufts in the sand. Along the early part of this road were large grids of stones, each a metre square, planted to prevent wind erosion.

After two hours, as the light is growing warm with the waning afternoon, we stop to photograph two groups of kiang near the road. Sweeping the open surrounds, I count forty-eight kiang within view and even more goa. There are many small ponds in depressions, and aggregations of bar-headed geese and ruddy shelduck roost on the sandbars of the larger ones. The road climbs steadily higher, our view to the south and west expands, and I feel happy to bursting. This is what we came to find. I want to stand on one of the small knolls and throw my arms wide and shout 'Yes!' to the sky. I tell Tenzin and Gongpo of my joy, how this surpasses all that we have seen.

My response to being alone in real wilderness, apart from the excitement and anticipation, is a joy I can feel physically in my chest. It leans against my ribs like some gentle hand that holds them open and I feel as if I can take the deepest breaths. It feels good. Sometimes so good that tears well in my eyes. And that is what happens when, coming around a lone, tall knoll, we run into a Tibetan wolf.

He is close, and, stopping in his tracks, holds his head low as he regards us. He is a surprisingly large animal, almost twice the height and weight of a German shepherd, and his coat is a blend of tawny, grey and black. The tail is dense and the mane full without being long. For a minute, he stands and then trots on parallel to the road. He stops, turns around and sniffs the ground, then moves on again. He is in no hurry. He seems to be searching for something as he circles ahead of us, then doubles back. Suddenly, he smells something and veers sharply to the side to inspect it. At the dried carcass of a yak calf he makes a thorough inspection of the surroundings, scent-marks a grass tuft, and then heads off on a certain and direct course to the west. We follow until a maze of ponds makes it too boggy to continue. As the wolf emerges on the far side, three other wolves come into view, further on, also moving away. They cover the high savanna at an easy pace and we watch until we cannot see them any more.

We would stop, it is enough for one day, but the clear evening light is so lovely that we decide to go on just a little longer. In a narrow valley, I climb out to try to catch three falcons jousting high above. As they drop lower, in their stoop-and-climb dispute, I see that they are almost white, peregrines perhaps, but as they whistle down into the narrow valley I can then see clearly that they are gyrfalcons. I have only ever seen two before in my entire life. They sweep close beside the road in a blur of speed, rise almost vertically 90 metres up, and then tilt lazily over the hilltop without once flapping their wings. I stare for almost a minute into the electric buzz of the empty sky they leave behind.

On our way out of the valley to a lower elevation where it will not be so cold to camp, we startle a small cat beside the road. Its turns on the vehicle, flattening itself on its belly, hissing and spitting, and chirruping a high-pitched, aggressive threat. When we do not move, it turns and slinks close to the ground into the cover of loose rocks. Tenzin calls it a yuca, and we later discover it is a manul or Pallas's cat. It is slightly smaller than a domestic cat but has dense, pale-grey fur that makes it seem chunky and full. It has the proportions of a miniature snow leopard. Its body, however, is almost uniformly pale charcoal-grey, with two distinct, pale tear stripes set between black outlines that run between the eye and the lower jaw. There are also thin, black stripes that run across its back towards the rump. The tail is grey, with paired thin black lines encircling it every 10 centimetres or so. Its eyes are a greening yellow. I watch it climb nimbly into the rock debris, and then above its head two more appear. They seem to be almost adult kittens, for they are the same size as the cat we disturbed, but when it reaches their height, they gambol with the playfulness of the young, while it remains aloof. We sit quietly and watch them until dark.

28 SEPTEMBER

The dawn is bitterly cold. Out tent is covered in a sheen of ice both inside and out, and although we do not have a thermometer, it is well below zero. The sky is clear, and as we head out from our camp, nestled in a hollow, the sun is hitting the top of the distant snowcapped mountains in a sharp line. There are goa close to the road and a herd of kiang in the far distance. We stop to photograph a flock of geese and duck on a large ice-edged pond, but they take fright as soon as we start walking out from the vehicle. The shopkeeper we ate with yesterday told me that mining companies are responsible for most of the illegal hunting in the Hoh Xil, but that in recent years it has decreased dramatically. The wild animals' behaviour argues otherwise. Everything runs or flies as soon as we stop. The goa are sometimes in such a panic that they run into each other. If we walk out into the land, the game runs when we are 300 metres away. There is the fact that there is very little traffic into these remote regions, but wild animals learn fast. People are hunting in the Hoh Xil, Buddhist doctrine and reserve status aside.

We climb towards the sun-kissed heights, and in the shadowed valleys we find woolly hares still about and chiwa poking their noses out of their burrows. We stop where we saw the manul last night, and a careful search of the rocks reveals one lying on a flat ledge looking at us with half-closed eyes. We move no further, and for the next few hours we remain absorbed by the remarkable nature of these cats as the kittens materialise from between the rocks to join their mother. She is a superb hunter and brings four pika to her kittens in a slow, fluid crouch. Though I watch her carefully with binoculars, her ability is so accomplished that not once do I follow her successfully through the rocks. Only when she emerges onto the hunting ground of the open slopes where the pika burrow do I pick her up again. She runs over the open ground in quick, abrupt bursts, her belly scraping the short grass. The tip of her tail twitches like a nervous bird every time she stops, as if a ratchet is wound up every time she runs and released when she stops. I do not see her make a kill, as she is wary of our presence and hunts away from us, but everything about her is the perfect predator. Her colouring is a perfect match of her environment, among the rocks and on the grass slope, and when she freezes she becomes a rock. She can creep with infinite patience, stretching one leg out at a time so slowly it almost seems to be growing, and she can in an instant be in a full sprint, bounding across the land. Watching her is addictive. The kittens wait in the safety of the rock jumble, and when they become used to our presence they venture further afield. Only occasionally do they stalk each other or play, but when the female returns they become quite boisterous. When the sun is finally warm on the slopes, all three retire into the crevices and we return to our camp. We have an extra day at our disposal and decide that there is no place that we would rather be. The camp is pitched, the day warm, and we are in the remotest place we have found in Tibet with wildlife literally all around.

After a breakfast of noodles and freshly baked bread we take a drive towards a distant group of ponds. We are only a little way down the track when we spot something crossing a distant skyline. Three wolves. They are trotting leisurely across the land, pausing often to smell the ground, but they are moving away from us, deeper into the untrammelled country. On our map, even the track we are on does not exist, and the entire block of country, in a rectangle between marked rural and principal roads, is at least 150 kilometres square, and this is the small eastern portion of the Hoh Xil. To the west of the north–south highway between Golmud and Lhasa, the country is roadless for at least 700 kilometres and perhaps a thousand kilometres north and south.

In the afternoon Beverly and I take our cameras and hike to the place where the manul inhabit the rocks. This time I try climbing to a higher vantage, hoping the cats will relax, given time, and we can get closer images of their remarkable features and behaviour. Beverly remains below to signal their move-ment between the rocks. Eventually the cats emerge, although by their nature remaining secretive, never quite coming into the open except when slinking along a ledge to sun themselves. A pair of gyrfalcons, white and fast against the sky, chatter excitedly at my intrusion into their territory.

29 SEPTEMBER

The night was warmer and I hike back before sunrise to the cats. It is a cool, overcast morning with a thin, chill wind and the cats are mostly content to sit still and seek what little warmth there is. I do not see the manul mother, and a fox trotting across the skyline of the hill is the only excitement in my observa-tions. The kittens see or sense it before I do – the chiwa alarm-chirr, alerting me. They retreat to the sanctuary of the deep clefts of rock and, stretching their necks slowly, crane their heads to watch the oblivious foe. The fox trots past high above, without a pause.

The others collect me and we drive into an even higher country of gentle hills that eventually become flatter. We have left the last nomads behind. The country here is all above 4 600 metres, and at its highest nearly 5 000 metres, but there are no mountains or even tall hills. Only to the far west do the moun-tains of the Kunlun Shan rise, snowcapped, to over 6 000 metres. The land is featureless and almost desert. Vegetation grows in tight, low tufts but for the most part the ground is gravel flats that are soft underfoot. I have known similar landscapes in Namibia, and remember the fragility of that ecosystem, where vehicle tracks remain for decades.

We see not a single other soul or vehicle during our whole traverse of this part of the country. The game is sparse but we still come across small herds of goa and the occasional kiang. The chiru, however, are nowhere to be seen. We stop frequently and scan the wide horizon, but to no avail.

As we drop imperceptibly lower, the short, stiff-bristled grass returns in patches to the gentle valley floors. Beside a wide, gravel riverbed we find the first wild yaks we have seen, known in Tibet as drong. They graze singly on a distant hillside. There are five altogether, spread out over two-or-so kilometres, but they are so far away that we need binoculars to see them. The weather has been clos-ing in all morning, now a low haze shuts out the horizon. Five kilometres further on, when we find a lone wild yak bull grazing shortly off the road, it begins to snow. The wild yak is distinctly different from the domestic. It is pitch-black from its nose to its hoofs, and larger by a third or a half than the domestic animal. This bull's coat is only long across the lower edge of the belly and shanks and does not hang to the ground, as does that of a domestic yak, but stands 20 centimetres clear. The horns, however, set it dramatically apart. They are thick at the base, tapering in a beautiful outward-curving harp shape that ends in two sharp, widespread points. They are fuller, more curved and narrower between the tips than those horns of fighting bulls, but the impression is the same. They are lethal. The wild yak has a reputation for aggression, so I approach this bull with extreme caution, staying downwind, proceeding unseen from behind a knoll and making the final approach on my belly so as to present as small a disturbance to the animal's view of the land as possible. Some 120 metres away, it grazes unaware. Gongpo does not exercise the same care, and 50 metres to my right walks upright directly towards the yak, his only caution an exaggerated stalking step. At 150 metres, the yak lifts its head towards him. He freezes, but the yak has seen him and snorts into the air. It looks very dangerous. He clicks a photograph and, dropping his hands, turns halfway around to retreat. The yak bucks, paws the dust, and lowers its horns. Gongpo runs, stumbling in panic, his arms flailing for balance. He scrambles into the vehicle just in time to hear our relieved laughter: the yak has stayed entirely still. The nuisance gone, the yak grazes on peacefully, moving slowly in my direction. At 30 metres, I back away on my belly until I am out of sight.

Snow is blowing across the land. We cross a river and find another single wild yak. This one, a cow, is wary of the vehicle, and although we stop she stands for only a few seconds before moving off. I am surprised by her gait: it has nothing of the shamble of the domestic animal. It is light and nimble, a trot like a large antelope, quick on the ground, strong and swift. I am glad the bull did not charge. This animal would catch a vehicle over the first 40 metres, and the thought of those horns has me arching my shoulders, and a quick shiver runs down my spine.

The snow starts to fall heavily, the wind gusting to 25 knots. Vultures, buzzards and falcons are driven to the ground and we find them perched beside the road, facing hunched into the wind. For hours, as the storm grows, the close environment of a circle of 150 metres of whiteout is all we get. There are cold raptors, goa with snow frosted to their coats, muzzles and the slits of their eyes, and distant groups of kiang standing huddled with their haunches to the wind. We stop to photograph flocks of pastel-coloured Tibetan sandgrouse. The wind stings the icy grit against our faces, and when we are done the track ahead is almost invisible under the snow. We press on. We have not seen sign of any human all day. I look at Beverly. We love this sort of unpredictable adventure, but an unplanned night stuck out in the middle of an invisible country in below-zero conditions, in a snowstorm, would tax her stamina immensely. At 4 pm we find a fence, and half an hour later a set of prayer flags strung between two steel posts across the road marks the unsignposted boundary of the Hoh Xil Reserve.

The Golmud–Lhasa Highway, eight kilometres to the west, is a sharp return to reality. Army camps line the highway outside Xidi Tan, the peaked tents and lines of trucks and artillery iced with white. We drive to Golmud through a desert landscape of tall, bare peaks shrouded by low cloud. The highway valley is dirty, untidy and abused. Power lines, a railway, hydroelectric dams and the work of bulldozers scar every inch of this desert place. There is not a nomad in sight.

Emerging from the close hold of the mountains, we enter an endlessly flat desert of brown sand, where Golmud rises in the distance. Golmud is an ideal Chinese city, in the sense that it is occupied. Garage-sized small industries, offering everything from yoghurt to Caterpillar crane repairs, stand dimly lit in the twilight, ready for business. Broad six-lane boulevards lead into the city. There is little traffic. Solar-powered streetlights shine on rows of casuarina and poplar trees that hold back the sand and the reality of the bleak desert surround. At a five-way intersection that is 250 metres across, vehicles ignore the distant traffic signals and follow their own set of rules. Downtown there are treed centre islands, fanciful streetlights and standard, square, bland-faced city buildings, made individual by a festival of neon lighting that runs a river of colour in the dark night.

30 SEPTEMBER

Our hotel breakfast buffet offers a 20-metre-long selection of foods, but no fruit. There are four choices of soup, six hot noodle dishes, including spicy shrimp, and a variety that I can only guess at. Eggs are deep-fried in oil until hard, and beside them is what looks like jellied pigs' trotters, sliced thinly. Perhaps the most peculiar offering is a sweet apple juice served as hot tea.

The morning is cloudless as we head back into the mountains of the Kunlun

Shan, whose high presence is today obscured by a dusty haze. Some 80 kilometres into the mountains we turn in to an isolated valley, where the road climbs steadily as it follows the valley west. There is no traffic bar one nomad roping a sheep, and several military vehicles. For such an undisturbed, remote place, I am surprised by the lack of life. Except for a few circling vultures we see nothing at all until the altimeter passes the 3 900-metre mark and stiff, bristly grass starts to replace the other shrubs. Pika appear, and with them there are raptors once more. The grass becomes more and more luxuriant as we move higher, and there are goa and herds of kiang. At 4 100 metres, grass covers the land and we count a herd of fifty-two kiang on a pretty slope beneath a rounded peak. Soon we spot a single wild yak on a far hill and then there is another and another. We stop and glass the mountain. There are wild yak everywhere, spread thinly over the steep land right to the rocky foot of the serrated peaks. A closer, clustered group has young calves that race each other in gay abandon, their tails held high like shaggy flags. There are more than seventy wild yak from the valley floor to the peak of this single buttress, moving like slow black ants across the land.

Beside the river, the goa and kiang move off at our approach but are tamer than those in other remote regions of the Hoh Xil. While out walking across the country, I am surprised to hear what sounds like four or five shots in quick succession, too quick to be the crack of a nomad's sling or whip. We find no people other than two nomads unloading their vehicle beside their white-tented, riverside camp, but far up the flank of one of the side valleys is the scar of a jade mine. Perhaps I was mistaken, thinking mine blasting to be rifle shots. High in the valley it once again becomes too cold for grass and the land is snow-covered and barren, the mountains pure white. The road ends at a lake, Naktso, where a great black-headed gull carves a divide in the mirror-calm reflection of the mountain surround. When we head back down Drong Long (Wild Yak Valley), the light is warm and cumulus clouds are gathering in great billows above the peaks.

Walking out to photograph a drong in the riverbed, I disturb two pale-chested saker falcons perched on the top edge of a low riverside cliff. The yak lifts its head and, seeing me, raises its tail, its feet planted. I am above it, the cliff preventing its charge. After a time, it trots upstream and, climbing a less steep bank, comes back towards me. I am contemplating a descent of the cliff when Beverly has Gongpo bring the vehicle down off the road towards the yak. For a long minute the yak stands still, weighing the situation, and then turns and heads towards the higher ground. The great horned black beast, walking alone against the silhouette of sharp-ridged peaks, and my own adrenaline are like the bow and string of a violin. My soul hovers on a high, searing sustained note.

While we are photographing a large flock of sheep returning to a single, white yurt-style tent in a scene lit by evening storm light, their nomad owner rides out on his elaborately decorated motorbike to greet us. He is a garrulous, happy character who talks with his hands. 'One thousand, one hundred and fifty sheep,' he tells me. 'Do you want to buy skulls?'

'Sheep skulls?'

'No, wild animal skulls.'

From beneath a tarpaulin beside his tent he hauls out the skull of an enormous Tibetan argali, or wild sheep, the huge, heavy boss coming almost a full circle. He wants 3 000 yuan. Tenzin thinks it would sell for 10 000 yuan in town, but is fearful of the consequences of carrying it out. The nomad shows us other horned skulls, of bharal and goa, that he has picked up in his wanderings with his flock. I photograph his father sitting on a low chair inside his tent, warm and smoky from the stove, the walls decorated with a cheap fabric printed in man-sized images of Snoopy.

We drive alone through this singular place, cupped in a valley that I have never heard spoken of, scenically outstanding by any worldwide comparison, with the setting sun igniting bare mountains and towering cloud. When it grows dark, Beverly's hand finds mine across the camera case that separates us and we share a look without speaking for the depth of where our choice takes us.

1 OCTOBER

We wake at 4:30 am to make the 6 am train to Lhasa. At the station we must present tickets, passports and permits to travel in the Tibet Autonomous Region (TAR), which includes Lhasa and the country to the west. Our cabin is defined as a hard sleeper. Inside, the carriage is dark. Door-less compartments reveal six sleeper bunks, where the sleepers turn at the disturbance of the station. The train is packed, we must manhandle our luggage down the narrow passageway, and when we finally establish which is our compartment, the small luggage space is full and our bags occupy the whole bunk. We decide to sleep in shifts on our one available bunk. Beverly lets me go first.

The Chinese passengers are smiling and kind, and it is a relief to find a few in our immediate vicinity who speak English. I wake after an hour and a half to discover that Beverly has found another bunk and is sleeping. I peel an apple and chat to my neighbours on the small convenience seats in the passage. Beverly and I are the only foreigners on the train. I am offered dates, bread and noodles, and must repeat a brief history of my origin several times. I draw a map of Africa with a cross at the bottom for the loving, middle-aged couple on the bunk across from me.

'Kay Taa!' he says.

I nod, patting my chest. 'Cape Town,' I say.

The train rolls steadily, smoothly through the country. Inside, the temperature is balmy, with extra oxygen provided for the passengers' comfort. Outside, the country is the cold, high steppe of the Hoh Xil. Wide plains run to far-distant hills. There is snow on the ground. The road that runs alongside the railway line is almost empty, save for trucks carrying loads of small red tractors. Game is everywhere. Accustomed to the rush of trains along the line, the animals do not run away, and for several hours I see many kiang and large herds of goa. When we pass scattered groups of drong and then a lone Tibetan wolf close to the tracks, I shout with such enthusiasm that all my neighbours rush to the window.

Later, there are small towns and army convoys on the road, and the periodic herds of nomads replace the concentrations of wild animals. A pretty young woman in black tights and a lace top comes to sit in the passageway two seats away from me. Every time I look up she is looking at me. It distracts me from my writing.

After a catch-up sleep for our early start, we find that the country grows empty again. It is still high-altitude steppe of long, gentle inclines, but at over 5 000 metres the animals and the people are gone. The country seems faded by the extremes of high-altitude weather: pale and washed out, the colours half of their lower-altitude counterparts.

We descend towards Lhasa, the landscape growing steeper and taller as we progress. A storm billows over the peaks, the clouds pierced from behind by sharp shafts of sun, the rain and snow drifting in grey veils over the darkened land. Small rivers twist lazily in the wide, flat valleys, where thousands of yak and sheep mow the grass to within millimetres of the soil. There are many double-storey buildings, painted white except for the cornice around the flat roof parapet, which is bright in reds, yellows and blues. Terraced fields and large stone-walled compounds indicate permanent cultivation rather than the presence of nomads. The man-height perimeter walls are painted white, and the effect is of an ancient, calm permanence. Some homes are clearly those of the wealthy, with tiled façades and shiny, high-end SUVs outside.

The fairy-tale Lhasa, with the Potala Dzong (Royal Palace) set in hilly country that I imagined, does not exist. Lhasa is a city of 260 000 people, divided between the old Tibetan city and its new Chinese counterpart. Tibetans now represent only 45% of Lhasa's population. The officials at the station are polite and helpful, guiding us and helping us with our bags on the long walk to the public security barrier.

It is an extended drive through extraordinarily chaotic traffic to the Lhasa Yak Hotel, whose brass reception plaque proclaims it 'The Best Tourist Hotel in Lhasa'. Upstairs, two mascara-ed ladies in high heels and short skirts stand sheepishly in the passage outside our room. The two porter women bending beneath our bags, which they carry on their backs with the straps across their foreheads, have a loud, harried conversation with them. The room has been recently slept in, and the bed and bathroom are in disarray. We move to a new room where the toilet flushes without emptying onto the floor and hot water comes out of the hot tap.

2 OCTOBER

In the morning, Bhuchung, the guide who will replace Tenzin as neither Tenzin nor Gongpo may enter the TAR, tells us there will be a delay in our departure because our driver has had to take a client to the Nepalese border. He is uncertain as to how long this will take, so we set out to explore Lhasa.

The Jokhang Temple, first built in the 7th-Century and rebuilt in 1419, is a short walk from our hotel through a part of old Lhasa, where squat buildings of stone and wood rise to three storeys, with tall, narrow windows, each topped by a coloured cornice. The stone walls are painted white, with narrow lanes running between them and, except for the chaos and jamboree of the streets, the effect is of old Europe. Tourist shops vie with traditional dress shops, jewellery shops, butcheries and hole-in-the-wall bakeries for the attention of the human river that negotiates the often-narrow pavements.

Today is the first day of the week-long Chinese national holiday and tourists are stopped everywhere with cellphones lifted to the sights. Two large urns stand in the plaza with fires burning in their open, head-height hearths, and pilgrims toss a stick-like herb and the trimmings of dense, short-needled pine branches into the maw of the fire. A thick smoke issues from the tall spout of the urns, filling the plaza with fragrant incense. It was here in 2008 that a more acrid smoke filled the air, when four monks performed self-immolation in protest at the Chinese occupation of Tibet. The accompanying protests shut Lhasa and the Tibetan Autonomous Region to all travel.

Buddhist pilgrims, now allowed back into the region, walk clockwise circuits around the temple, spinning ornate, handheld prayer wheels and reciting mantras to complete what is called a kora, a special prayer set or pilgrimage. Before the temple walls the especially devout prostrate themselves flat on their bellies, stand, raise their arms wide to the sky, bring them down into a position of prayer held close below the chin, and then prostrate themselves again. Many will keep this up all day. The greater the effort, the more worthy the kora. Some have come from far. We have seen them beside the road near Yushu and Nangchen, more than a thousand kilometres away: pray, prostrate, one step forward, pray, prostrate again. The pilgrimage can take three to four months, and families camp beside them, providing food and a place to sleep as they walk towards this most respected kora. To arrive in Lhasa today, at the beginning of a week of special prayers, is auspicious.

Inside the temple, photography is not allowed. Each chapel holds a special deity and the pilgrims carry brown flasks of molten yak butter that they pour into huge, brass candle urns before each deity. In a slot in the glass, before each representative god's statue, money is shoved to stand in a pyramid at the feet of the gold-painted, brocaded figurine. There are flat, wooden trays where farmers place money and small offerings from their crops. The sense of spirituality is overwhelming. Monks pray in small unobtrusive corners, their eyes somewhere far beyond the melee.

In the afternoon, we time our visit to the Sera Gompa monastery to coincide with the afternoon debate. It is a lively event held in a walled garden beneath old trees. Monks sit on cushions on the gravel in groups of two or three, while another, standing, fires questions at them about the traditional philosophies of Buddhism. The debate is rigorous, often heated, and the standing protagonist claps his hands and stamps his feet in a gesture of attack at his seated debatees. I find myself compelled, wishing I could understand. It is clearly a university of thought.

3 OCTOBER

On the broad sidewalk skirting the perimeter of the Potala Dzong a sea of Tibetans is flowing clockwise as they make their kora. Beyond the entrance the grounds are lavishly arranged with potted marigolds, statice and other plants. The stone steps up the steep façade take time to negotiate and present a view over the city. It has snowed in the night, and as the clouds recede they hang in trails between the white-dusted hilltops. One enters the Dzong itself from a large plaza set on top of the hill, high above the city.

There are halls and chambers: places for worship, council and debate. We see some of the finest examples of ornate Tibetan interiors, with beautiful tapestries, rows of huge statues of deities, gold-foiled sculptures of the rites of life and ornately painted alcoves of contemplation, but the passages of the Dzong were made for small groups of monks, not masses of tourists. The stairways can take only two abreast and the many aged or short of breath cause bottlenecks, with literally no space to move. In this place of extreme wonder, the quick, the rich, the raw, the poor and even the sincerely devout become swept along by the inexorable flow of humanity rather than pause for spirituality.

In mid-afternoon, our vehicle and driver arrive. There is much explanation about having to drive to the border of Nepal to deliver an Indian client suffering from chronic altitude sickness. Our driver does not return our greeting. Climbing inside the vehicle, I introduce myself, but his eyes only flick to my chest and he nods. I must ask Bhuchung for his name.

On our way out of town, we discover that Bhuchung and Khama, our driver, do not have a tent. Bhuchung is quite sincere when he says they will leave us camping and drive to the nearest village. He is at a loss for words when I ask him, 'What if the closest settlement is four or five hours away?' I try to impress upon him that the express purpose of our journey here is to travel into the wilderness, that we want to be in a place where there is not a single other person, that we are determined in our desire to see the deserted lands and wildlife of Tibet. This seems to convince him and we detour back into the city where he procures a tent and a mass of blankets borrowed from a relation.

The road west from Lhasa follows the wide, treed valley of the Kyichu (Lhasa River) to the southwest. Where this joins the Brahmaputra River, known locally as the Yarlung Tsangpo, it turns west into a narrow, steep-sided valley where towering peaks are glimpsed between other heights, their white-capped sentinels standing between the clouds. The river is brown and fast. Far to the east, though the whole valley expands to kilometres across, it cannot escape the hold of the mountains that serrate even the distant horizon beneath radiating white clouds.

The land here is heavily agricultural. A few cows have replaced the vast yak herds. Sugar beets lie piled to dry atop courtyard walls and house parapets, and in the myriad small fields is the stubble of cut barley and neatly stacked bushels of reaped crops. I am perplexed by the severely eroded hillsides, as I see almost no stock and the sparse grass is quite tall. Bhuchung tells me that during the harvest all stock is kept contained, usually indoors, to prevent the animals eating the harvest, as all the people are busy and the herds would be unattended.

I note too that along the valley all the Tibetan homes are flying the Chinese national flag. It transpires that, by law, every Tibetan household must fly the national flag and, further, that the flag must be replaced with a new one twice a year.

4 OCTOBER

The morning starts badly with a confrontation with Bhuchung and Khama, who, as we pore over maps to indicate side roads we would like to explore, refuse to consider any deviation from our itinerary. I call the local office of our tour company in Lhasa on our satellite phone, to ensure that our guide and driver grasp our purpose to explore the wilderness and approach it willingly, but their assurances ring hollow. It is only when I call the head of the tour company in Xining, who has hired the Lhasa company to handle our trip in the Autonomous Region, because they are not permitted to guide here, that our point is understood. He promises to rectify matters. There are further delays as we wait for Bhuchung to clear our itinerary, travel permits, visas and passports with the police in Shigatse.

Eventually on the road at 11 am, we leave the Brahmaputra Valley behind and enter the Yongdungling Valley, which carries us through a landscape of arid hills. Here the barley harvest is still under way. The fields are larger, and the ubiquitous red Dong Fang Hong tractor, with its single-cylinder motor, proves to be the incredibly versatile workhorse of the community. I see it with a narrow mower attached to the front, felling the tall barley, dragging flatbed loads of the cut crop that tower to twice the tiny tractor's height and width.

Crossing the Brahmaputra again for a short time outside Lhatse, the road leaves it once more as it climbs higher into a country of low, jagged hills that rise individually in high steps. Snow lies on the hills, where yaks and nomad encampments are once again part of the scenery. Being back among nomads makes me feel as though I am among friends. The land climbs continually until it is too arid and harsh to support stock. The country here is defined by mountains, and in their repetition, one has a sense of vastness. There are no fences, and as it grows too dark to photograph I daydream about taking a horse and riding far into this open place until I reach some other side.

5 OCTOBER

It is still dark when we leave the guesthouse in Saga. I have slept well in the tiny room, with its colourful fish-and-bird graphic motifs and low blue-and-yellow wood-beam ceiling. As we pack to leave the room, a young woman I assume to be housekeeping waits outside our door. When I am gone she comes into the room, lies down on one of the beds and invites Beverly to join her, patting the bed beside her, and smiling. Beverly is uncertain as to her intention or where she has come from, so thanks her kindly and leaves.

It is raining. As the weather abates and the clouds drift low, we find ourselves in a fantastic landscape, with peaks appearing and vanishing as if conjured before our eyes. Snow is thick on the ground, as are a worrying number of dog packs. I count one pack at thirteen dogs running at a trot through the snow halfway up a hillside. They set the chained guard dogs around the still-sleeping nomad camps to furious barking as they pass.

Smoke is just starting to issue from the chimney of a nomad's tent when we stop to photograph a small owl perched on snow-covered clothing hung on the fence. It is the size of a dove, but round as it squats cold on the wire. It is mottled, pale brown and white, with faint yellow false eyes on the back of its head. Its movements are quick: a swivel of the head, a sudden standing upright, then back to a hunched perch. In the valley below the encampment yak bulls are tussling on the snow-crusted ground. They ignore our close presence and horn the ground and butt heads, pushing each other into the cold, fast-flowing stream. Up the valley, there are others doing the same, and I wonder if there are still traces of a wild rut in the domestic yak or if it is just the first cold weather that brings on this vigorous behaviour.

We climb through several small passes. Khama has the terrifying habit of taking hairpin bends on the inside, into the face of unseen, oncoming traffic. Already yesterday I asked Bhuchung, who seems rather to have a mediocre comprehension of English than an unwillingness to cooperate, if he would please ask Khama to take these bends on the outside for our sake. For a time, it went better, but today he has relapsed. We don't know how to communicate with Khama, who sits taciturn behind the wheel, his eyes permanently hidden behind dark glasses. We do not understand Khama's conversation; however, there is little ambiguity in his reluctance to stop or, even worse, turn around for a photograph.

The country we are travelling through is extraordinary and we stop often and walk out, the sun burnishing low drifts of cloud silver against others that are a heavy, dark grey. Sunlight, shooting down in narrow shafts, illuminates a varied earth of foothills, rough rocky peaks and valleys, made lovelier yet by the meander of brisk, crystal streams and freestone rivers.

The town of Dongpa is divided into three sections: the original village, where officials at another checkpoint verify our passports, visas, travel permits and itinerary; a new, smaller adobe nomad settlement on the far side of the same hill; then, several kilometres out onto the plain beyond, the brand-new Chinese town, all square and modern, with fresh paint and shiny, steel-palisade perimeters, and only windblown litter, dogs and a single new, high-end SUV in the streets. Nothing sums up the current status of Tibet better than this high, remote town in the lap of a spectacular mountain country.

Slowly, the signs of nomads fall behind as the altitude reveals a more arid land. We see a herd of kiang and later a lone chiwa. The steppe is seldom less than 4 600 metres in elevation. The Mayunga Pass is at 5 217 metres, and on the far side the country is without people. Kiang are suddenly everywhere, and we start to see goa scattered across the valleys. The valleys are sweeping and remarkably flat, and those where snowmelt streams feed into them become marshlands. Using binoculars, we see a pair of black-necked cranes, and then another, and suddenly there are pairs to be found dotted throughout the valleys.

At a small pass into the next valley, Khama careens into a blind corner, on completely the wrong side of the road, touching the inside dirt. Both Beverly and I gasp simultaneously in fright. The incident seems to pass, but 10 kilometres on, when I call out, 'Stop, stop please,' to photograph a shrine on top of a hill against the white peaks of the distant Himalaya, Khama hits the brakes so hard that we leave 25-metre skid marks on the tar. When I raise my eyebrows in amazed query and Beverly asks Bhuchung what this demonstration is about, there is an outburst of vitriol from the driver that must have been brewing for a long time. Even Bhuchung is left speechless and wide-eyed. When we ask Khama, with Bhuchung translating, what it is that we have done, his anger just grows. Both of us try to get across that we need to know the cause of his upset before we can make amends, but he just shouts at us in a monologue that gives us no opportunity to reason. We stare at him in stunned silence until he fishes out his cellphone, storms over to the deserted highway, and calls the agent in Lhasa. Once again, he spits his vitriol into the phone without pause until Bhuchung intrudes and asks if he may speak to the person. Shortly afterwards, he hands the phone to me. I am met with an irate tirade from the Lhasa office's principal, who accuses us of not treating the driver correctly. It takes thirty-five minutes of reasoning for her to see the point that we have directed all our requests through Bhuchung, as Khama speaks no English, and that all of these have had to do with stopping or driving back to photograph a scene or event. That is the only communication we have had, and we always thank him when we are done. She tells me that we are abusing her staff by not sticking to the schedule. I explain as calmly as possible that there is no strict schedule, only an itinerary guideline, as the sole purpose of our trip is to explore and photograph the wild places of Tibet, and we were quite clear that we would need flexibility and time. I suggest that perhaps we should change drivers, as Gongpo had found enthusiasm, enjoyment and real fascination in what we were doing. We became friends, and he told us that he saw his beautiful country and the wild animals with new eyes. She suggests that both she and the driver would be happy to cancel the tour.

We are 250 kilometres from any town in either direction, three days out of Lhasa, on the wide plain of the high steppe. A freezing wind is blowing, Beverly is in tears and I am walking down the deserted highway seeking to calm myself. I am incredulous that we could be delivered to such an unrealistic place by the fragility of an ego, and filled with frustration that we had no way of gauging that fragility.

To his immense credit, Bhuchung initiates a resolution. I speak to his principal again and outline our simple request and promise. Khama must be made to clearly understand our purpose, and if it is against his preference then the agency should replace him. Our request about not cutting corners on blind rises is not a criticism of his ability, but rather a sincere request made entirely for our own comfort and safety. For this I will shake hands, apologise for any upset we have caused him, wipe the slate, and move on. Finally, after two frozen hours standing on the highway, I request that Khama look me in the eyes when I shake his hand.

Near Lake Manasarovar we find three black-necked cranes close by the road. Khama turns the vehicle around, approaches slowly, and for close on an hour we watch and photograph these fragile, elegant birds feed and preen 30 metres from our station on the side of the road. Both Khama and Bhuchung sit with their heads slumped against the dashboard.

Our guesthouse, in the saddle between two lakes, has no other guests and no toilet. Our room is spartan and rudimentary, with hand-made beds, two water bowls and a water tank on the veranda outside. We eat instant noodles beside a stove whose issue of warmth is a welcome comfort. We are shattered and emotionally fragile, and at an altitude of nearly 4 800 metres, we are finding it difficult to breathe.

6 OCTOBER

We leave our simple room before dawn. An icy wind is blowing across the lake but the sky is clear, the occasional star still resisting the coming day. We explore a nearby hill overlooking the tiny village where we have slept. There are two stupas, prayer wheels and a yard piled to the height of my waist with stones made beautiful with inscribed prayers. Prayer flags flap in the wind that chills my fingertips until they ache, but it cannot detract from the sweeping view of the Himalaya with the lake at its foot. The mountains dominate the southern horizon. They form a marked physical barrier, mountains rising from a land of mountains until they close with the sky. To the northeast, the sacred mountain of both Hindus and Tibetan Buddhists – the classic, triangular peak of Mount Kailash, at 6 638 metres – is capped with a pure-white monk's cowl made golden by the early sun.

Around the stupa a pack of dogs is baying after a female in heat. In the midst of this place of fluttering prayers, a savage dogfight breaks out. There are too many in the pack to intrude, and we walk to the southern side to photograph the Chiu Gompa, perched precariously on the side of a small, steep, rocky hill. Prayer flags hang like streamers down its side, and in the far distance the peak of Nemo Nanyi, at 7 728 metres, peeps above the hills. As the dogfight subsides I can hear the monks chanting.

Our journey west is into an arid steppe where grass is tightly tussocked and sparse. The day is clear and the sunlight is reflected off the thinly covered ground until all is bleached and the land seems somehow exhausted, waiting for some respite. There are nomads, but they are few and they walk with their sheep out into the country, which I take to be a sign that there are predators about. There is the occasional fence, but it runs only a few hundred metres and then ends and the land runs on and on. There is the road and its parallel of power lines, but walk a hundred metres either side and you stand in the same place as the next valley and a hundred valleys beyond that. It is open country, and even though it has clearly in its time been hunted there are still kiang and goa to be seen. I am certain that, as one moves in any direction away from the road and the influence of man, the numbers of wild animals will increase. Even beside the road, in the places where the nomads have left this high summer grazing, moving their herds lower at the onset of winter, the wild animals have returned.

The settlements here are not big enough to be considered towns. A few nomads gather at a lower, warmer place, where there is water and wide plains for winter grazing, and perhaps a shopkeeper serving the road traffic. I am interested that although many of the houses we see at these places and in the small towns are new, they seem to be traditional Tibetan buildings, and not the new settlements that are being constructed in the north on such a vast scale. Bhuchung tells me that here the houses are built by the Tibetans themselves, often with a government subsidy. The government provides a 30% subsidy to any person building a home in this region. I ask what other support is given to Tibetans and learn that both education and healthcare are state-provided, and that any attempt at enterprise, such as acquiring a tractor or a truck, will receive a 30% subsidy. It is clearly an attempt at genuine development, but the contrast with the harsher attempts at governmental control of people's lives leaves me confused. How can this seeming philanthropy be forthcoming from the same government that applies a rule of a maximum of two children to the nomads? To those children who come after, the authorities simply refuse to issue a birth certificate. Without a birth certificate, you do not exist and cannot register as a citizen, nor receive the benefits of a citizen.

In conversation with a Himalayan guide at a hotel in Shigatse, we were told that the issue of government control of the number of children is not fixed across the whole of China. In big Chinese centres such as Beijing it remains at two, and it is two for Tibetans outside the TAR, and three for Tibetans registered inside the TAR. There is, however, one notable exception: the Shigatse Region, where an old custom persists that is, to my knowledge, almost unique in the world. The women of Shigatse have multiple husbands, up to four on occasion, and they are allowed seven children by the government.

At the 5 400-metre Dartala Pass, the bare mountains are multihued, as if some broad rainbow had fallen upon the soil and stained it. There are golden ochres, rust-reds that bleach to a faded maroon, pale-blacks and veins the colour of oxidised copper. At the watershed divide of the pass, the rivers to the north flow east and west on their long journey around the Himalaya to the sea. Incredibly, the rivers to the south cut directly through the barrier of the Himalaya to proceed almost due south into India. Between the watershed of the Dartala Pass and the Himalaya to the south lies a vast valley more than a hundred kilometres across whose centre is divided by the Lanengq Zangbo River, also known as the Macha Kapo. The river and vicious summer thunderstorms have cut a canyon into the base of the valley that forms a landscape entirely of its own. The eroded earth is pale-ochre and grey and has been shaped into a thousand castle-like buttresses, rounded and sheer, standing alone or as a winding wall, kilometres long. Some are incredibly tall, towering over the crumbling earth around them. Gullies run outlined like fish skeletons from their tops and in the later afternoon are deep with shadow.

The land we drive through is all but deserted. In the three-hour drive to the sleepy rural town of Tsada I count a total of five nomad encampments, three of which in the dry country are clustered in the valley of a small stream. Tsada lies on the banks of the Macha Kapo and, leaving town, we follow the river downstream for 10 kilometres, as it cuts sharply into the eroded canyon valleys. In a tall buttress are the caves and remaining buildings of the Guge Kingdom. They are the remnants of a 10th-Century city of several thousand who followed an exiled Lhasa king to this place. Several hundred metres above the gentle foothills, a building called the Summer Palace sits at the precipitous, narrow point of the buttress. Cut into the buttress underground below it is the Winter Palace, beneath which are a multitude of caves and tall walls, teetering on cliff edges that formed the heart of what was clearly a defensive position. The rebel king, a believer in an alternative non-pacifist Buddhism, survived here for twenty years before the army of Lhasa caught up with him and executed him.

Our Toling Gompa guesthouse is neat and clean, offering pleasant small rooms with tapestry-covered couch beds and a communal flushing squat-toilet. I like the touch of incense sticks and matches beside the dipper in the water bucket for hand washing. We eat in a tiny Chinese restaurant of three tables. A family of four sitting beside us speaks some English, and as the evening wears on and the proprietor and his wife join in an animated and often funny conversation, they take some trouble to translate and include us. When we walk home down the broad, deserted street, I am feeling the happiest I have felt since we left Lhasa.

7 OCTOBER

The myriad eroded canyon walls are warm in the early sun, with the Himalaya in the haze of distance behind. There are kiang in tall, yellow grass on the canyon rim and a few goa against the slopes of the closer mountains, but for kilometre upon kilometre all is still except for the flitting flocks of sparrow-type birds in their drab, earth-coloured browns. Then, near the top of a pass, above a low ridge, a sand fox appears, the first predator we have seen in the TAR. Our walk up to a higher vantage point to find it again though, is painfully slow, with the thin air rasping in our lungs, and by the time we crest the ridge the land beyond is empty. It is, however, striking in the striated colours of the earth: yellow grass runs like spilt liquid over a meandering red valley, and a tiny blue pond is fringed by white salts in a depression that is the colour of faded jade. I stand and catch my breath. Far, far below, a nomad is walking his flock across a place of maroon earth in front of a hill that is a rich mustard-ochre. It is a fine place to pause, and I linger a little before turning back to the vehicle.

In the late morning, we descend a short pass into an arid place. The valley of the Gar Zangbo, which flows into the Senge Tsangpo, is cup-shaped and very wide, the mountains to the south hovering at around 6 300 metres, their peaks etched by snow. The ground is grey-white and I must squint against the glare.

In the far distance, the town of Gar occupies a short, sheltered strip against the rise of the bleached earth hills. We drop lower and lower into the desert valley, our first taste of the Chang Tang Plateau. Gar is under construction, with scaffolding covering buildings and sidewalk paving stones being laid beside wide, dusty six-lane boulevards. Khama and Bhuchung get lost in the maze of new streets and cannot find the police station to report our arrival. Once that is done, we take the road out of town to Gakyi in the afternoon, travelling with the sun behind us, Gar being the western terminus of our travel. This is the road that, in our planning, had us filled with anticipation, as it runs through an otherwise-roadless tract of Asia that extends more than a thousand kilometres to the north, beyond the Kunlun Shan, and from the barrier of the Karakoram and the Himalaya 600 kilometres to the east.

We search the wide vistas of the colour-leached desert hills and valleys for any sign of life, but the land is still. There has not been a single tree, except for those planted on the streets of Gar, for hundreds of kilometres. Only when the road joins the Senge Tsangpo does life return. The river is a slow, oxbowing meander where the drift of aquatic plants is the only indication of its flow. The nomad winter homes here are deserted, and we find large flocks of waterbirds feeding in ponds, backwaters and small lakes. There are geese, pin-tailed ducks and teal, while far out in the marshy flatlands where the valley grows wide we can see several pairs of black-necked cranes.

Each time we stop Khama and Bhuchung slump over the dashboard in a display of boredom and tedium. To come face to face with such an attitude is astounding. There is no rising to the day, just an overt submission to the drudgery of life. Even Bhuchung, who is young, in his mid-twenties, is lacking in curiosity for the places we explore. Their listless exasperation at any pause or deviation from as fast as possible a transit from A to B is like a ball and chain shackled to our enthusiasm. I stare out of the window, wishing with every cell of me that it could be otherwise.

8 OCTOBER

The road east is dirt. We encounter a large herd of goa and, with the sun just touching the land, they are lovely, backlit against the yet-dark mountains, standing in the yellow grass made incandescent by the sun. There are kiang too, and as we move into the desert the game is prolific, if not varied. There are great concentrations of waterbirds on the occasional backwaters or ponds. The presence of water seems bizarre in this completely arid place, except for the snow, but even at midday it is hardly warm enough to cause evaporation.

We walk out to a wide isolated pond where the water is black with birds. They must be gleaning the last riches before migrating, for even at mid-morning the ice is solid, 15 to 20 metres in from the shore. They fuss away to the far edge, some of the Eurasian teal taking flight, their silhouettes black darts against the mountains.

Every time I step away from the vehicle into this wide place my heart quickens and my tread feels light. We walk far across a plain after kiang, and they and the goa keep a long, safe distance, but the land feels close, an intimate garment that warms me inside. The air is chill and the tip of my nose burns in the breeze.

We stop in the tiny town of Xungba to repair a puncture. The town is dusty, the people entirely traditional. There is little that is Western here, except for what is associated with motor vehicles and, incongruously, pool tables. Nomads arrive crowded up to three on the back of motorcycles and jostling on the heavily corrugated roads in the back of trucks of bright orange and red. All are wearing their finest. The men wear a broad-rimmed, round-domed cream hat made of stiff, finely woven straw. The hat sits perched on the crown of their head, well above the ears, and is held there by a black chinstrap. Their strong black hair is plaited into long ponytails. The women wear unusual dark red, blue and green vertically striped chupas that sweep the ground. They mask their faces below the eyes with lined silver-grey lace and wear hats of rich gold brocade on green or maroon, trimmed around the brim with fur. The brim is cut into four or five rounded sections and then stiffened, angled slightly upwards so that the hat resembles the petals of a flower opening. Everyone is drinking; some are drinking and shopping, grinning broadly as they look at the portraits I have taken. Others are drinking in the sun, watching the pool players. The women play pool with the men and are as accomplished, also drinking as they play. As the women walk, bells jingle from their waists, held there by braided cloth belts that are fastened by ornate work in studded silver, 20 centimetres across and eight centimetres high. The atmosphere is festive and everyone is friendly. We are given hot water to make tea while we wait for the repair, and Khama is charged an extravagant fee for the work.

Towards midafternoon we come upon a large lake, Chakatso, its shore and middle encrusted in white. We stop and photograph a shepherdess with her mixed flock of goats and sheep against the backdrop of the white-rimmed lake where sheer black mountains and snowy peaks lie on the distant shore. The shepherdess is spinning wool in a small, handheld wooden spindle.

9 OCTOBER

I wake in the crisp predawn to a star-cast sky. The plain is dark, the mountains ink-black against cobalt-blue. I guess the temperature to be close to -6°c, because the water left outside our tent is frozen solid. I make coffee as the sky pales and definition begins to emerge from the land. It is Marco Polo country, the landscape I imagined as a boy when I read of his adventures. It is not difficult to picture a camel caravan in this twilight landscape before the sun reveals its harsh austerity. I picture a company of them as I sip at the steaming coffee: knelt down in a small throng, held by their tethers, their loads of silk, spices and gold bundled together off to the side, a scimitar-wielding guard dozing against their bulk. I am the only one awake.

We break camp before sunrise and a pair of ravens land close by, strutting nonchalantly through the places we have deserted, looking for leavings. Further down the valley, lakes shimmer in the high steppe air with large flocks of waterbirds rafted in the centres and roosting on the shore. While we are photographing the scene, a single yak walks down to drink. The country here expands and contracts. Nomads are sparse, but there is little country save the high dry valleys where they are not present. Goa and kiang, though in small groups, are more numerous than yak or sheep. They run from our approach on the dusty track, but on the hillsides and out on the far plain they are lying down in the mid-morning warmth.

Predictably, there are fences intermittently throughout the landscape here, bisecting the valleys or cutting a random quadrant out of some long plain. We have read of China's mandatory fencing of nomad-allotted land, ostensibly to increase stock numbers through modern methods of animal husbandry, promote rotational grazing, and keep stock out of areas reserved for wildlife. This policy is, however, in direct conflict with the thousand-year-old nomadic pastoralist tradition of wild and domestic herds living together sustainably. The fences are without exception in poor repair, and I am confounded at the magnitude of effort to erect them in the first place. They are not used by the nomads to contain stock, and their placement seems far too arbitrary to be part of a successful government-regulated plan. Most distressing, though, are the remains of wild animals caught in the fences.

At the frontier town of Gertse, beyond which lies only plateau and the Kunlun Shan, we are swamped in the street by curious locals. I call my mother from our satellite phone to wish her happy birthday. Her complete surprise and delight at hearing my voice is a wonderful tonic. I talk until the connection ends in static, because I can feel the presence of home, of the familiar, filling reserves that have become severely depleted by Bhuchung and Khama's complete disinterest in the country, ourselves and what we are trying to achieve.

We drive out of Gertse, on a back street headed north from the main road, in search of Drakpotso (Drakpo Lake), inside the Chang Tang Nature Reserve, where neither Khama nor Bhuchung have ever been. The road is washed away within the first kilometre and we take to a track that follows the unusable road, stopping two passing nomads on motorcycles to confirm we are on the right track. At a fork in the road we leave the small, dry river course and head into the deserted hills. The track is rough. We have 140 kilometres to go. We work our way up through gentle hills until all other land falls away, and a high undulating country gives us a 360° view to the distant tops of far peaks all around.

Kiang are plentiful, but as we descend into a long chute that extends its flat hills for 30 kilometres on either side there are suddenly as many as the herds of Africa. It is sight for our souls. Such country: pale-yellow plains alive with wildlife, and rugged hills beyond, stretching to the very limit of the earth. It is as it was made, and for me no monastery or cathedral can compare. When I walk out into this place and stop, I stand before it as one does before fire. It comes to me then that this land is, for the most part, not protected by exclusionist decrees, such as national parks or reserves are. Even the truly gigantic 334 000-square-kilometre Chang Tang Reserve seems, from our brief experience, to be a reserve in name more than the land's use or protection on the ground.

There are no people, just a yellow land with hundreds upon hundreds of kiang and goa dotting the gentle undulations of the stony ground. We camp on a flat, where distant, sharply eroded hills in the long shadows of early evening look like the backbone of the earth pushed through a parchment skin.

10 OCTOBER

I am fifty-six years old and I feel this morning the same unbridled vivacity in this place as I did at fifteen wandering the banks of the Chobe River in Botswana on safari with my parents. I am glad that the feeling is still with me. I have devoted my life to the places that take me there, to the awareness of them, to their tenure in the face of human encroachment, and to their conservation. It has made my life full, and the battles I have fought have seemed worthwhile. There have been times when, standing opposed to the inexorable tide of human expansion, I have felt minute and helpless, but I have come to understand that no matter what I see as the final place of human destiny, what the ultimate nemesis will be of our overrunning of the earth, it does not excuse my duty to stand for what I know to be magnificent and irreplaceable. I will pass on to others only half-won battles, but there will be the sense in them that has given my own life a purpose beyond myself. And standing in this frozen, wild dawn, I encourage all to know that, no matter what the odds, there is a remarkable power in one voice raised in sincere objection.

My own voice is quiet this morning, stilled by the cold and the sound of the kiang across the valley. I have heard them close in the night and in the black of the land. Now they call from all around. The cold is a vice on every exposed part of me. The water bottle in our tent is frozen solid and the tent's outer cover is so stiff with frost that it almost holds its shape as we lift it free to shake the ice from it. The average altitude here is 5 000 metres, with the peaks that divide the plains reaching 6 000 metres. Beverly says it is the coldest night she has spent camping.

The first sun casts mountain shadows long across the plain, and the kiang are caught in its illumination, standing bright against the dark of the hills. As we walk out with our cameras, the goa appear as cold as I am, and only move from their position broadside to the sun as we approach. Trotting off, neither too far nor too fast, their plump white rumps, with the short, black exclamation mark of a tail, face towards us. They stop and peer back to see if we are following. Yesterday, I noticed bands of horses roaming far from any settlement. Today there are more. Their herds number from a few to twenty-five or thirty, and they are as far out on the land as the kiang. The horse and the yak were once the load-bearers of the nomads' households as they trekked, but they have now been replaced by trucks and motorcycles. The herds of horses I see have clearly not been subject to the human hand for a considerable time and seem entirely feral. It is interesting that they remain separate from the kiang.

On the road back to Gertse, we pass several trucks piled high with belongings, as the nomads retreat from the high country to the village to see through the harsh months of winter. On the far hillsides, during the course of the morning, I count three motorcycle-shepherds moving their stock down out of the high hills and valleys.

II OCTOBER

Two ravens are once more in attendance as we break camp. They are remarkably bold, strutting within a few metres of us in the twilight. This morning there are a few clouds advancing over the mountains from the south. We stop to photograph a plain thronged by kiang, against the backdrop of cloud-shadowed hills. Their numbers are wonderful to see, and their small groups spread from close by the road to the far distance against the dark hills. The goa, too, are in abundance. The intensity of the morning, its glorious views of hundreds of kiang and goa, Tongtso (Tong Lake) and snowy mountains, is enhanced, raised up, made even more singular by the harsh, freezing environment.

Bhuchung and Khama, I think, are finally resigned to our slow, stop-start progress when there is wildlife or beautiful light on the land. Our relationship has gradually improved, and although both still show no spark of enthusiasm for the country or our work, the open hostility has dissipated. It makes our work easier and the days more pleasant.

Shortly within the hills we come upon a point where two valleys meet and its inclines are full of grazing game. We walk out to what we think are goa on the side of a small hill, with the darkly shadowed mountains behind. The first time I frame them in the lens I see that they are bharal. We follow them at a weary trudge, enforced by the altitude, to where they climb nimbly into the rough, crumbling, ochre-red cliffs of a tall hill. There are four, and they stand silhouetted against the blue sky. Up the valley we find more, grazing along the base of jagged cliffs, and as they come into the sunlight we see that there are eighty or more, the biggest herd of bharal we see. They take to the cliffs at our approach, but they have little to fear from our advance, which is mechanical and laboured. When I stop to photograph, I must pause for fifteen seconds to catch my breath before raising the lens. The bharal seem to sense this. Although they keep their distance, they stand looking down at us from their precarious perches. They are nimble across the rock and precipice faces. When caught in the occasional sunlight between the shadows, their grey coats stand illuminated against the dark ochre background.

At the foot of the snow-covered 6 822-metre Mount Xiakangjian, two hawkers' trucks have set up a market beside a tiny cluster of nomad adobe and rock buildings. Tents have been erected and the air is festive. The hawkers have set out their wares on the ground beside the trucks, and it is a treasure trove of the bizarre, the glittering and the everyday. Out of a petite, gold-plated casket with brass hinges and staples, a woman fishes ornate bracelets one by one, as if they were each created by magic. She places them on a plastic-wrapped carton of small containers of skin cream, with a blue label showing a blonde European woman. A nomad woman, with a pink scarf and an ornate dress with a long, embroidered, multihued bib, lays them on a pile of blue scarves to see them better. There is a whole sack of dehusked, dried coconuts, and a crumpled old cardboard box filled with something dried and salted that looks like mushrooms but whose musty odour does not smell like mushrooms. In one of the tents an entire extended family is sitting cross-legged around the dung-fired stove and drinking hot yak-butter tea. Two of the men wear hand-tanned sheepskin chupas. They are pure-white, hang to below the knee, and have a single Tibetan character inscribed with a razor and then stencilled red on the back. One of the most beautiful women I have seen uses a huge bone-handled knife to slice meat into cardboard bowls of steaming instant noodles and gnaws the last meat from the bone with her teeth.

The vehicle's altimeter jams at 5 000 metres and we are still climbing. Each stop to photograph the scenery that requires a walk to an edge takes twice as long as usual, and Khama once again shows thinly veiled displeasure at our interruptions to the journey.

The pass finally descends and on a tall cutting Beverly spots a lone, charcoal-black wolf directly below the road. It jumps nimbly across a small stream, and when I howl it stops and looks for a long moment at where we stand high above. It is an older, adult male, its coat full and heavy, and it runs in that wolf trot that seems so light and insouciant, yet carries them across the country so fast. Twice more it stops and looks back, then runs over a ridge beside a stream, turning upvalley and out of view. I spot a small road a little further down the pass that crosses the stream on a shallow ford, then follows the same stream upriver, on the bank opposite the wolf. I rush back to the vehicle, pointing out the road excitedly to Bhuchung as we bundle into the back seats. We will be directly opposite the wolf, just 60 metres away.

Khama refuses to move. I am too stunned to speak. It is a distance of no more than half a kilometre off the main road, and it is not a track but a proper minor road. I ask Bhuchung again. Again, Khama refuses. I don't ask why; it is beyond our understanding. Beverly, frustrated and helpless, is close to tears. Through the binoculars the wolf is receding and there is no way we could keep abreast of it on foot, even at normal altitude. The rest of the afternoon passes in a numb blur. I am bewildered, and although Beverly bravely tries to comfort me, she cannot hide the fact that her spirit is defeated. We have come so far, loved the country with a passion, and found the animal that so embodies its wild heart, only to have it taken away.

12 OCTOBER

The dawn is overcast, cloud closing in and the wind searing cold. Perhaps it will snow. No sooner do we leave the outskirts of Tsochen than we hit construction traffic. Khama weaves between the precariously leaning loads as trucks creep over the very rough surfaces. For several hours, we grind through the chaos and dust of roadworks. We watch with longing a road that branches off to the right, going south to Saga.

For the past five days, we have tried to persuade Bhuchung and Khama to take this alternative route south. 'Impossible,' we are told. Then it seems that Khama has never driven it before. 'We can only drive the main road,' we are told several times. As we drive slowly past the road we would rather take to get to Saga, I see it is everything we had hoped it would be: a quiet country back road into the mountains, away from the fences and power lines, with little or no traffic. Instead we grind our way through dust and detours into the hills.

By mid-morning the sun is breaking through in patches and we stop to photograph flocks of bar-headed geese, Eurasian teal and two black-necked cranes on a litter-strewn lake a short distance from a construction camp. A little further on we have just begun the first stage of the ascent of Sangmola Pass when Beverly stops us for a small herd of animals. Through the binoculars they are the mythical chiru, the straight-horned Tibetan antelope, finally come to life. The chiru has in recent years been brought to the brink of extinction due to illegal hunting for the soft underbelly fur, used in the making of a much-prized luxurious shawl: the shahtoosh. At every place we have visited here that is wild, we have been told that we could expect to see these famed, rare creatures, but we never have. I had begun to believe that they were gone from this land. They are a small group: five females, one juvenile male and one adult male. The adult male has long, oryx-like horns, perhaps three-quarters the length of his body. His upper muzzle is marked with strong black stripes, but he is somehow less than impressive. Perhaps it is that his head is small and his gait more light and tripling than assertive. The females are the most sheep-like animal I have seen in the wild; the only thing setting them apart is their longer legs. The whole head and muzzle are that of a merino ewe, even down to the rounded ears. The coat is fluffy and full, a lovely, pale-russet tan, the tail a plump rounded bun. They do not linger, and instead trot over the open hillside, climbing rapidly up and away over the open slope.

We are still talking animatedly about the chiru when, in a narrow part of the climbing pass, a Tibetan sand fox crosses the road ahead. It is the second fox we have seen in the whole of the TAR and it is once again in a place where we have not seen any evidence of nomads for quite some time. There is a pretty river running beside the road, and the fox, ignoring our presence, comes trotting closer to its favourite crossing place. It is paused on the far side, looking back, when another comes into view and crosses the stream a little higher up. It is aware of our presence and, faced with the dilemma of a short pool or approaching us more closely, it chooses to swim. On the far side, each pauses briefly and scent-marks before the entrance to a den at the base of a narrow scree slope. They are climbing away into the rocky hills when, 200 metres upstream, the first of a herd of bharal come skipping out of the shelter of the cliffs onto the short grass verges of the river. Our standing still results in a bonanza of wildlife as nearly fifty bharal drop down out of the shale cliffs and come to feed so close to where we sit that we can hear their chewing as they mow the already short, dense grass with a quick chopping action, using the side of their fore-jaws. I wonder, as we watch them graze, how much their coming out into the open has to do with the midday pause of the construction gangs. Far up the valley, I can see that the construction lorries and earth-moving machines are still beside the road while the workers take their lunch break. It does not seem to be entirely coincidental that both where we are watching the bharal and where we saw the chiru are, for the moment, construction-free areas. I notice the bharal raising their heads, and, looking towards where they are staring, I see the foxes returning. Most of the bharal ignore them, but the females with young lambs make their way unhurriedly back into the shelter of the cliffs. The foxes stop and watch, trot forward, and watch again, but their focus seems more on the pika than the sheep.

The Sangmola Pass is, in truth, no more than an uphill climb through hills that are neither steep nor particularly tall, to the watershed apex of this range of

5 600 metres. The hills and ground are covered in light snow. On the descent, we find a group of five chiru, all females, close by the road and then, lower down, beside a small stream, two groups close together that make a total of seventeen animals. There is only one young male in the group. Every chiru that we have encountered flees as soon as the vehicle stops, and the almost complete absence of mature males adds to the suspicion that guns are brought to bear on these fragile, rare creatures. They run without stopping until we can see them only through the big lenses of our cameras. Bhuchung tells me that a single decent-sized chiru horn will fetch 600 to 1 000 yuan (US$100 to 150). Shortly after seeing the last group, the construction work resumes and we see no other wild creature.

All the chiru we have seen are in high-altitude, open-hill country where the vegetation is sparse and ground-hugging. It seems that each species has its niche: the bharal choose the more rugged steep hills and rocky cliffs, leaving the flatter valley floors and savanna to the goa and kiang, which are the only species that seem to overlap. The wolves like the mountains and distance from humans; the foxes fringe the steppe and the valleys leading down to it.

My breathing grows easier as we descend, but there is no respite from the constant dust and bumping. Every 10 kilometres there is a new construction camp working the new road in either direction. The Chinese do not undertake construction by halves. They will complete the almost-250 kilometres of mountain road in the space of two short summer seasons, the winters being too cold for work.

By 5:30 pm, we have been cramped in the close confines of the vehicle for nine and a half hours and have covered 200 kilometres. We are deeply saddened and chafing with frustration that our repeated requests to take the quieter, wilder side roads have been ignored, compromising what is for us a once-in-a-lifetime journey of exploration for wild land.

When we ask Khama to stop so that we can make something to drink, as the road is too bumpy to pour anything or drink from a cup, he sighs a sigh to bring down the Pillars of Hercules and slumps forward over the steering wheel as he comes to a standstill. This is the last straw; I step from the vehicle, pick up a stone, and, in my exasperation, hurl it with all my force at a steel drainpipe awaiting placement at the side of the road. I turn with my arms and head raised to the sky and then walk away down the road. My outburst upsets everyone.

As we drive into the night, still three days' drive from Lhasa, I reflect on our time in Tibet, the wonderful first few weeks with Tenzin and Gongpo, who shared so much, and how the circumstances of the last few weeks could so easily have been happily otherwise.

As the dark outside my window shrinks the country until it is no more, and my reflection stares back at me from the glass, my focus turns inward until I see with a different sense, my eyes glazed and fixed, my mind turning to another world, the thoughts within my head.

I have been wrong to allow the foibles of men to distract my focus, my seeing, but I did, and I have lost something precious because of it. The land is not beholden to men, nor to the follies of men. It lies beyond that: permanent, solid, enduring change in the gentle transition of aeons, of geological time. I know that, and it brings a calm to my distress.

I wander down the avenues of my memory, through the mountains and valleys of Tibet, the dawns, the cold and the expanse of the wilderness we have seen. Of plains wider than the Serengeti, populated by wild animals: free, without restraint. Of the power of a sudden vista, sufficient to cause one to draw breath in exclamation. And to stand there before it, reduced to silence by reverence. I remember laughter and kindness, but most especially wonder, and it is this that I lift to my mind's eye. We found what we came to seek.

I feel the eyes of the manul fixed upon me, only just discernible between the jumbled scree in which they shelter. They challenge me to remember what is worth remembering, what I have gained, not the measure of my loss. I see a jet-black wolf stop and turn to stare as I howl into the lonely hills, and although its eyes are too distant to see, I can feel their pale intensity bore into my chest. It floods me with warmth. I recall walking the steep slopes with game stretching to every horizon, and realise then that the Tibetan Plateau is wild land in its purest sense. What is here is not managed by humankind: what people there are here are part of the system, bound to it, their lives turning in the most intimate sense with its seasons. This is an open land, filled with free-roaming, wild animals and birds for thousands of kilometres in every direction. It crosses roads, railways, power lines and pipelines. It encompasses whole towns, its tide swinging around them like the current of a river around an island. It is remarkable and utterly unique in the world. I know that if we had the time to trek far away from the roads, we would find the tracks of the shy snow leopard and bears. Only yesterday, walking in the sand of a dry river course, right beside the gravel road, we saw the tracks of a fox and I am glad that he is here.

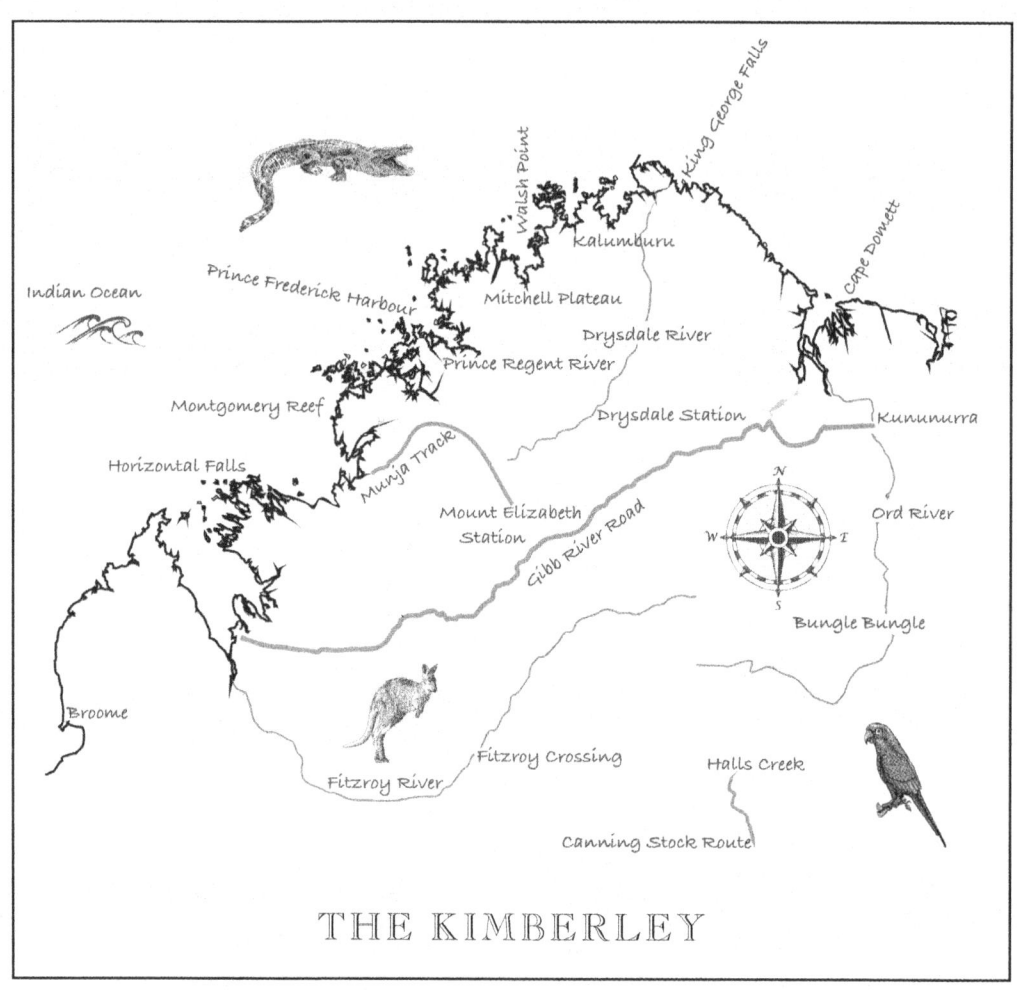

THE KIMBERLEY

VI

AUSTRALIA
The Kimberley

WINTER 2014

The sun is cresting the horizon and I am here, cold and enthralled, above the earth, undisturbed, in a place so singular that it tightens my chest to see it.

20 APRIL 2014

'An not with a reefall.'

I am standing in the evening twilight at 3 Mile Campsite on Gnaraloo Station, a full day's drive north of Perth in Western Australia. In front of me a young surfer with long, bleached-white hair is slowly turning the handle of a spit on which a whole goat is browning over the low coals. Behind him the waves shatter in a spray of white against the low cliffs.

'A reefall?' I ask the dapper Frenchman who is the cook for the gathering of an extended family, come together for the Easter weekend.

'No, an endgoon.'

'Endgoon?' I am at a complete loss as to what we are talking about. I've never heard of this kind of goat.

'Yes, an endgoon,' he repeats, making the classic symbol of a pistol with his hand: two fingers pointed straight out and the thumb cocked high.

'Aah! A handgun.'

'Yes. It is difficult with ah endgoon, eye ave to run.'

'You shot it with a handgun?' I ask, indicating the goat.

'Won shot! In the ed.'

Gnaraloo Station ('station' being the Australian term for ranch) occupies a staggering 1 080 square kilometres. About 18 kilometres wide, it runs for 60 kilometres along some of Western Australia's most picturesque and pristine coastline, the waters offshore and out to sea falling within the Ningaloo Marine Park. Along its northern beaches three species of turtle come to lay their eggs: loggerhead, green and hawksbill. We are here to try to photograph turtle hatchlings as they make their way to the sea.

We had not intended to begin our work this far south of the Kimberley, the desert region of northern Western Australia, which by general concensus starts at Broome, but the coastal waters we explored from Shark Bay as we moved north were in such pristine condition that to ignore them would run counter to the very purpose of our quest. Seldom, if ever, have we seen an expanse of coast in such condition. It is some of the most intact, unexploited marine environment, with fish of every description occupying, in their full complement, every niche of the coastal ocean.

With goggles and snorkel, I lay on my belly in a wash of shallow water off the shore-side plate of rock at False Entrance and marvelled at the myriad tiny lives. At Steep Point, I endured only a partial regret as a 300-kilogram giant grouper stole the gamefish off the end of my line as I struggled to haul my morning catch up the sheer fall of the red seaward cliffs. The sun was peering over the dark horizon of the land, the sea cast a fragmented gold, and the grouper a gigantic dark shape that rose to the surface through the blue and orange to claim my prize as its own. What a fish. What a way to start a day.

Further north, we had swum over the edge of the reef into deep water and followed three turtles as they grazed on the vegetation of the steep sides. Schools of trevally and darts had trailed us, their curiosity piqued by our splashing fins. Hundreds of multihued fish swept and darted through the contoured fall of the undersea cliff, and it was only when the movement of our swimming raised the interest of a shark somewhat bigger than myself that we relocated to shore. I had heard and read much of Australia's Great Barrier Reef, but nothing of the seas off Western Australia. What we found was like being given a whole parcel of Christmas gifts at Easter.

21–26 APRIL

It is very late in the season for turtles still to be hatching, and as we wait in the airy, high-ceilinged main room of the Gnaraloo homestead, I wonder if luck could have sway on our slim chances. The sight of turtles hatching would be the cherry on the top.

Colleen, the unflappable, gravel-voiced overseer of the Gnaraloo Station tourist operation, has gone to raise the boss from his nap, and is heating him a cup of tea in a buzzing microwave, while Beverly and I sip iced water in the 38°c heat. I hear footsteps approaching down the passage when a giant lizard with a bright-yellow eye steps through the rear door and walks in a wide, swaying gait to drink from the dog's water bowl beside one of the kitchen cupboards. There is a mad scramble from behind me as a tiny, curly-haired mongrel careens after the lizard, which bolts down the passage, knocking over a Hoover and a cardboard box as it goes. Colleen attaches herself to the tail of the chaos, shouting, 'No! No!' and flapping her hands at the dog as the bungarra retreats into the depths of the homestead.

The man who walks into the room in the pitch of this commotion is not the image that 'the boss' had conjured in my mind. Paul Richardson wears threadbare shorts and an even more worn Camel T-shirt. He is sleep-tousled, with sandy-coloured hair and eyes that sparkle with an impish enthusiasm for life.

'Does Colleen need a hand?' I ask.

'Ah! Noo. The bungarra has hidden under the bed. It'll come out in a while.' His accent is thick with his native Irish brogue.

There is a quality that I have always been drawn to in frontierspeople. Perhaps it is their unassuming resilience, their knowledge that there is no backup, just them. It may be something deeper, the spirit that seeks more out of life, that is more inclined to see the cup as half full. Whatever it is, I am quick to like Paul. We talk for hours about wilderness and wild places, of lodges and tourism and fixing pumps and generators, his eyes bright behind a pair of broken-rimmed glasses that sit skew on his nose, with lenses so smeared with grease I am surprised he can see me at all.

In the evening, he drives us down to the turtle-nesting beaches and shows us where we can park behind the dunes. For the next few mornings, from 3:30 am to 8 am, beneath a waning moon, we walk for kilometres through the soft sand of the shore and pant up and down the low coastal dunes. But the turtles are gone and we find only ghost crabs, each of which raises its single large yellow pincer in threat at our approach.

To the far north a cyclone has struck the coast, and on our last afternoon the sky grows ominous with low, heavy cloud. The rain falls in a wall when it comes and the roads become wild, discoloured puddles. The young station staff dance to loud music in the torrent, and outside the office one strips down to his boxers and showers in the rush of a broken downpipe. At night, the storm sounds like a river raging over the tin roof.

27 APRIL

We leave Gnaraloo early in a cool morning. The rutted dirt road is a patchwork of deep puddles, but the rain has mercifully smoothed out the corrugations to some extent. Talk is impossible above the din, and I fall into a reverie on wilderness. Over my dealings with Paul I have learnt much about land in Australia. Paul has owned – he says 'owed' – Gnaraloo Station for eight years and I am surprised to learn that he does not have title to the land, but only a lease granted by the state of Western Australia.

Virtually all rural land along the West Coast is subject to an exclusion area, reaching 40 metres above the high-water mark, to which the state has historically retained title, but now specific exclusion zones are to be included in the lease renewals of 2015. These exclusion zones exorcise some of the most exceptional natural features along the coast from the areas granted under pastoral leases, which would then be available for lease separately to developers. It seems that practically no government anywhere has yet realised the worth of a wild place because it is wild. Western Australia seems bent on taking the same path as most other First-World countries: that of taking a place of outstanding natural beauty and ruining it with a plethora of money-making enterprises.

On Gnaraloo Station, Gnaraloo Bay is slated for exclusion under the terms of the new leases. Paul is an eccentric, but he is also in earnest, and he knows that any kind of development at Gnaraloo Bay will endanger the several hundred turtles that nest along its northern beaches. It is the reason he so readily agreed to our request to photograph hatchlings, for he knows that to sway the politicians he needs not only science but also public opinion.

The science is quite plain. Hundreds of turtles use the northern beaches of Gnaraloo Bay for nesting, making it one of the top three turtle nesting sites in

Western Australia, and for loggerheads it is significant on a worldwide basis. It has been proven that lights at night confuse turtles coming onto a beach to nest, disorientating them. Within a relatively short period they simply stop returning and the beach becomes deserted. To anyone with sense, this should be enough to negate the possibility of any development, but money talks, and if there is little or no public participation, money often wins. By the time this book is published, Gnaraloo Bay's fate and that of its turtles may well be sealed. There is no doubt in my mind that it will not be the government of Western Australia that does right by the turtles, it will be Paul Richardson, an untidy Irish immigrant to this harsh, arid land who sees value not in dollars but in the wonderful richness of life.

I MAY

We have seen no one for two days. Driving through the serene sunrise with the windows down, I feel close to the land. I drive slowly so that I can hear the birds call. The track is muddy but drying. At Boat Harbour we enter the Cape Range National Park.

Using our inflatable-kayak repair kit, I help the campsite host repair an inflatable roller for launching his boat and over tea he tells us that 400 millimetres of rain fell in one night in the storm here. It caused massive flooding, closing roads, flooding buildings, and sweeping vehicles and two sleeping campers into the sea. Yardie Creek, just to the north, broke through the sandbar at the mouth for the first time in years. Crossing it now can only be contemplated at low tide.

When we arrive at the crossing, the tide is high. The estuary is 50 metres across and the water comes almost to my chest. While we wait the six hours to low tide we unpack the kayak and paddle up the creek into a strong headwind. The estuary follows a gorge into the coastal ridge that parallels the sea, a cut through dark-red cliffs where green-leaved figs issue from cracks in the rock, their shaggy roots trailing like thin, grey beards in the wind. The bird life is suddenly prolific and we see herons, gulls, parrots and an assortment of smaller birds busy in the shrubby bush. Along the course of the gorge we see four ospreys and a single white-bellied sea eagle. Small raptors, black against the grey sky, patrol the cliff tops, with birds mobbing and fussing at them as they pass. Throughout the cliffs we find black-flanked rock wallaby perched on ledges or peering sleepily from shallow caves. They are diminutive, standing the height of a small terrier, and sit upright on their haunches preening their short black-and-grey coats and long, brush-bristled black tails. Our field guide notes that their numbers have been seriously depleted by fox predation, often making them rare over sections of their range. When they rouse themselves towards evening, they are remarkably agile over the tumbled boulders and steep faces of the cliffs.

The gorge is an oasis of life, and we spend a contented afternoon alone, watching and photographing its wild residents. In the evening, the spring tide pulls the ocean far back down the beach and only ankle-deep water runs over the sand. We are apprehensive, though. Traditionally, the Yardie Creek mouth is closed and one faces only a heavy sand crossing, but our guidebook warns that 'only experienced 4WD-ers should contemplate the hazardous Yardie Creek crossing'. Now it is soft sand and running water. Too often in the past I have seen vehicles bogged down in river-mouth beach crossings. Stop and the water digs around the mired tyres and the vehicle sinks quickly lower, becoming more and more stuck. As the tide returns the drivers can only stand hopelessly and watch.

I walk the river crossing seven or eight times, seeking the hardest sand beneath the flow. I let my tyres down to the point of almost flat so that they bulge like sumo wrestlers, giving me as wide a purchase area as possible on the soft surface. On a shelf of flat rock that protrudes into the mouth I stop and walk the crossing, then back up and head into the water at a moderate speed. The sickening surge to a stop as one hits soft ground does not come and I make the crossing easily and drive off the beach on the far side. While I stop and wait for Beverly, I count my pulse. It is 104. On the far side of the creek we hit civilisation again. There is a tar road, a sprawling car park, an ablution block and an attractive campsite, but no one else in sight.

At night, a fox visits our camp; I do not see it, but I hear it in the quiet night and in the morning its tracks are all around. Originally, I am told, six foxes were released for the hunting pleasure of some long-ago English nobleman. The foxes multiplied and became a scourge as widespread and prolific as the rabbits. With dingoes persecuted and locally wiped out by settlement, the fox entered a niche with no competition, and on a continent with not a single small animal predator the damage they caused was immeasurable. Today, fox numbers are quite effectively curbed by poison-baiting. 'Warning 1080' poison-area signs are well heeded by dog owners, for it will kill dogs and cats as quickly as it does foxes.

2 MAY

Euro, a strangely named species of kangaroo, are everywhere and they watch us nonchalantly from where they doze in the shade of bushes. I watch an Australian hobby stoop high out of the sky onto the flocks of small birds. It flies like a jet on narrow wings and, selecting a single bird taken fright out of the flock, pursues it through a steep climb. After several rapid jinking turns it explodes in a puff of feathers as it connects hard with its hapless prey. I utter an impulsive 'Phewaah!'

At Oyster Stacks to the north we snorkel on a close inshore reef, where fish ignore our presence and go about their busy lives. In the shadowed caverns beneath the oyster stacks, shoals of many-lined sweetlips, at a size that makes my fisherman's heart beat faster, circle unperturbed a metre from my nose. I particularly like the tiny zebra-striped humbug dascyllus that inhabit the green staghorn corals, and the blue-green chromis whose shoals of pale electric-blue would be not be out of place in the garden pools of the Taj Mahal.

As the afternoon softens to early evening we find an echidna crossing the road. The tiny, hedgehog-like creature, with short black and yellow-white spines and a long narrow snout like a beak, shuffles along on short legs with enormous, splayed backward-facing nails that make it seem quite deformed. I am fascinated to learn from our field guide that it lays a single egg, which it incubates in a pouch. When its young is hatched and old enough, the adult leaves it in a nest of brush and leaves, returning to feed it. The creature itself seems permanently lost in its slow, shambling shuffle, and it is hard to imagine that its small, deep-set eyes provide a clear enough picture of the country to know where it is, let alone find its way back to a nest with food.

15 MAY

Leaving Broome, the western boundary of the Kimberley, we begin our planned exploration of the extensive 420 000 square kilometres of the Kimberley.

On the spur of the moment I turn south onto a dirt road that parallels the Fitzroy River and drive into a more secluded country. Cockatoo Creek is filled with water, and we stop and walk slowly down the bank, photographing bee-eaters, honeyeaters and friar-birds that are not shy at our presence and come to us if we sit still. Further south, we turn northeast onto a 4x4 track that follows the river. It issues us, eventually, into a cluster of houses scattered beneath shady trees. A tall old man who walks without any stiffness comes over to greet us.

'What are you doing here?' he asks by way of introduction.

'Exploring. What are you doing here?' I return.

'What do you mean?'

'Here in the bush, in the middle of nowhere?'

'This is my land. I've been here since the year dot. This is my place. I've seen on television what goes on in the rest of the world, crazy people. I stay here where it's safe and sane.' Ernie Hunter introduces himself. I explain our purpose, the road we seek.

'The road ends here.' I bring out the map, but Ernie's map is in his head, not on paper. He tries to explain using names of places I cannot find, places you can walk to.

'Ask my brother: he drives around more than I do. He's been to Africa.' Neville is driving an overland tourist truck on the narrow, sand road between the houses. He is the youngest of Ernie's eleven siblings. He also wants to know what we are doing here and I explain our purpose and our work.

'I believe you've been to Africa.'

'Yeah! I'm chairman of the Aboriginal Tourism Board and they invited me over to Namibia to give a talk on indigenous tourism. But when I got there, I realised I had nothing to teach them – you're way ahead of us in Africa in community tourism.'

Neville pinpoints the road we can use to get further down the river and invites us to have a look at the river just beyond the homestead. 'Just follow the car tracks, but watch out for the salty; it bloody well ate my dog.'

The Fitzroy is 50 metres across, running clear but lightly discoloured over yellow sand between high banks lined with tall, mature, thick-trunked trees. The vegetation is lush and riotous beneath the high shade canopy. I see no tracks of any crocodile, but I know from home that the deep, opaque green of the slower pools is not to be trusted. Back at Neville's home, we eat cold watermelon in a shaded outside dining area. The whole kitchen is set beneath the trees: the cupboards, the sink, the stove and the refrigerator stand in the leafy respite of the garden, where birds undeterred by our presence feed in the low branches. We drink water as it melts from a bottle frozen into a solid block.

'Pure spring water,' Neville tells me. 'Had some guy come out here and he put a bottle of Evian, a bottle of Aussie Natural Spring Water and another bottle of this water they charge for on the table with ours, and he tested them all right here with this whole big Geiger counter kind of apparatus. Our water came out tops. Are you carrying water?'

'Yep.'

'We'll empty your tank and fill up with the garden hose.'

'I carry 220 litres.'

'I got it by the 10 000 litres.'

I pull the car closer and run the hose into the tank, while a young Australian couple working as part of the World Wide Opportunities on Organic Farms project, universally known as 'wwoofers' throughout the outback, fill us in on the dog and the saltwater crocodile. Apparently, only three days ago, both Neville and Ernie's entire families were at the swimming hole in the Fitzroy River when Matthew, the woofer, had spotted a three-metre 'salty', or saltwater crocodile, about a hundred metres upstream. One of the young boys was swimming close to the far bank when Matthew put out the call, which meant the child was stranded alone away from the family with a crocodile lurking in between. Someone went up to the house to call Neville, with the thought to run the small boat across the river to fetch the child. When Neville arrived to assess the situation, his dog, which came with him, saw the boy on the far bank. Before anyone could stop it, the dog jumped into the river and headed across to the boy. With every single member of the extended family watching, the crocodile closed in.

Neville must go to a meeting in Kununurra so our time is cut short, but I find I like this man. As he shakes my hand he says, 'We met for a reason.' I feel the same and know that I will make the effort to see him again when the time comes.

For the rest of the day, we drive on roads where we do not see another vehicle or person. Towards evening we drive through the yard of the abandoned Luluigai homestead. Wild Brahman cattle scatter at our coming. The house, set on an open copse surrounded by boab trees, stands gutted and forlorn, the paint still white beneath a roof open to the sky. I am told that this is happening all over the outback, farmers moving off the land. The reasons are many. Some stations are bought and amalgamated by large farming corporations. In others, the farmer moves to town and operates the station through a manager, or by visiting only to perform the vital functions. Some simply fold, or grow too lonely. Whatever the reason, an abandoned station always seems a desolate place. We stop in the darkening twilight in a forest of boab trees. A barking owl calls from the riverine forest and later the resonant song of a rufous owl drifts through the leafless limbs and shiny, bulbous trunks. We make images of the moon rising over the deserted track, lined by the distinctive trees.

16 MAY

There is an abundance of birds in the still wet floodplains of the Fitzroy River. A pair of brolga, their pale-grey crane heads encircled by a band of distinctive red, and a brown falcon are new to us. At the river, there are egrets and herons fishing on the wide causeway, and black-necked storks chasing fish over the shallow sandbars downstream. Every time we pause some other life arrives and we are kept busy. We make camp on the far bank beneath the deep shade of the riverbank trees that tower 15 metres overhead. We bathe in the shallows of a wide sandbar where we can see the bottom clearly for a long way around.

A group of four Aboriginal women with a young girl stop close by our camp and, walking barefoot out onto the causeway, throw a cast net into the cascade downstream, catching palm-sized fish that they empty into a large green bucket. One of them plunges fully clothed into the deeper water of the upstream side of the causeway, kicking and splashing and calling her dog to follow. I scan the river from my high vantage for movement, but see none. The dog is reluctant. I know the Hambukushu people of Botswana believe that the crocodiles do not take them because they know them as the 'River People'. Perhaps these people are the same. The girl swims while the rest of the group fishes, the child running up and down in the shallows of the causeway between them.

The women have returned to their car from fishing and are ferrying armloads of clothes, blankets, fishing and cooking utensils past our camp to a grassy spot about a hundred metres down the bank when two smart 4x4s pull up on the verge of the causeway opposite their car. There is a pause and then three well-dressed women, one carrying a child, and an older man and a youth get out and walk gingerly in their clean shoes to stand cautiously in a line on the edge of the causeway gazing across the river. After a time, some return to their car to fetch cameras. They do not acknowledge either us or the Aboriginal women and I notice that even the young tousle-haired girl, standing beside their car gazing at them, in her wet underpants, does not elicit a wave or greeting. The fisherwomen take their tackle, and kicking water so that their clothes get drenched and cling to their skin, they cross the causeway to the far bank and walk upstream. The visitors seem as shocked as I am by this blatant disregard of crocodiles, but it emboldens them and, placing their shoes square by the roadside, they hitch up their skirts to the modesty of their calves and wade out into ankle-deep water. The older man is digging in his car for something and when I look again he is standing with a sturdy, leather-bound book in his hand and talking to the matriarch of the local family.

'Devil dodgers,' I say to Beverly, using a term a devout friend of mine uses to describe himself. 'Bible people,' I explain when she seems confused.

Before they leave, the youth comes awkwardly over to our camp with a pamphlet in his hand.

'Beautiful, isn't it?' he asks.

I nod.

'Ever wondered who created it?' He is stiff, uncomfortable, the pamphlet held forward like a peace offering.

'Was life created?' I read the title aloud. He nods vigorously. 'Thanks,' I say, taking the pamphlet, and he raises his palm toward me and then turns and escapes back to the comfort of his known place. The security of faith is not for me. I prefer the paradise that I have right here, and I decide right then to celebrate it by spending the afternoon fishing.

I take my fly rod and walk downriver. A storm is threatening and a low, heavy cloud has mushroomed across the sky, making the heat stifling. The riverside forest is a tangled mat of undergrowth. Creepers hang from the trees and crisscross the ground, and I trip often over their thin, rope-like stems. The foliage is dark-green and close, and birds call in lethargic 'whoos' from places hidden in the branches. It is as humid as it is hot, and sweat runs freely down my chest and cheeks. There are locusts the size of small birds that whirr up before my feet, and kites wheeling slowly over the water, but little else moves. The water is still. Fallen trees stained the colour of the red earth reach out of the opaque, emerald-green depths. I catch three bright-eyed catfish with yellow irises and short white whiskers. They grunt and squeak as I turn the hook free. The barramundi, the large, silvery predatory fish that I seek, have either grown wary of hooks or have been pursued too long, and I find none.

In the evening, we feed small fish that the women gave us to the brahminy kites, photographing them as they stoop down to the water's edge. In the dark, the Aboriginal women return again and, lit from behind by the headlights of their car, throw their cast net after small fish. They are etched in silhouette in the pitch black, the net flaring briefly like a silver spider's web before splashing into the dark water. In the light of a small spotlight I count a total of nine sets of crocodile eyes. The dog does not seem to be with the women, or perhaps it is not as certain of its aura of protection and is hanging back.

17 MAY

I am attempting to photograph kites with a remote camera when a road train, an articulated truck with three empty 20-metre cattle trailers in tow, comes grinding in low gear down onto the causeway. It throws a wake out from its multiple axles that runs as waves onto the sandy bank. An hour later, the kites have just begun to show interest in my bait when a helicopter comes buzzing slowly right at treetop height along the far bank. It is a small two-seater, with just the pilot in the open-sided cockpit. Cattle flow onto the roadway like water and a red dust cloud rises from their panicked flight through the trees.

This is the modern Australian cowboy, or jackaroo. On stations that often extend to hundreds of thousands of hectares, the cattle roaming almost as freely as wild game, the horse has been replaced by the helicopter, and herding cattle to market now resembles more game capture than droving. The helicopter worries the cattle out of the trees, hanging low so that the branches shake tormentedly beneath the buzzing rotors. It pendulums back and forth along the bank, sweeping the herd before it like wheat kernels in a threshing yard. Flocks of white corella flee the din and perch on the outer branches of our side of the riverine forest, protesting loudly. The kites take to the air.

Three hours later, when quiet has been restored to the river, the road train returns to the ford, the trailers full, the cattle lowing. It slows for the water, the engine sounding deep and full-throttled as it pulls its live load, nearly 80 metres long, up the far bank. The dust from its passing leaves a wake of red ribbons across the green water and I can smell the dung of the fearful animals waft over me in the still river air.

18 MAY

Much of the country we drive through has suffered burning. In places of scrubby trees, the effect is most devastating, leaving stark blackened skeletons. Taller trees seem to withstand fire better, especially if the land is grassed or grown with spinifex rather than dense bush. At times a fire's scar runs for 50 kilometres or more. How or why they stop is a mystery; they just seem to fizzle out and end quite abruptly in places that seem as fire-prone as any.

At the Fitzroy Crossing petrol station, the only store that is open on a Sunday, an Aboriginal man approaches and asks if I am interested in buying art. He is as tall as me, with wild, curly hair and heavy hands, and his manner is so reticent, with a tangible gentleness, that although I have little place to carry paintings I feel moved to see his work. He unrolls a single canvas of curving, wandering graphics in bright lime green, with lines of yellow and red dots outlining the bold design. At the centre is a crudely drawn man's face opposite the head of an eagle, held within a circle. Between them, on the palm of a hand, lies a golden stone. In his soft voice, he tells me that it is the story of a salvation. The man was lost wandering in the wilderness and became thirsty to the point that he thought he might die. It was then that an eagle came to him carrying a golden stone. The eagle placed the stone in the man's hand. He took the stone and placed it on the ground, and as he stood watching, water came out of the ground and he was saved. I am tempted to buy the work just for the story because I would like to unroll the canvas slowly, as he did, and retell the tale simply as it was told to me. It is the closest I have felt to touching something more than just the superficial of this place. I thank him and wish him luck with selling his work and he bids me a safe journey as he slowly rolls up the canvas.

19 MAY

By sheer chance, I spot the Mangkaja Arts Resource Agency Aboriginal Corporation. I had in Perth seen several books on Aboriginal art and had been taken by the power of the naive-seeming graphics, with their vivid use of colour and unusual patterns and line. We stop beneath the shade tree outside and walk into an unassuming hall. Massive works hang on the walls. A disabled woman sits beside her wheelchair on the tiled floor painting a huge canvas of tree-like forms with rose-blush-coloured foliage perched on the top of exaggeratedly long stems. On a child's sleeping mattress an old man with rheumy blue eyes is painting the story of a dingo and a kangaroo, in blue and red and a pale, almost black purple. We are riveted. Art is everywhere. More is being made, not in the hushed, studied ambience of a famous studio but on the floor with the telephone ringing, a baby sleeping on a mat with arms and legs akimbo, and two men slowly sipping hot tea on a shaded veranda beyond the airy, glass windows.

One of the project coordinators leads a large, plump, older woman shuffling through the door. Sonia Kurarra is wearing a thin blue-and-white dress and a fire-engine-red cowboy hat. She is vociferous, waving her walking stick in the air. Sitting heavily in a chair at a painting table, she folds her arms and announces loudly that she refuses to paint. No one takes any notice. Another of the project coordinators places a canvas on the table in front of her. Her tirade is high-pitched, sharp and incessant. She picks up a brush. 'Sarah,' she yells. The coordinator comes over after the fourth shout.

'What colours do you want?'

'Blue.' The blue is delivered in a shallow plastic bowl.

'Too blue.'

'You want some white?'

At one point, Sonia comes across to the rheumy-eyed man, who has now moved his work to a table, and begins a keening weeping on his shoulder. The man sheds real tears, which he wipes away with the knuckles of his fist. Ngarralja Tommy May lost his wife six weeks ago and it is his first day back at painting. Of all the work in the gallery, I am most taken by his unfinished work and offer to buy it. He tells me that he might finish it today, or maybe tomorrow.

20 MAY

The 4x4-only track to Purnululu National Park turns out to be a lovely, meandering, slow drive through forested hills, between clear running creeks. We drive the final few kilometres to the Bungle Bungle Mountain Range through a land flushed green after fire, to arrive shortly before sunset in the most singular scenery we have encountered thus far in Australia.

A tabletop limestone massif, kilometres across, has eroded along its edges into a series of towers and beehive domes striated by broad, alternating bands of burnt orange and charcoal black. It is so structured that it has the feeling of ruins, but humankind has had no influence here. Although long known to the local Aboriginal people, it only came to outside attention in the 1980s. Now every brochure of the Kimberley carries images of the unique dome-like congregations of rock, but all fall short of delivering its most impressive aspect: its scale. I walk through the ancient, silent sentinels of rock in the fast-fading twilight half expecting, half hoping, to hear someone singing opera.

21 MAY

The land is cold in the dawn as we walk alone through the tall cliffs. The first sun ignites the inclines as if they are lit from inside, the shades of orange bright against the black. As the day warms and the full light of the sun bleaches the first flush of colour from the rocks, I find in my silent wandering through this place that I am craving something, something that is missing here. It comes to me suddenly: life! Animal life would make this place entire – something that matches the majesty here. But there are only a few birds and the tracks of night animals. I am amazed that in all the time life has had, nothing has colonised this place, that there is no ape or rock rabbit or wallaby that found use for the abundant water and hideaways from predators and heat. Even the Aboriginal people seem to have passed this place by. In Cathedral Gorge, a vaulted, narrow chasm that terminates in a hollow cavern amphitheatre, complete with a circular pool in white sand, there is the sense of spirituality, of a place of worship, the sense that one is drawn closer to God, but its still walls echo the silence that is the absence of life.

In the night, I wake to the drawn-out 'auwwwwu' of dingoes calling. They are close and there are many. It is late, the night quite still, and I listen to the life that I have craved to find. The eerie chorus changes the way I see the stars as I lie on my back, and raises in me something instinctive, like a garment fitting to the land.

22 MAY

The day begins with a helicopter flight over the Bungle Bungle. The doors are open to the wind, and for a change the air is cool. Impressive from the ground, the eroded sandstone is breathtaking from the air, like the turrets and ramparts of some gigantic castle, reaching out into the landscape with the extended paws of a giant dragon. Orange and black are vividly offset by the green of the land, where termite mounds and the white trunks of trees appear like the graphic, bold stripes of Aboriginal paintings.

Back at our camp, I try to catch a single cane toad that hops out from the cover of tall, matted grass to where the water from my shower is pooling, but I am wary of the poison glands on its shoulders, and it hops heavily away from my clumsy attempts, back into the safety of the grass. The campsite has cane toad boxes where visitors can dispose of any of the slow-hopping toads they catch. The warden told me that last year they euthanised forty-four thousand toads.

The cane toad was introduced to eastern Australia in 1935 by the Bureau of Sugar Experiment Stations, which saw it as the perfect remedy to control the native grey-backed cane beetle and the French's beetle, which were a threat to successful sugar-cane farming. It was a strange decision to release an exotic toad that can neither jump nor climb trees to eradicate a beetle that spends its entire life in the branches high off the ground, coming down only briefly to mate and bury its eggs and larvae underground. The cane toad, however, is a robust amphibian, and firmly routed the competition for food, often eating other amphibians, helped by the absence of successful predators. They multiplied, began marching across the continent, and today are believed to number more than two hundred million. Purnululu is presently at the forefront of their westward advance.

One instance of the ill-considered repercussions of releasing the cane toad into Australia is the case of the northern quoll. This small, kitten-sized predator has a short life, particularly the male, as it needs to mate within a year of birth, after which, in one of the strangest quirks of evolution, it dies. The quoll is an aggressive nocturnal predator, and the slow, fat cane toad should be an easy meal for it. But the cane toad's poison glands, which release sufficient poison to cause discomfort, nausea and extreme irritation to humans, are deadly to the quoll. Northern quoll populations have plummeted with the advance of the toad. We must hope that, perhaps, like the wily, long-lived crows, the quoll will learn that the toad can be flipped onto its back and the stomach eaten without ill effect.

23 MAY

South of Purnululu we turn northwest onto the Springvale–Lansdowne Road in the hope of negotiating a 4x4 crossing to Mornington Station and the Gibb River Road. The road has recently been graded and the going easier than we had hoped.

At Springvale Station, the welcome sign proclaims that owner Peter Burton is 'full of piss and wind, somewhat tough, but inclined to bullshit'. The main house is finished above and below with elaborate cast-iron filigree. There is washing on the line, an open-lidded cooler full of meat standing in the sun in the yard, but no one at home. There is a pick-up under the trees and a fan whir-ring in the office, but no response to my 'hello' and no barking dogs.

We drive on into picturesque countryside, with tall rolling hills of good grass beneath widespread trees, not quite thick enough to be called a forest, yet in suf-ficient numbers to cut off the long view. We cross many small creeks. Some are still flowing, others already dried to a series of billabongs, the Australian word for isolated, standing pools. The billabongs are a feature of the country here: secreted beneath groves of tall trees, with thick grassy fringes, they usually lie in the catch-ment of rocky layers, the rock worn to a polished smoothness by the water's passage. The water is clear and cool and there are legions of small fish. Throw a pebble into the water and the fish rush to align themselves around the source of the plop, like iron filings around a magnetic field. Dragonflies are busy low over the water, and birds call in the trees. Often in the sand one can find the snake-like trailings of a perentie lizard as it prowls the bank in its swaggering walk.

At dusk, the first dingo we have seen runs off down the road ahead of us. At first I mistake it for a feral domestic dog, because it lacks any of the attributes of wild canids I have seen anywhere else in the world. Although it has the upright pointed ears of a wolf, it lacks the wolf's broad facial disc. Its stature is closest to that of a coyote, but its fur is short, making it seem lighter bodied. The coat is a light tan that blends to white on the chest and lower jaw, and the tail is short-haired above, growing longer and denser below, as it tapers. The dingo's origins have been traced back to the Asian variety of the grey wolf. It is thought to have come to Australia about four thousand years ago as the domestic hunting dog of Asian seafarers, and then to have evolved from those that escaped their custody, becoming in the process a free-roaming predator.

24 MAY

'How would you describe their call?'

Beverly is listening to the fluted chiming of a flock of hundreds of woodswal-lows sallying back and forth over our camping place in the forest.

'Like a chandelier in the wind.'

'Write that down, that's exactly how it sounds.'

The country is rich with birds. A flock of galah passing overhead spot us, circle down and land in the nearby trees for a closer look, discussing our pres-ence in harsh, guttural screeches. The night has been cool and the smoke from a distant wildfire lies like mist between the trees.

Past Bedford Downs Station we turn north, and almost immediately the road becomes a track, defined only by two ruts running through tall grass. I am encouraged, as it is still a road, no matter how little used. Beyond, the track worsens and we crawl along in first gear into a narrow, rocky gorge between two hills. I can see the open land we are seeking beyond when we are halted by the boulder-strewn crossing at the gorge bottom. What passing place there might have been has been tumbled into chaos by floodwater. We walk the crossing twice. It will require a half a day or more of road-building to make a passable track. I decide against it.

Lansdowne Station is deserted, the cattle fleeing the shade trees of a garden gone wild as we arrive. I walk through the decaying buildings, intrigued by what people have left behind. There are bats circling in the base floor area and mud wasp nests on old moth-eaten easy chairs stacked in a pile beside a grey-faced television set, all herded into the centre of the room. There is a single, faded photograph in the small living room of two toddlers holding an enormous green frog with sticky yellow toes. A dog lies in the background. Linen is piled in the bedroom cupboards and a mirror with blue-painted shells glued to its wooden frame still hangs on the wall.

In the large kitchen, there is an industrial gas range and a full set of crockery and cutlery. In the pantry, there are rows of basic supplies, canned food, detergent, mosquito coils and matches, suggesting that this place still provides shelter from time to time. On the counter lies a map of the station, which is 331 271 hectares in extent. On the floor, fallen leaves eddy in the breeze from the door I leave open.

Three pairs of old shoes hang from a tree and the vice in the workshop stands open, its handle poised at ten past as if expecting to be used again any moment. I wind it closed, leaving the handle hanging straight down. It feels like putting pennies on the eyes of the dead.

27 MAY

On our return journey we come across a broken-down grader. The driver and the Springvale Station caretaker are working on it. Both are dirty and dusty, and do not introduce themselves, but I learn from their conversation that the driver's name is Chris. The other refers to himself only as 'Ringer'. He wears a broad, fancy copper bracelet, has a broken nose, and half his front teeth are missing. He is genial and full of information this morning, but I suspect that on a Friday night he likes to fight. Ringer tells us that the jackaroos are mustering at Mabel Downs Station at the moment.

We stop at the muster on our way back to the Great Northern Highway, and meet Dave Burton, the station owner, a taciturn, brusque individual who nonetheless agrees to our making images of the drovers as they work. There are several hundred cattle crammed into the multi-penned race. The din is continuous and the red dust rises like smoke from a fire. Six hands work the cattle, three men and three women. All of them, bar one, are young. The cattle are almost exclusively Brahman and as wild as they come. When cornered, they often charge, and several times during the afternoon I photograph drovers sprinting for the fence and leaping up and over, head first, an angry cow tossing its head at their heels. A bull charges a man on the other side of a gate, crashing at full tilt into the steel-pipe fencing. It comes away shaking its head looking for more, a broad gash across its muzzle. 'Sale!' shouts Dave, as it enters the small sorting enclosure, the drovers sheltering behind the heavy steel gates as it spins, dusts and snorts at its confinement, before charging for the freedom of the open gate leading to the sale holding pen. The older drover, a hand-rolled smoke dangling from his lips, grins at me.

Dave tells me that they will sell between eighteen and twenty-two thousand animals this year, gathered from the four stations: Mabel Downs, Texas Downs, Alice Downs and Springvale, with a total area of more than one million hectares. They are still working when we leave them after sunset. I walk out through the 'bush' pen, which contains the lucky cattle selected as still suitable for breeding to be turned loose again. The hair on their tails has been cut short to identify them should they be herded up again in the next round of mustering. Some have bloody heads, where disfigured horns growing into the skull have been docked off. Thinner horns are sheared off with a massive bolt cutter, heavier horns cut through with a saw, the animal's head clamped between two heavy metal gates, its body kicking behind.

30 MAY

A short distance down the road to Cape Domett we find a dingo, and then another. They are not wary of the vehicle and stand and watch us. The first is breathing hard, as if it has been running, and moisture drips from its open lips. After a time, they move off, stopping to scent-mark here and there beside the track, then they turn in to the bush and vanish in the thick grass that grows higher than their heads. Their reaction to our presence reminds me that for much of the year this reach of the country is impenetrable, except perhaps to the very determined on horseback. The wet season makes the ground a bog and renders the roads impassable. We are here not much more than a week after the road has become viable.

Nearing the northern coast, the road breaks suddenly out into the open onto vast expansive flats. Single boabs punctuate unbroken vistas clear to the horizon. There are shallow lagoons of water, grass-fringed, where hundreds of waterbirds feed. We stop at the edge of a pan dominated by two giant boabs standing leaf-less beside the water. A black-necked stork hunts fish in the knee-deep water by flushing them with the flapping of its wings and stabbing after them with its long, strong beak. It is a jester's dance, full of fanciful flaring of its wings and crazy pirouettes. There are brolga and spoonbills, herons and stilts, plovers and ducks, and in the copses huge flocks of little corella fuss and take to the air like white clouds. We cross salt flats that look like the tide may occasionally push this far into the land, and twice we must drive a circuit through the grass to avoid boggy runnels. The bush closes and opens like the curtains of a magician's show, and we come into an opening where a dense stand of boabs grows through the grassy carpet of a clay pan. There are hundreds of them, some old and fat-waisted, the younger ones slim and tall in tight groups. If this were Africa it would be a favourite place of elephants. There are small agile wallabies that lift their heads to our passing, peering just over the tops of the grass with their ears cocked forward, and forepaws held hand-like, raised, and slightly crossed before their chests. A final stretch of dense overhanging forest, with only just sufficient space for the vehicle to push through, brings us out onto a point just before Cape Domett.

The view to the west is arresting. Across a vast tidal flat, where swathes of short mangrove trees follow the lines of the rocky shore, a peninsula of massive, red eroded cliffs rises tall from the sea. Beyond its tip, two islands of rock withstand the erosive force of the sea and its dramatic tides. The sea is far way, but the tide is coming in. The ocean is a dirty, turbid brown.

I watch the tide advance, rising up the rocks like water in a bath. An Aboriginal family comes barefoot over the rocky ridge of the reef to fish close to where I stand. They use hand lines and, unlike any other fishermen I have seen, they sit down to fish cross-legged on the flat rocks. It is strange how this alters the mood. It removes the urgency of anticipation that one senses in all fishermen standing with one finger touching the line. It ignores the time pressure of the tide's steady advance and makes of the atmosphere something more festive, like a picnic. The woman suckles an infant cradled in the crook of her left arm. She casts her line with her right hand, swinging it in a fast circle to the side of her head and then casting it into the sea. She never stands. The infant falls asleep in her lap with the nipple still in its mouth. She catches more fish than either her husband or I.

1 JUNE

'The Needles', as the locals know the rocky peninsula that juts out into the sea towards Cape Domett, glows like a coal in the first light. The green tops of the mangrove trees stand surrounded by a mirrored sea of silvery powder-blue. The tide is returning and wavelets run up the slow shelving sand of the beach. We are captivated by the design and colour of the huge flat sandstone rocks of the shore. The sandstone is formed in layers of varying colours, and as the ocean has worn them smooth, different layers have become exposed to create an abstraction of forms and colours that rivals Aboriginal art. There are blood reds, dark maroons, burnt oranges, yellows and deep mustards, just in the red spectrum. There are whites that go through all the hues to grey, and a purple that when wet is the colour of the skin of a grape.

Returning to the wide salt flats, we make camp in an area of deep shade beside a long lagoon, where myriad birds and northern nail-tail wallabies, half seen through the luxuriant grass, keep us company.

Sitting beside the Land Rover, with only the stars and the sheen of the water visible in the dark, hundreds of whistling duck descend from the blackness to land with a patter of little splashes. Their piping whistling draws more. Their plaintive calling suits this place: it has an edge of longing, of being lost.

2 JUNE

Like the ducks, the little corella arrive beside the lagoon before it is light. They come in one vast raucous flock of hundreds of birds, a dark cloud against the twilight sky. They land in the bare branches of the boab trees, like sudden white flower buds sprouted from the stems, and shatter the silence with screeching chatter. The lagoon holds us captive for the day. I'm on my belly on the edge of the mud in the predawn when a masked lapwing sidles closer to inspect me. This northern race of the species has a yellow wattle that covers the entire face to above the forehead. From head-on, its face is like a bald-skinned yellow barn owl. It pretends to feed, dipping here and there into the shallow weeds and then turning once again to stare at me. The whistling ducks have gone, but the royal spoonbills are hunting the pool in a dense flock spread over a wide front, advancing like a net. Little egrets run ahead of their flank with quick steps, striking fast after the small fish that dash away. White-headed stilts pick at small morsels on the surface, like ladies at a tea party, while the lapwings dive-bomb them, forcing them to duck into the shallow water. The brolga are gathered in a large flock, feeding in the reed-fringed edge of the shallow water. The pure red band that runs from their bill, through the eye, to wrap around the back of the head like a bandana gives a roguish aspect to the otherwise dignified, crane-like bearing of these large all-grey birds.

The birds grow used to our stillness and venture close, and the sun is well up when I finally stand up from my awkward position. My clothes are wet and I smell of mud. Walking back around the drying mud fringe of the lagoon, I cross the tracks of a large crocodile that has abandoned the shining water: a flattened, narrow python-like smear over the surface of the mud. It is headed towards the distant sea.

5 JUNE

Turning west on an indistinct track, we go in search of old Aboriginal sites in hill caves that we were told about. The track is unused, the grass taller than the bonnet of the Land Rover, and the unseen ground eroded and rough. It comes to a place of long, still, deep pools of clear, jade-coloured water between a narrow margin of trees. We walk out into the limestone kopjes, shrouded by dense growth. There are numerous caves but no human artefacts or art, only congregations of butterflies sitting with folded wings, like thousands of petals hung from the yellow-stone cave walls.

Crocodiles lie as still as the rocks as we roll quietly, motor off, down onto the causeway of the Keep River, lit with the first sun. I count two, then three, then four of them, all large. They look identical to the crocodiles we see in Africa, but by some curious twist of evolution these have become adapted to a saltwater environment. They will penetrate far up into the freshwater reaches of rivers, but essentially they are crocodiles of the estuaries and sea. The Keep River estuary pushes deep inland, and at the crossing it is almost 55 kilometres as the crow flies to the open sea. The crocodiles slowly drift away from our presence, only their eyes and nostrils showing as they move without ripples beneath high banks overhung with green fronds.

Following the river northward, we turn off onto a sidetrack, and find a chain of small lagoons with tall, purple-flowered water lilies. Waterbirds, heron and ibis move in the shadowy retreat of mature trees. The river here is still a meander between rocks, with the occasional crocodile basking in the sun. They are shy of human presence and slide into the water when we are still a hundred metres away. Coming stealthily through the bush, I am able to approach a 4.5-metre individual from the cover of a high bank until I am quite close. I make a few images and then descend through an erosion gully to get a water-level view. Halfway down, a dry branch snaps beneath my weight. The crocodile swivels its large head to the side. There is no distinction between its neck and body; its large girth begins immediately after the large scaly plates at the back of its skull. For an interminable moment in time we stare at one another, unmoving, over eight metres of water, and then in one broad slash of its tail it disappears into the muddy water, leaving only a swirl on the surface.

Beneath a half-moon we camp out on the open riverbank, at the edge of a plain where we had come across a dingo in the twilight of evening. The dingo calls and is answered from across the river and then by another further upstream. Their call is thin, not as full as a coyote, more like that of a jackal. Later, when they call again, the dingoes are joined by what is clearly the baying of feral dogs. The calls are distinct, that of the dingoes a thin one- or two-note wail, pitched high and long to announce their presence on the land. The dogs have an overture that seems to come from a different source, something deeper, more primal, more sinister.

7 JUNE

The picturesque Ivanhoe Crossing of the Ord River has a lovely grassy area beneath a spreading shade tree. Cormorants and pelicans fish on the downstream side of the causeway. Further down the river we see crocodiles, but all of them are freshwater – or Johnstone's – crocodiles: a much slighter-jawed, smaller reptile than the estuarine crocodile that does not present a danger to humans. Through the camera lens I can see that although the head of the bigger ones is quite large around the eyes, after the gape, the jaw narrows to a thin snout. The teeth are narrower and perhaps longer than that of their saltwater relatives, more suited to snatching fish than to wrestling larger prey.

Along the Ord, I talk to an Aboriginal fisherman who has caught a barramundi. He tells me that it is exciting to catch, a furious fighter and excellent to eat. I do some research and learn that the barramundi, the iconic predatory fish of the estuaries, can even be found in deep billabongs far upriver, well away from the sea. It can grow to up to 40 kilograms. Barramundi undergo an unusual midlife crisis. All barramundi are born male and become sexually mature at three to four years of age. Then, between five and six years, they undergo a remarkable physical change, losing their male organs and becoming female. They remain female for the rest of their lives, which can be as long as twenty years.

Marlgu Billabong, on the western flatlands of the Ord River, is massive in extent, stretching out for kilometres in either direction. It forms a vast, shallow floodplain, and all day the sky is busy with the coming and going of birds. A flock of fifty or so pelicans seems to fish in time to some unheard internal music, a slow choreography of glide, lift and dip. The only element that is inelegant is the splashing of their webbed feet as they tip, tail skyward, to reach deep into the water with their pouch bills. There are flocks of hundreds of plumed whistling duck on the shore, darters, cormorants, pygmy geese and six species of heron and egret immediately visible. The world is alive with birds. It is pitch-dark when we finally stop photographing and retreat to the height of nearby Telegraph Hill to escape the hordes of mosquitoes. The waxing moon burnishes the silvery, smooth bark of the fat boabs that stand etched against the dark land.

8 JUNE

In the mid-20th Century Marlgu Billabong was a different place. The only abattoir in these northern districts was in Wyndham, and cattle were marched overland from the stations to Wyndham for sale. After the long overland crossings, Marlgu Billabong was a pause point outside town where the cattle could be fattened before being driven the last leg to market. Thousands of cattle, year after year, feasted on the rich grasses of the Marlgu floodplains, trampling the vegetation and ripping out the water plants, until there was little of this remarkable environment left intact. The birds were all but gone. Finally, in 1971, the importance of the billabong was recognised and an era of protection began.

A wind that has been blowing for the past few days is sweeping hard across the flat country of the wetlands. It keeps the mosquitoes down, and on the water the pelicans lift to flight with half their usual heavy flapping. The pelicans are fishing early: there are a hundred or more, which from time to time break into two separate armadas. I see one individual catch a large catfish. It breaks from the group and sits off on its own wiggling its gular flap to manipulate the fish into position. Catfish die hard, and the pelican drops it to the point of its bill and bites down with the sharp nib at the point of its beak to try to kill it, but after fifteen minutes I can still see the catfish's lateral fins, spread wide in defiance, pointing sharp through the pink skin of the pelican's bill-pouch.

9 JUNE

On the corrugated Old Karunjie Road south, we climb to a high cliff face to view Aboriginal paintings made on the shallow caves there. Most are weathered and hard to see until we discover more on the low roofs of the overhangs. Lying on my back in the dirt, the paintings an arm's stretch from my face, I feel strangely close to the artist who made them, who must have lain down exactly as I am. There are men and animals I do not recognise, and what I think are boomerangs. It is strange that all people in the paintings are depicted with hands and feet with more than five fingers and toes. The most common theme is that of the outline of hands, over which the ochre-clay pigments of red, yellow and white have been spat in a fine spray. There are small hands and large, thin and fat-fingered ones, as if a whole extended family had left its signature.

10 JUNE

At Diggers Rest Station, Roderick Woodland is offloading horses after a five-day ride. With him is Daryl, who introduces himself, and Monica, who ignores us. I help unload the tack and food and water barrels into the shed. It is the same as all farmyard tack sheds: a chaos of saddles, blankets, bridles and bins of oddments, with pieces of leather, buckles, bolts and assorted tools strewn across the few work surfaces. Daryl is a professional horse-breaker.

I watch him handling the horses. He is calm, slow-moving and gentle. He touches them as though placing his hand on the shoulder of a friend. Both he and Roderick have the unhurried air of men of the country. They take their time even when they are busy, and when they stop to talk you feel that your conversation should slow down to let your thoughts settle and gel, to hear the birds, to give a kind word to the dog.

We sit in worn, thick-cushioned chairs in the shade of the wide, stone-walled veranda. Roderick talks of the backcountry without flair or fancy, like someone who has spent his life in remote landscapes. He tells me of the places we should try to see, of the tracks that lead to wilderness, and he gives me the names of people I should talk to. We mark our map with circles and notes and cross out marked tracks that have long grown over. Eventually, Roderick gets up and we shake hands. He bids me the greeting that I have only had so far from Aboriginal people: 'Enjoy the country.'

The Old Karunjie Road becomes a smooth track that runs at the toe of the foothills of the Cockburn Range to the east and the wide mudflats of the deltas of the King and Pentecost rivers to the west. It leads suddenly, in the middle of nowhere, to the edge of the mudflats and then disappears. Although the surface looks hard and caked dry, their soggy mire can lie a few centimetres below the surface. Break through and you become properly stuck.

Today, we stick to the longer but more certain route following the stony edge of the hills. The going is slow and, after a time, I see an old track leading into the hills and we cross the stony ground to join it. There is no need to drive fast. We are alone and the country is vast and lovely. From the blonde-grassed hills with their forests of pale-stemmed trees, the tall hills of the Cockburn rise to a fringe of sheer red-rock cliffs that encircle the entire range. In the afternoon light, the colours are amplified to a saturated richness. Cattle are coming up in long columns from the flats into the hills, raising tails of dust that grow golden in the sun. A few euro kangaroos move just out of our path to stop and turn to watch us, their forepaws held loosely before their narrow chests.

On a ridge with an open view to the west, we stop for the night. We can see 60 kilometres or more over the flats, the riverine trees a distant, dense forest that collides with faraway hills and the horizon. When the last light is a crease of orange over the western horizon, I notice a flock of birds flying low over the close trees, coming into the hills. At first, I think they are night herons, but their numbers increase from a few to fives and tens and I realise suddenly that they are flying foxes. Steadily, like a river swelling in flood, their numbers grow until they are flying low over our heads in their hundreds and thousands. A flood of black bodies on whooshing wings, ephemeral against the purple moonlit sky.

12 JUNE

Scrutinising the map, I see that the Carson River track, which leads to the now-deserted Aboriginal settlement of Oombulgurri, begins at the Home Valley Station and crosses the Durack River to the west of the island where I suspect the flying foxes have their roosting place. I have been told by several sources that since the closure of the Oombulgurri settlement due to violence and the abuse of alcohol, the track has fallen into disrepair. The only internet-based information I can find on it was by someone who, coming at it from the north, spent four days bashing his way through the wilderness, at the end of which he and his party had travelled only 35 kilometres. They gave up and went back. It has the allure of wilderness and may also lead us to where we can find the flying foxes. We decide to give it a try.

On the track north, the very first obstacle we encounter is a wide riverbed crossing over a jumble of round black boulders, 70 metres across. Beverly is uncertain. I can see a gap in the far bank where a vehicle could climb out. I get out and walk the crossing. There are two channels of water with a dry ridge in the middle. I hop from rock to rock, trying to examine the bottom for any differential-crunching boulders. I persuade Beverly that we have to try, that there is unmarked wilderness beyond. We grind across in low-range first gear, the Land Rover lurching over the uneven bottom. Climbing out at the far bank we cross a small ridge of heavy sand straight into another boulder crossing. This one is shorter, but the boulders are all the size of soccer balls. I spend ten minutes scrutinising the water from the boulders heaped on either side. I commit and enter the water. The wheels climb over the invisible obstacles, spraying water as they fall down their far sides, but we cross without incident.

The track is two indistinct paths through thick grass that stands half a metre higher than the top of the vehicle. We push through 300 metres and are rewarded with yet another boulder-strewn riverbed. I hope this is the last one as I edge the Land Rover forward into the water. We make our way through kilometres of tall grass and ford two more streambeds, dry with heavy sand, before we come out into the clear. There is a sparse forest of leafless, brown-barked trees adorned with bright yellow flowers and green fruit.

The road finally emerges onto floodplains. The vista is vast, broken only by a single boab tree. Close by the edge of riverine trees a herd of wild horses turns and gallops away at the strange apparition of our vehicle. We stop and follow them on foot. We walk slowly, headed past them to the side. They grow inquisitive and come forward. They are all chestnut-brown, some with a white blaze on the forehead, a few with blonde manes. The two stallions are dark-brown. When we sit down in the grass they walk forward to see us better, their coats shining like polished metal in the setting sun. Behind them the Cockburn Range is shadowed beneath a horizon-wide pall of wildfire smoke. They stand etched bright against the shadow, stamping their hooves from time to time. Wild horses are so much more thrilling than wild cows.

13 JUNE

All night we hear the horses from our camp beside the boab tree. In the first light, we find a herd of sixty or seventy animals, and when the sun rises behind them it turns the dust they raise into a curtain of gold. We walk far across the plains, the horses turning to run, circling wide, and then coming back in for another look, closer than before. They stand, nostrils flared, heads raised high. Each of them is perfect, fat, with the muscles showing clearly defined beneath the short sheen of their coat. But it is their attitude, wild, defiant, regent, that lights a flame in my chest.

Their presence accentuates the lack of animal life in these remote stretches of the land. It is a clear void and I wonder how it came into being. Most of the native animals are nocturnal but the smooth sand of the road has revealed only the tracks of a lone kangaroo.

Australia has the dubious distinction of being the continent with the greatest number of recent extinctions. The mammal guidebook I am using lists a few species together with their descriptions, habitats and habits, with a big bold 'EX' to indicate extinct. Perhaps they remain hopeful that some lost remnant will still be found. There certainly is enough remote land for the possibility. It is strange, however, that not a single species in all the millennia of this ancient continent evolved to fill what the horses, cattle, goats and pigs are now filling.

The track leaves the flats and climbs into the hills towards the Durack

River jump-up, the Australian term for a saddle or ridge crossing. The track is extremely rough. In places, it has been eroded into deep gullies and we must fill in sections with rock or find a way to drive around. On a steep descent into the Durack River we finally concede defeat. We could risk pushing on, but the going is murderous on the vehicle and every few metres there is the danger of major damage. Both of us are dirty and cut from manhandling rocks. At the Durack crossing the tide is in and, being full moon, it is extremely high. It feels strange to be thwarted by the sea this far inland, but the tides in this part of Australia can reach more than 10 metres and the flood of brown salty water is 100 metres across.

15 JUNE

The sun sinks below the horizon and the mosquitoes buzz around our faces as we sit on the stony ground between the spiky spinifex tufts, where we have returned to the Old Karunjie track to wait for the flying foxes. The sky changes from pink to purple to black, and when there is only an orange crease over the western horizon they come.

In three minutes, there is a flood of black bodies on whooshing wings. We are prepared this time, our equipment chosen and set for photography in the dark. The flying foxes show black and heavy against the slightly paler sky. They fly fast and appear suddenly above us, out of the dark background of the land against which they are invisible. They sense us if we are standing and veer away. For twenty minutes, they cross the sky above us and then their numbers slow to a trickle.

I am elated about the images I have shot, the bats close, the landscape a thin row of pale-trunked trees, before a thread of orange light beneath a star-filled sky. When I review the images, however, I am stunned by my lack of success. The images are there, but each has its faults that render it useless. Beverly has fared much better than I. My failure falls heavily on my mood. The experience I have had has been remarkable but my photography is the voice with which I speak of it to those who will never come here, and tonight that voice has failed me.

16 JUNE

We opt to spend another day here to try again with photographing the flying foxes in the evening. We find a shaded camping place beneath the low, dense trees of a tidal creek. There is a tree without leaves that is flowering bright-red blooms in abundance below our camp. Birds of all sizes and kinds are busy in its branches. The locals call it a kurrajong tree. On a shady spit on the terrace of the riverbank a squadron of small, striped Australian digger wasps are frenetically busy digging tiny tunnels, making nests into the sand. They dig like dogs but at three times the speed and disappear rapidly underground, a spray of sand spurting out the mouth of their holes behind them. They feed on nectar and honeydew and will hunt for flies to feed the larvae in their nests.

In the evening, we return to our hill in the flight path of the flying foxes and set up once more for the brief phenomenon. I am doubly studious about my preparations, but there are so many variables when photographing a fast-moving animal in the darkness that there is still a huge margin for error. I sit on an uncomfortably hard termite mound among the spinifex and wait. The mosquitoes find me as soon as the flies have finally settled for the night.

A flight of red-tailed black cockatoos passes along the hillside, calling mournfully. Of all the birds that we have encountered, this large, slow, almost entirely black parrot has become my favourite. They are the largest of the parrots, cockatoos and corellas we have seen, but are benign and retiring in their behaviour. Their whole manner is gentle, from the laboured way they fly to their sorrowful call. They occur in small groups; only very occasionally have we seen flocks, and invariably it is in the wildest lands that we find them. They are shy, and it takes patience to approach them closely. Their only extravagance is the flash of scarlet-red in the tails of the males as they take off.

The waiting makes it seem as though the bats are late, but the sky still has its rind of orange on the horizon when they arrive in a trickle that within minutes becomes a sky-covering rush. It is hard to choose one in the melee, and even harder to focus as they fly directly at us, low and fast, their bodies a dark silhouette against a dark sky. After twenty-five minutes, it is all over and too dark to see any stragglers. We return to the Land Rover to review our results. I am flooded with relief when the very first image I check is what I have been looking for.

20 JUNE

The Gibb River Road is an endless corrugation. Whatever is not tied down or clipped closed falls open, spilling its contents, or shimmies across the floor and smashes into something else. At mid-morning, we turn north on the Kalumburu Road towards the coast.

It has always fascinated me how thoroughly and quickly one's view of the country can change depending on how you are seeing it. Driving along, I am looking at the land as a whole, able to give it only some of my attention as I must constantly focus on the road, avoiding rocks and ruts and finding the areas of least corrugation. The continual passage of vehicles in both directions makes me edgy. They shower us in dust, racing through the country from destination to destination, very few slowing or pausing to absorb the country in between. It drives away any sense of wildness. But, walking upstream along the sandy banks of Plain Creek, their rumbling presence recedes between the trees until we are alone again, the forest quiet. The same land that had me disgruntled becomes a clear stream, birds flying from the trees, and small fish occasionally dashing into the shadowed depths as I approach. There is a tranquillity here because I am living at the same pace as the land, and I see all the myriad features of its kaleidoscope that an hour earlier I had passed by. Perhaps when fossil fuels run out we will ride horses again and our perception of the earth will change because we will see it so differently from a saddle. As though our own garden, we will notice its every aspect and value and love it and will give worth once more where it belongs. Our rushing, it seems to me as I walk slowly upstream, has brought us little other than more of a need to rush, an anxiety made and consummated by our own actions whose ultimate purpose remains elusive.

21 JUNE

At Drysdale River Station, we wait on the dining-room veranda for Anne Koeyers, the station owner. She wears a light floral dress, keeps a tissue tucked into the top of her bra, and smokes hand-rolled cigarettes, fishing stringy red tobacco from an old round, flat tin. She is almost laid-back, but possessed of too dynamic an energy to actually give in to it. She is wary at first, a little guarded from a world that for half the year is filled with the beck and call of tourists often out of their depth. After an hour of talking, however, she warms to our project and I can see that my honesty appeals to her. I do not fit the crush of people who fill her yard with their 4x4s standing in line to take on fuel and water, or beer at the bar, or iced soft drinks in the shop. I am seeking wilderness. From what she tells me, I suspect that her husband, John, is the same. He is away for a month grading the roads to the north.

'He likes it quieter.'

Quiet must have been what the Koeyerses found when they came here twenty-eight years ago. Drysdale River Station today is at the frontier of some of Australia's most remote land; twenty-eight years ago, it was outback in the extreme.

'It wasn't much more than a goat track to get out here. Not as bad as the days when it took a mule train and thirteen days to get to town, but almost.'

Anne knows the country intimately and we draw dotted lines and mark Xs on our map at her suggestions. She is excited, and I feel a strange weight in the trust she has extended to us on her gut impression of what we do and who we are. I find myself for a moment removed, knowing that I will strive to keep sacred the remoteness and untouched nature of what she loves of this land. But I am also aware that by our work we will make it known. I assure her that I will not recite directions or give GPS points in any published work. She shows us her own photographs of the country; she is clearly a photographer in her own right. It is mid-afternoon when I suddenly realise that she has given us four hours of her time while people have constantly been coming to the office door for her attention. It is rare and special when you find someone to talk with about land and places, and you know from the first that what you have known is different but how you feel is the same.

In the forest, shorthorn cattle graze on the fire-burnt areas that are now turning green. There are no fences any more, and I wonder how the station knows its cattle, but I suspect that it is merely a case of catch what is on your land when the time to muster comes. Anne has told me that the shorthorn cattle that we see are the stock that was originally brought out here when cattle ranching first began in the Kimberley. Further south, they have been replaced by Brahman, which are more tick-resistant, but to phase one cattle type out and another in has thus far proved too difficult in this place where many stations are a million acres or more.

22 JUNE

The road to Mitchell River National Park is so deeply corrugated across its whole space that there is no escape and the vehicle dances to the side as the bouncing causes it to lose contact. The turnoff to Mitchell Falls brings no relief, except that the road is narrower, winding and the surface rutted and rocky, making it impossible to travel at speed. The King Edward River crossing is a deep ford over a loose-stone bottom. I drive slowly; the water is as translucent as gin and I am looking for fish.

On the far side, we walk between scattered clumps of sandstone boulders where the overhangs and recesses have Wanjina-Wunggurr paintings made by the ancient Aboriginal peoples who lived here. The paintings are of people, animals, faces, and the outline of hands. The proportions are stretched, perhaps to represent something beyond the literal. The most striking are the faces, white discs haloed in a broad band of maroon with hollow, staring, lifeless eyes. They seem commemorative rather than celebratory.

Alone at the site, we make our way along the narrow footpaths between the rocky outcroppings, and I try to imagine the ancient people walking as I am between the trees, looking up to the light, translucent green through the leaves, but I cannot. The paintings remain rock art, their artists too distant to reconjure in the imagination. There seems no modern counterpart who lives their tradition. It is strange because I have a strong sense that the Aboriginal people I have encountered are people of the land. Being on the land, in wide country, suits their manner. It is in keeping with how they see time and how they are, both in repose and in action. They seem most comfortable when they are in contact with something natural: the ground or a tree. Oddly, however, I have met only a single Aboriginal person living out in the land. Every single other lives in a community where dysfunction blinds one to the truth of their being. Everywhere else we have been in the world, we have found people who have maintained a strong connection with the land. They live it today as their ancestors did and it marks them and their claim to the country with a powerful identity. The Aboriginal people lay as strong a claim to the lands of Australia, but somehow they seem lost as to how to return, to reassert what the colonial governments here tried to quash. They seem to me a people adrift. As if they know the land to be their place but lack the fortitude to return.

Along the road, the transition into the Mitchell Plateau is marked by the palms. Suddenly they are everywhere, until their stands crowd out almost all the other trees. They grow tall, on thin stems, and the leaves of the young are emerald-green in the shadowed understorey. The road becomes a cavernous tunnel through their tight ranks, like driving through some exotic botanical garden.

23 JUNE

Walking to the edge of the plateau, we photograph out over the long view of the valleys below. Mist lingers in all the depressions, and mixed with the smoke of the fires it is the palest blue. We drive enchanted through the trees, the soft light falling in shafts as the mist rises from the dew-wet ground. The scene is reminiscent of a close, smoky theatre.

We finally catch up with John Koeyers. He is moving camp, driving his grader and towing a long trailer with a container and bulk fuel tank, behind which his pick-up is hitched, its white body showing only vaguely through a thick carpet of red dust. He climbs down shirtless. He, like Anne, is reticent at first, unaccustomed to being stopped and talked to. But he warms quickly to our anecdotes of our time with Anne, laughing a little incredulously at all she has told us. It is clear that not many people are given the privilege of what she has shared with us.

John has lived all the country here, and there is not a single one of our questions that he cannot answer. He is, however, under pressure to complete the grading here so that he can turn his attention to the Kalumburu Road. The settlement at Kalumburu is running out of fuel and the fuel truck cannot get through. We agree to get together tomorrow evening. He draws with a stick in the dirt a map of where we can find him.

The road descends sharply off the ridge and then climbs another lower jump-up. We cross a clear, fast-running creek and on the far side take a small track that turns upstream. It delivers us to a long, deep pool of crystal-clear water where small fish fuss along the edge. We are in washing, naked, when John arrives in his pick-up, the one working windscreen wiper having carved a half circle of vision through the encrusted red dust. He places two twenty-litre water containers on the ground and, dropping his shorts, climbs naked into the pool.

'What about crocs?' asks Beverly.

'Naah! Only freshies here. Never had one even think about it and I've swum in all the rivers. We used to play stalk the croc. Get in the water, get as low as you can, you know, with just the eyes showing. You can see the croc away on the far side of the pool and you try and get as close to him like that, but you never can. They always move off, come up behind you somewhere.'

I help him fill the containers and then dive in for a swim. John joins me and we stand in the chest-deep water in the middle of the pool, talking more about land and wilderness and change. He gives me a short overview of the geology of the region when I ask him about a pale green stone I found beside the river yesterday. It is an epidote crystal. Incredibly, this flat place with its few ridges is the eroded remains of heights that in aeons past towered up to 5 000 metres. As the land eroded, the laterites, which naturally attract each other, started to form in concentrated pebbles as the rest of the minerals and soils were leached away. Today the rough stony surface, with its layer of rock-marbled, unevenly faced pebbles, is the concentration of lateritic bauxite formed into rock. It is what the Mitchell Plateau palms and the mining companies like.

While we are talking, the small fish are nibbling at my skin, biting me on the cheeks of my backside. John tells me they are a species of rainbow fish. They are small, with translucent bodies and red dorsal fins and tails. We both have goose bumps and wrinkled fingertips when we climb from the water. I haven't noticed the time pass. It feels like John is a friend I haven't seen in years.

24 JUNE

At first light, we take off on a helicopter flight to the coast from Mitchell River National Park. The rush of air with the doors open is cold. Below us, the land lies under a shroud of grey-blue smoke, giving definition to the ridges that stand through it. It is mostly flat, with a few hills, and from above one has a real perspective of its remoteness. Not a single path or road, nor any kind of structure is visible as far as the horizon in every direction. No one is out here. There is no way out except on foot, and the terrain is hostile to walking. John Hayward, the ranger at Mitchell Falls, who stays here year-round, enduring the torrential rain and heat of the wet season, is a bushwalker and has covered much of the country. He classes it as hard walking. 'The sandstone country is much easier, but it lacks the forest and the geographical interest.'

Towards the coast, the land suddenly changes. Rough, red-rock hills rise between deep valleys, growing more and more rugged until at the coast it is a dramatically pleated landscape of red cliffs, dark gorges and mountains of stone. We follow Hunter Creek on a twisting course to the sea. The tide is in and the sea lies a sheeny blue between red cliffs, fringed at their bases by the bright green of mangroves. In the upper reaches, there are wide circular mudflats and the mangroves trace the watercourse across their surface like veins, bright-green against the flat-grey. Seaward, the bight of Prince Frederick Harbour is rimmed by red cliff heights and dotted with islands. We work the helicopter slowly back and forth over the drama of the rugged land and its meeting with the sea.

The sun is cresting the horizon and I am here, cold and enthralled, above the earth, undisturbed, in a place so singular that it tightens my chest to see it. Many have told me I am lucky. Perhaps I am, but you also make your own luck.

We turn back as the sun warms to a white light, the land starting to show through the gauze of the smoke. The wind slaps my hair into my eyes, as if they must pay penance for what they have seen. We land in the bush away from the camping area and walk back through the wet grass. People are sitting around their campfires sipping coffee, waking up. I feel like I have lived a whole life already today.

25 JUNE

On a hike to Mitchell Falls, in the pool beneath a wall of faded red-ochre paintings, we are preparing for a swim when I notice a small perentie sunning itself between white and yellow flowers, right on the edge of the water. Lying on my belly on the uneven rock, I edge closer with the camera. Several times it starts as if to move and then relaxes again and closes its eyes as it flattens its body against the warm rock to absorb the heat. When it finally does move, it steps forward into the shadows, licking at the water with its long, forked tongue. I assume that it will swim to the far side and so lift myself to my knees for a better perspective when, to my astonishment, it dives beneath the surface and, swimming just like a crocodile, hugs the bottom, disappearing into the dark crevices between two submerged boulders. Small schools of fish spook out into the open water from the shadows. I stand on a rock jutting into the pool, waiting for it to resurface, interested to see if it has caught anything. I wait a full ten minutes but do not see it again.

Later I read in the guidebook that it is a Mertens' water monitor, which 'can remain submerged for a considerable time'. The pool we are swimming in is the Mertens River, and I wonder whether the species was discovered by the same person for whom the river is named.

John Hayward is a rock art enthusiast and has told me of art sites upriver from the falls. I have his hand-sketched map crumpled in my pocket. The river is a series of huge pools, hundreds of metres long and fifty or more wide, the water held cupped in the shallow valley of rock. We find the place John has suggested, a low sandstone ridge just back from the river. There are numerous paintings on the front wall, but there are caves beneath the stone and, bending low into their darkness, we walk through the eerie low cavern, the roof supported on tapered pedestals that give the feeling of columns, as if the place was constructed. There are plates of flat rock on the cave floor, worn to a slippery smooth patina, with holes, varying in size from tiny to that of an emu egg, ground into the surface. On the far side, we find a broad and well-preserved wall of ancient art. Beverly likes the huge red-ochre fish. I like the dark, mud-brown figures drawn in fine detail, depicting elaborate costumes and energetic ceremony.

Without cattle, without fences, away from the press of people, the land feels wild here, and we linger as long as we can in its solitude before heading back in the dusk.

26 JUNE

Mitchell River National Park and its surrounds are one of the few areas in the Kimberley where burnt land does not exceed the unburnt. As John Koeyers said to me, 'We all chuck a match in for our own reasons.' The cattle ranchers set fires to create fresh pasture for their stock. Many fires are lit as firebreaks to prevent other fires from burning infrastructure and homes. People set fires to entice birds and wildlife into areas for hunting. Fires are set as 'conservation management'. Some fires are even set by arsonists angry with government or land occupation, or just personal feuds. Many fires are lit for fun, just to watch it burn.

As I write this, I have a wide vista of the land before me and I can count eight separate fires raising their columns of smoke to the overall haze that makes sunset come early and sunrise take longer. The fires that ravage the Kimberley country, until more than half of it is burnt annually, are the result of men with their own agendas and they know not what the final effect on the land will be.

Leaving the park, the road northwards along the Mitchell Plateau to Port Warrender becomes a track. At the edge of a low escarpment, a view out over the sea to Walsh Point appears suddenly between the trees. The sea – it seems so strange to find it here in this place of remote solitude, forest and dry earth. It is here that we decide we will have to explore the Kimberley coast by boat. The very few rough tracks that lead to the sea reveal very little of the gem that lies out there: 3 500 kilometres of convoluted coastline, where 12-metre tides push as far as 50 kilometres into secret river estuaries.

Very soon, what was a track becomes a rough, boulder-strewn passage through the hills, and for the last six kilometres to Walsh Point, which takes one hour and fifteen minutes, we are reduced to a low-range 4x4 crawl. Beverly, walking ahead to point out boulders and holes, walks faster than I can drive. We arrive in the late afternoon at a black-stone beach with low mangroves and a far view across the calm sea, held within the hills of a massive inlet.

We make camp on the beach. Across the water, in inlets and coves, there are white-sand beaches and, between them, valleys that terminate in deltas of mangrove trees. I stand well back from the opaque water that laps the shore, for the crocodiles that might be patrolling and the box jellyfish that may be lurking there. The box jellyfish – or stinger, as it is locally known – is finely constructed with long trailing tentacles, and often occurs in large schools that are notoriously difficult to see. Their sting causes extreme pain, paralysis and a fairly quick death in humans. It is a sad compromise not to be able to swim, but the consolation of the wildness here is ample. Like the tsetse fly in the Kafue of Zambia, the presence of crocodiles and stingers here defines the frontier of wilderness more effectively than any human-imposed boundary or law.

Twice, as the sky darkens and turns to pink, I must hurl a stone into the water where a three-metre crocodile stalks us from between the green camouflage of the mangroves. Sitting in the darkness, safely back from the water, with just the glow of coals in my fire, I wonder at the value or worth of things. I have known the luxuries that money can buy, but little has ever made me as happy or satisfied as to sit, like this, alone on a beach on a dark night and feel like all the world is mine for a time. It comes as a feeling, like a coal growing brighter somewhere inside.

29 JUNE

There are glimpses of the long view between the trees as we follow the ridge out, and in all of them there is a haze of smoke from fires. Beyond the turnoff to Mitchell River an entire valley to the south is burning, the smoke rising like a curtain, obscuring the valley behind. The smell of smoke is strong on the road and then suddenly there is fire. The forest is alight, the grass crackling as the flames creep through its thin carpet. The dry fronds of a young Mitchell Plateau fan palm are consumed in a rush of flame, the stalks popping in the sudden heat. An old tree, its trunk scarred by past fires, crashes into the road, its branches shattering like bones, its base burnt through and unable to support its weight. It makes me distraught. I can see no reason that this forest should be burning. The palms are still marked with blackened stems from last year's fire, the understorey and grass are thin, barely covering the ground, and are nowhere near the kind of fuel load that a natural fire would require to start.

The research that I read about fire in the Kimberley stated that fires in Northern Australia release 80 million tonnes of carbon into the atmosphere each year. But to stand before the pall of smoke and flame as the bush burns and know that the whole land burns with an almost annual frequency, I cannot help but feel that that is anything but a very conservative estimate. John Hayward said to me that a fire will eventually go out, but the question is at what point and at what expense to the natural cycle of this place? A fire will go out when it hits the sea, perhaps when it arrives at a river and the wind is not behind it, occasionally roads will stop it, but for the most part fires burn until they find a place where another fire has already been. It consumes the whole country. In the Kimberley, when you throw a match in, you have no idea where or when the flame will finally die.

In my discussions with the people of this land, the regime of fire in the Kimberley strikes me to be more habit than considered purpose. Only the cattle farmers, with their desire for new grass for their stock, have a sound reason. Whatever the purpose, the result is extreme, and to witness it is disturbing. Kites twist in the thermals of heat in great numbers as they stoop to snatch burnt insects that flutter to the ground, their dark bodies in sharp silhouette against the blue smoke. I wonder what happens to the reptiles and small marsupials that cannot move far or fast.

30 JUNE

Past Theda Station, we cross the Carson River, a boundary I hope will lead us into the wilderness we seek, for to the north is Aboriginal Land. We find John Koeyers camped to the west of the road, about to set out with his grader for a day's work repairing the road into Kalumburu. Our quick chat turns once again into a yarn for hours that slowly shifts from the fireside to the shade, and then follows the shade as the sun moves across the sky. It is a pleasure to find someone so informed, so passionate about the land, and we ply John with questions about everything from roads, maps and history to kangaroo meat, fire and culling donkeys. Donkeys gone wild, John tells me, had quite recently become such a plague that during the muster, if he gathered 1 500 cattle he would at the same time have herded 1 000 donkeys or more. It has such a dramatic effect on the grazing that the farmers and the government have settled finally on an eradication programme.

At midday, I am sitting on a small campstool in the shade of the Land Rover. John is shirtless, sitting in his cut-off denim shorts on the red dirt ground. We are talking of friendships, the important lasting ones that, if you are lucky, you will be able to list on the fingers of one hand, maybe one or two digits more. He says that it has been important in his life that his wife has been one of his best friends. We have been fortunate, him and I, for our wives have come with us. We are silent for a time after he tells me of some men whose partners have not. I think this makes both of us realise how much Beverly and Anne being along has changed the scope of our lives. Our map is full of John's notes, 'burnt out six-wheel Mack truck', 'road indiscernible, follows the creek bed of flat plates of rock', and he lends us a hundred litres of diesel to see us through the extra distances we will now cover in our search for wilderness.

We turn south, a vague spoor through the trees. The cattle spook at our appearance and then turn to stare. The occasional bull stares us down, stamping the ground. One bull takes issue with possession of the track but gives way at the last moment and dances and pirouettes on his heel and swipes at our passing with his sharp horns. John has asked whether we carry a firearm, because he says that the cattle where we are going are completely wild and have never seen people. The bulls, he says, will often charge. The bulls are stocky and compact, and their short horns, curved forward and upward, are sharp and tipped with black. We do not have a gun, so we will carry bear spray and hope that if we are charged, it is from downwind.

The road twists finally through a thick forest of thin, tall trees to deliver us onto the banks of the Drysdale River, with the high sandstone hills of the Drysdale River National Park warm in the last sun on the far side. At night, the new moon sets quickly through the dark-etched branches of the high canopy. I hear a dingo call once and not again.

1 JULY

Wading barelegged through the knee-deep water, we cross the Drysdale River. The wet sand of the edge is cold beneath my feet. There are no roads, no designated trails, no official access whatsoever into the park. Bushwalking is the only way in. The river is more than a physical boundary, and that knowledge is enthralling. It is a steep climb into the crevices of the sandstone bluff that begins immediately at the riverbank. We push our way through a tangle of undergrowth, breaking the branches of dead bushes to make our way around the rocks. The sun has yet to hit the river, but we are warm.

At the top of the bluff there is a short, sheer buttress, and it takes us a while working around the foot of its face to find a way up. The view from the top is more than we had hoped. With the rising sun at our backs, we look down, along the edge of the bluff, which runs north–south, on the whole country far to the west. The river is a wide run of blue that is swallowed at its extremities by the tall green riverine canopy. There are yellow sandbars dividing its flow directly below us before the river stills into a bank-to-bank pool that stretches for kilometres to the north. White-barked trees, 15 to 20 metres tall, along the bank merge into an endless canopy of green as the land rises away from the water and runs flat and unbroken to the western horizon. A dingo emerges from the forest and crosses the river almost directly below us, shaking as it leaves the water and then trotting off down the sandbars, headed north.

From vantage to vantage we clamber across the eroded, deeply creviced sandstone until the sun is high above the horizon, and then we begin our clutching, scrambling descent. A quarter of the way down I grab at a sapling, releasing a shower of ants with striking translucent emerald abdomens. They are fire ants and they are soon down my shirt, up my sleeves and in my hair. Their sting burns but fortunately does not linger like that of a bee or a wasp. Halfway down the hill, I am still frantically fishing them out of my sleeves and from my chest while Beverly picks them out of my hair, she too getting stung for her trouble. At the river, I wash in the cool water to soothe the bites.

We are driving towards a fire and at sunset we encounter the flaming front. Without wind, it is eating its way slowly through the grass with a soft crackling. We break off branches of a leafy tree and beat out the flames beside and in the middle of the road and drive through. On the far side, we stop to photograph a gathering of brown falcons that are following the advance of the flames, perching in the low trees and watching the ground intently for any life that might scurry or limp away. They are undisturbed by our approach, letting us get to within five or eight metres. We camp for the night on a knoll of bare rock, the fire burning away from us and turning the night sky orange all around.

Before we turn in, I call my father on the satellite phone to tell him we are safe and well. He tells me that our pointer, Blaid, who he is looking after while we are away, is dying. She was seventeen years and eight months old yesterday. Every time we have gone away to work on our project for the past three years, we have said goodbye to her in my father's small garden in the belief that we will never see her again. Each time we have returned, she has been there. We were hoping against hope that this time it might be the same. Our conversation is short, our choked voices distorted by the delay and echo of the satellite connection. The moon, just in its first quarter, is setting as I finish the call and I close the phone and send her my love via the moon. It hurts, and more so that we cannot be there. I will from now on refer to the first quarter of the moon to myself as 'the Blaid moon'. She has been the finest companion one could ever have wished for. As I lie in bed, awake and sad, I let my memories of the past, almost eighteen years, flow past my eyes. It is her animated eyes, her unbridled excitement at the adventure of our lives together that marks her history. I remember how she would stand with her hind paws on the floor of the car, her chest and forepaws on the camera compartment between our seats, and the sound of her tail wagging against the fridge was like an unstoppable drum roll as we drove through the wilds. Of all the dogs I have known, she could run. Not the fastest, but forever. When we built a house in the mountains she would run the seven kilometres from the road to the house and back every day, racing me in the old Land Rover if I ever tried to overtake. She never had much time for people; wilderness and the outdoors were her business.

2 JULY

I follow John's directions and find the disused track that turns to the north. The going is dead slow as we bounce and shudder over boulders, and trees and bushes drag screeching down the side of the vehicle. After an hour or more the track finally delivers us onto the banks of the King George River. John told me that the track followed the bank down until it came to an end at a series of bouldered sandstone ridges about five to six kilometres downriver. But the track is overgrown, a forest of trees and deep, eroded gullies, blocking any way forward.

When I return to the Land Rover after trying unsuccessfully to find a way through, Beverly is grim-faced. She does not want to do this any more. She is finished with driving endless rough tracks into the trees and doesn't have the stamina or stomach for more. I have known that this slow exploration deep into the country has not thrilled her as it has me. She misses the walking and animation and excitement of encountering life. I do too, but that has not moved my goal of going as deep into the wilds as I can. I feel utterly flat. When I finally do find purpose, it is to find us a suitable campsite, but I am confounded by indecision.

As it grows dark, Beverly comes to stand beside me. She holds my head against her and apologises and tells me she will stick by me and see it through. She runs her fingers through my hair, inside my collar against my neck. When we do speak it is to admit that we have both had Blaid on our minds on and off all day.

3 JULY

The dawn is bitingly a cold 3.4°C, the coldest we have experienced in the Kimberley. We are walking before sunrise, our goal to reach King George Falls. No one we have spoken to has ever walked to the falls so we do not know how far we have to go. Our guesstimate is between 10 and 14 kilometres one way. The cold is invigorating and we walk strongly for the first few hours, keeping to the higher banks between the trees where the ground is harder than the riverbed sands, and the going more direct.

We come into boulder country, where for the first kilometre our pace slows as we climb up, down and through the rocks. After that it is still rocky, but flatter, and cut by crevices we can walk and do not have to clamber much to get through. We are walking towards yet another fire, but the day is still cool, the wind not yet blowing, when we hit its front and we are able to get through without fuss or fear. Gradually, the river emerges onto the solid rock of the plateau and, following its course, where the wet-season floods have scoured the stone to a smooth patina, our pace increases.

In the mid-morning, we come upon the edge, the land vanishing into a gorge 80 to 90 metres deep and several hundred metres across. On the red cliffs, giant stacked stones lean out over the sea below. It takes us almost an hour to navigate the tangle of vines and tumbled stones of the cliff top to reach a point where we can look back to see both branches of the King George Falls plummet into the ocean. I take a self-portrait of us with fists raised. Although far from any record for height, depth, width or volume, this is a remote place that few others have reached on foot.

We do not linger long, for the fire we passed earlier is burning towards our vehicle. Even though it seems unlikely to reach it, it would be foolish to take the chance. Swimming briefly in one of the pools above the falls, we move on. We are tired, the sun hot, and our feet are hurting from the hard, uneven, boulder-hopping terrain. We stop finally for lunch in the early afternoon after we have caught up and passed the fire's front once more. The riverbed is wide and divided into several sandy channels and I come stealthily upon the main river, frightening a crocodile close below me into the clear stream. It is a Johnstone's or freshwater crocodile, and instead of swimming off, it settles on the bottom beside a rounded boulder, lying quite still, its slender snout pointing downstream. I strip off and wade in with a camera and get to within two metres as it lies motionless and hard to see. The water is cold, which I am hoping accounts for its apathy. Although not much more than 1.5 metres long, I can see the rows of needle-like teeth in its jaw. When I move again for a better angle, its nerve breaks and it swims off downstream.

When we finally return to the Land Rover, we have been walking for nine and a half hours and we lie with the cold water of the river soothing our feet. In the stillness, a group of three brolga feeding upriver walk within 12 metres of us, the closest we have come to them. With their grey plumage, and band of red around the head and eyes blending between the shadows and late afternoon sun as they stalk through the trees, they look like a posse of masked Mexican bandits headed into the shadow country.

4 JULY

Today is Independence Day for my American friends. As I drive back on the slow push down the overgrown track, I wonder about independence. I think the days of personal independence – in terms of self-reliance – are, strictly speaking, gone. Even out here, far from the presence of other people and the rigour of law and conformity, I am dependent. Dependent on the contribution of others towards my independence. I am dependent on the fuel I carry to bring me out. I am dependent on the last of our food stores to see us through the next few days or week until we can resupply. We are now 1 300 kilometres and three weeks distant from our last stop for food. We can last another two weeks, perhaps, at a real push, relying on being able to catch fish.

The San Bushmen of the Kalahari and the Nenets reindeer herders of Arctic Russia are perhaps the last few souls left on earth not dependent on others for their survival. I had expected the same of the Aboriginal peoples of Australia, but so far those I have met are so distant from any true sense of independence that it seems as if their whole identity has become unhinged. From the outside, it is hard to watch, for it seems fraught with tragedy and pain.

Reaching the banks of the Drysdale River at dusk, we make a fire in the sand, for the night is already cold. There is light from the moon and I sit for a time with my back to the fire looking onto the silvered landscape, listening to the dark. I hear no bird or animal sound except for the bellowing of the cattle. Their din has been a constant presence. I had hoped that we might have left the cattle behind, but they are as present in the wilds as they are anywhere. I realise, finally, the full impact of their running wild. They have colonised the whole country. I have been told that there are places they have not penetrated because the terrain is too hostile. Everywhere I have been, even in the tumbled boulder country around the King George River, or on the tops of the rough mesas we have climbed, there have been hoofprints in the sand, their paths through the grass, and their pats of dung.

The cattle change the country, change its very nature. Even though they are feral and have existed wild for a hundred years or more, they are domestic stock, the mark of humankind upon the land. They have in their time also changed the vegetation here, putting the land under such pressure that many of the original species have had to move or perish. I wonder how it was before the cattle came, how it might be if they, like the donkeys, were removed and the land became pristine wilds again?

5–6 JULY

Kalumburu is a tiny, square-grid town, five streets across, with the Catholic mission dominating the entrance. Everything, bar the single store, is closed. Kalumburu is a dry town and it is Sunday quiet. Outside the bright blue community centre, several young people are sitting on the ground against the wall playing with phones.

The store is surprisingly well supplied but there is almost nothing fresh. The manageress tells us that the barge will arrive tomorrow from Darwin with their restock of fresh food. I am curious as to the meaning of the signs posted all over the shop: 'No humbugging in the store at any time.' She explains that this is a term for begging. When you know someone has money, particularly family, it is the custom to follow them into the store and then place what you want in their shopping basket, expecting them to pay. In extended families, refusal can lead to a riot.

We visit the mission, where all the doors are closed but the church is open. It is a small, corrugated-iron building, with wooden pews and Aboriginal depictions of biblical events on the walls. There are two stained-glass windows and a life-size carving of Christ in a heavy, dark wood.

At McGowan Island campsite, we stop in to ask about the conditions of the roads to the coast.

'Got a fellah here wants to know what the road's like to Lull Bay.'

Out of the dark interior a heavy-browed man appears in a Metallica T-shirt and sunglasses. It looks like his nose has been broken a couple of times and he is nursing a hangover. He asks me why I want to know, and when I say we want to drive there I am taken aback by his reply.

'You can't do that. We don't want every Jack tourist driving all over the country. Besides, you gotta get permission.'

'I have all the permits.'

'What permits?'

'From the Department of Indigenous Affairs, plus the one you get in town to come out here.'

'I said permission, not permits.'

'Oh, you mean Matthew Waina. They told me in town that he wasn't around.'

'Well, they lied. People got enemies, so they'll lie to you. Matthew's around – you ask him. I'm not letting any old anybody go drive around out there.'

7 JULY

The two administrators we seek are away but the ranger for the district they represent is in. Jason Adams is a forthright man who listens to our story and is happy to have us explore his people's traditional lands. He speaks to me at length about the road, the country and the sights. When I tell him that I don't drive at night, he is pleased, because at one place there is a dreaming stone, and if one passes it at night something bad will happen. A dreaming stone, he tells me, is a flat rock with another rock like an egg placed on top of it. He gives me directions to Matthew Waina's home.

The house is shut, the unkempt yard littered with discarded cans. An old fire pit is being slowly overtaken by the tall grass. A dog sleeping on a sagging sofa on the veranda opens one eye at my knocking and goes back to sleep. No one answers my call. Walking away, a man approaches me on the street, a sleeping boy cradled to his shoulder. He is Matthew's son-in-law and tells me Matthew could be back anytime. He tells me that I might bump into him on the road driving his green Land Cruiser. Just as I am leaving, Matthew arrives in a Land Cruiser with no windscreen or windows, the body raised high above the chassis and a solid steel frame welded high around the back. It is white with a patch of pale green painted on the one side. Matthew is a genial man whose broad, round face, framed by extravagant sideburns, one feels is inclined to a smile or laughter. He lifts his sunglasses away from his eyes as we talk on his veranda. He gives us permission to drive out into his people's country and tells me of a few places to visit. I draw a complicated map in the dirt and he laughs a loud belly laugh and gives me the simplest directions. We drive out into the evening and are soon in lonely country. Crossing a few creeks where water pools in wide billabongs, we camp at no particular place out in the open, as darkness falls quickly between the trees.

8 JULY

Flat Rock is a picturesque meeting of the sea and stone. The water, as still and flat as the rock, is a silvered blue, through which single mangrove trees and a Zen garden of loose stones stand locked to their mirrored reflections. Fish move out in the sea, their fins breaking the surface. Occasionally, a bright splash sends a shower of tiny baitfish leaping across the surface. Of the original site of the Kalumburu Mission, only a concrete slab and some tall poles remain. A stone oven stands incongruous in the trees, the opening a dark maw and the door long rusted away.

At a tiny creek crossing where the ground looks muddy, I get out to check it and frighten some finches into the trees. On closer inspection, we discover them to be the rare Gouldian finch, and we back the Land Rover into the shade to sit between the trees on the creek edge for an hour to wait for their return. A myriad small birds come to the water: doves, finches and honeyeaters. And, twice, two small groups of Gouldian finches come down to drink, descending slowly through the branches, watching our strange presence. A lone kingfisher holds station from a high tree.

It is a long sandy road winding through a tight press of trees that finally delivers us to the coast of the Drysdale River delta. We stop at sandstone out-crops near the beach, where we find old Aboriginal art sites on low overhangs. Pictures of men and a dragonfly in red ochre on the white stone.

The coast is a scalloped shore of short beaches between sandstone outcrops and mangrove trees. We make camp beneath the shade of a cluster of long-leafed paperbarks, where the mouth of a tidal creek interrupts the shore. The fresh tracks of a large crocodile come all the way up the beach. Its foot is almost as big as Beverly's. It is nowhere to be seen but one can feel it watching and we walk the sand well back from the water's edge.

In the evening, the mouth of the estuary erupts with feeding fish, the black water spitting showers of tiny silver flashes as the predators hunt them from beneath. There are no recent tracks to this place and the crocodile's and ours are the only footprints on the beach. The night is without the stain of emitted light on any horizon and the only sound is the rush of small waves onto the sand.

9 JULY

The tide is high in the dark-purple of dawn, but the crocodile is gone and I am alone on the shore with a single brahminy kite, watching from the rocks at the end of the curve of the beach. I have been writing for a time when I hear boats approaching. I stand among the trees and wait. They come right to the mouth of the tidal creek and pull up onto the beach 30 metres from me. A bearded man jumps out into the knee-deep water and wades ashore. He is wearing a deep-brimmed cap with a neck flap that comes right the way around his ears. Sunglasses cover his eyes and he is wearing a long-sleeved camouflage jacket, long trousers and shoes. I walk out onto the beach. He shows no surprise at my intrusion at so remote a place and takes his time walking over, looking at the sign in the beach sand as he comes.

He introduces himself as Don. I have been told of Don MacLeod, who lives out here alone on the edge of the Drysdale River, where the tidal reach meets the fresh water. He has the slightly awkward manner of a recluse who is unaccustomed to conversation. He is taking some relatives fishing and talks to me over his shoulder as he walks the shallows throwing a cast net for baitfish. I tell him of the crocodile and he says, 'Yes. He's a black one, but he's alright. He survived the croc hunters.'

'Croc hunters?' I ask, knowing that estuarine crocodiles have been a protected species since the 1970s.

'Poachers wiped out most of the big ones around here. I told the Department, but they just ignored me. I think that they don't mind a few less crocs since that American got taken. Not good for business.'

The American he refers to was twenty-four-year-old model Ginger Faye Meadows, who was working as the assistant chef on a Kimberley coastal cruise boat. She had decided to visit Australia after seeing the film *Crocodile Dundee*. Although having been warned that the water was crocodile infested, she and her friend decided to have a swim while the rest of the crew went for a refreshing clamber up the rocks under the thin veil of fresh water of the Kings Cascade. When a crocodile approached the group, drifting closer on the sea, Ginger panicked. Instead of climbing higher up the rocks she dived in and tried to swim back to the boat, anchored a short way offshore. She never got there.

Don warms as I query him about the wilderness here, about the animals and the birds. He is a nature aficionado and knows the species that occur here like few others I have met. Don throws his cast net in a perfect arc every time, even though I am talking to him all the while. He collects the baitfish in a bucket to use them later out on the deep-water reefs. I am interested in his life in the wilds, far from any human infrastructure or contact. He tells me he is a little crazy, but it seems to me that he is one of the very few who is sane. He invites us to visit him at his camp once his relatives have gone. We exchange addresses sitting on a huge driftwood stump on the beach. He calls me Pete when he says goodbye, wades out to his boat, and, without a wave, points it out to sea.

10 JULY

I am excited by the prospect of my day. I have nothing planned, but we are on our own in the world in one of its last untouched corners, where it is as it was made, alone in wild space. There is the smell of last night's campfire, the faint freshness of the dew, the sound the wavelets of the tide breathing against the rocks, the glow of the dawn showing through the trees towards the east, and a single, small bird singing. It strikes me how rare this is, how precious.

Wild space lasts. Forever is not something we consider much, but protected wild space can be there forever and it will enrich the lives of millions to come. The less wild space there is, the more critical its existence, and it becomes more valuable in its natural state than anything that we would take from it. If this concept is difficult to grasp, think of Yosemite National Park, or the Kruger National Park: if a government proposed mining in either of these parks, there would be a national outcry. They are wild lands that belong to the people, tight in their hearts, and their sanctity is inviolate. In the first year of its proclamation, only thirteen adventurous travellers made the arduous journey to visit the Kruger National Park. Today nearly 1.8 million enter its gates each year. This shift took less than one hundred years. For it is a rare gift to know the earth as it was made. I am attached to it by some instinct, like an anchor in my soul.

The Drysdale Delta bay remains wild and quite pristine both above and below the sea. The flow of life on the outgoing tide is like a busy street. I fly-fish from a copse of rock where the water of the tidal creek passes close and deep. I see sharks, queenfish, threadfin and cod and the edges teem with shoals of small fish. Beverly photographs huge groups of tiny golden ghost crabs that march slowly like an army of tiny round aliens across the sand flats exposed at low tide. I am successful with my fly rod, but the barramundi remain hidden. A pair of white-breasted sea eagles eye the estuary cod I have kept for dinner, but I am wary of their thieving tactics and keep the fish close. An osprey hunts for itself in the stiff breeze, slowly coursing the contours of the shore. The tide falls all day until the morning's shore stands six metres clear of the sea. At dusk an almost full moon brightens the world with its colourless light, and a dingo hails its rising with its wailing call. It is standing 30 metres from our fire, head thrown upward, pale against the even paler sand.

13 JULY

I decide to call Don MacLeod on my satellite phone. Don is happily agreeable; he is coming down from his camp anyway to collect a group of Aboriginal rangers who are on a training and recce exercise. He will meet us on the beach at 11:30 am.

The day is muggy-hot despite a wind from the northeast. The sky is billowing layers of clouds. Don arrives on time, his yellow skiff appearing out of the silver haze long after we can hear the outboard's buzz carried ahead by the wind. The wind-driven chop slaps the boat against the sand. We take a long route back up to the mouth of the delta, following a passage through meandering mangrove creeks of calm water, avoiding the open sea. The tide is at its peak and there is little to see, the water driven far back through the dense forest of the trees. Milkfish, cruising the surface, explode away from our bow. They are come to feed on a plankton bloom, and across the 150-metre width of the estuarine canyon they send showers of spray arcing into the air as they make short rushes of speed, their broad tails and fins showing like knife blades through the slate-grey sea.

Don's camp is a ramshackle cluster of tin buildings and shade-cloth-draped structures of bush-cut mangrove poles. The ground is rock, the main area roughly paved with stones beneath an open-sided tarpaulin roof that looks out through an opening in the low cliff onto the sea, close below. When we climb from the boat onto the rock, a huge grouper drifts beneath the keel and edges gently onto the flat rock of the shallows until its dorsal fin and back stand 10 centimetres clear of the water.

'Coddly,' announces Donny Imberlong, Don MacLeod's companion in camp, as he strokes the nearly two-metre, 150-kilogram fish along its dark flank. 'He wants a fish.'

Donny came to live with Don when he was eight years old and has never left. Don treats him like a son and Donny calls him 'Gramps'. Donny is on vacation from university in Perth, where he is studying for an environmental degree. Donny tells me that the crocodiles also come for fish, edging up the rocks under cover of darkness.

Our quarters are simple, a mattress on a sagging wire-and-steel frame, on a concrete floor beneath a corrugated-iron roof, the surrounds hung with shade netting. The camp is a jumbled disarray, the discarded lying between the used and the new. Everything seems to be put down where it was last used, at the ready to be picked up again. A crocodile skull competes for space on the sacking-covered table with boxes of fishing lures, three rods, a pair of running shoes, solar-powered lights and two bird identification books. Lime, mango and coconut trees grow in shady profusion outside the kitchen enclosure, where its wastewater feeds a tiny vegetable garden. The fence is broken down, the work of nabarlek wallabies wanting to share in the food. Neither Don nor Donny remark on the breach with any concern. Don is more affectionate than angry at their intrusion.

Donny is the first Aboriginal person we have met who is openly vital in the bush, in tune with its rhythm. He sees the ground when we walk out the back to look for birds along the billabongs, he doesn't just look. He mimics the birds exactly, and although young, he has his stories for each place that reveal the country like a book. I go fishing with him after a lunch of locally hunted beef, which we eat on low stools beneath the lime tree. He catches a fish on his first cast, dispatching it cleanly with a stab of his broad-bladed knife in the top of the head.

In camp, until late in the evening, I listen to Don's deep interest in the lot of Aboriginal people. He is the first person I have encountered who argues from their point of view. He is well versed in their affairs and situation and lives here at the benefit of their grant to him of a lifetime right to live on their land.

'I'm so tired,' he says, 'of all these people who keep thinking that they can come in and make changes that in one year will fix everything. It won't. The Aboriginal situation is something that will turn at its own time and its resolution will evolve. How can we decide what it should be? It will be what it must be, no matter how good or how hard.'

A crocodile hovers stealthily on the edge of the water, its twin red eyes bright in the dark of the sea. As we talk, a small parade of nabarlek comes to the fringe of the light, nibbling the bits of Weet-Bix that Donny has put out for them. In the night, it starts to rain, soft at first but growing stronger until it thunders on the roof.

14 JULY

Rain. It drifts across our iron roof in varying intensity and I wriggle deeper under the single blanket I dug out in the middle of the night. The world is an opaque grey and dripping wet. It puddles in the rock and gurgles as it overflows down toward the sea. I raise myself reluctantly and hurry through the drizzle to make coffee.

Don and Beverly make a hearty breakfast of banana fritters, but afterwards there is little to do but stare out at the rain. Towards midday the rain softens to a light drizzle and I join Donny in the skiff, where he tries to jig one of the milk-fish that shoal on the surface in large schools. We have drifted about 200 metres when, grunting with the exertion, he is suddenly successful. The milkfish has a reputation as a fierce fighter, and this one takes off at such a blistering burst of speed that the line cutting through the water raises a wake 20 centimetres high that shatters into spray. It is a dogged fight. The fish is 1.5 metres long and weighs about 18 kilograms. It is built like a torpedo, with a giant scythe-like tail. The lateral fins have special indentations in the side of the body to make the fish more streamlined. The milkfish's large iridescent green-and-blue eye sits behind a clear casing of hard, nail-like substance so that nothing is exposed or can be damaged at speed. Because the milkfish is considered a delicacy, Donny will take the fish with him when he goes to town as a gift to relatives and friends.

The rain finally stops and there is just the dripping of the water gathered in the canvas of the roof and falling from the trees. The day remains grey beneath the close sky. We go out again and catch a queenfish, below where the river enters the sea, for our dinner. Once I have removed the fillets, Donny feeds the skeleton of the fish to Coddly, the giant grouper that hovers on the edge of their boat landing. It approaches slowly, opening its maw of a mouth just slightly so that I can see its definition in the white rows of small sharp teeth that line its jaw. When the fish is 20 centimetres or less away, it lunges suddenly with a snake-quick strike, opening its huge mouth and sucking the water and the fish with the unavoidable vacuum that it creates. I jump involuntarily. Its open mouth is almost big enough to swallow my head. I have kept the head of the milkfish that Donny caught in the hope of baiting a crocodile, and in the last light of the drab afternoon I rig a remote camera on the rocks beside the water with the fish head as bait.

As soon as it is too dark to see, we pick up a crocodile's eyes in the torch-light, approaching from out in the sea. It comes slowly until it is right against the rocks, and I wait with anticipation for it to emerge. My waiting drags on, becomes half an hour, and nothing happens. We can no longer see the croco-dile's eyes, but approaching the camera to see whether it has perhaps taken the bait unseen, we find it lying in the water tight against the rocks.

After a dinner of rissoles of beef and stir-fried vegetables from Don's small garden, we retire to bed and hope that the remote system will work. When I return after an hour the head is gone but the crocodile is still hovering at the edge. When I check the images, I find that it is a water rat that has taken the milkfish head and dragged it away. I search the surrounding rocks and find the head in a crevice. I retrieve it and replace it in front of the camera. Three more times I wake to the same situation and replace the head, but on the fourth occasion the crocodile is gone and when I check the images I find that this time the crocodile got to the bait first. The images are stirring: the crocodile has its snout almost on the camera.

15 JULY

Back in Kalumburu, Bevan Stott, the administrator who needs to give us permission to take the track west through Aboriginal Land, has still not returned. The office is closed, but I find Lilian at the community centre, where everyone is crowded around for what must be payment day. She tells me that Jason Adams is also gone for the day, hunting turtles and dugongs from a boat with a spear.

Beverly takes the Land Rover to find what fresh supplies remain in the store, while I go to the mission to collect our emails. The first one is from my father. Our dog has died. I sit stunned. We have expected this, but when the moment comes it is so final that the loss numbs my mind. As I sit staring unseeing into the mission grounds, a snake slithers onto the veranda, crosses it, and glides beneath a stone in the small garden that the veranda surrounds on three sides. A young girl raises the alarm, and three men and a woman drag the rock away and pursue the snake with a hoe, a broom and a stick. Pinned beneath the hoe, the broom holding it down towards the head, the snake raises the last 15 centimetres of its body into the air, snapping terrified with its wide-spread jaws at the open air. The hoe comes down against the concrete, severing its spine behind the head. It is too much death and the tears start from my eyes. I walk blindly into the garden and squeeze with my fingers at the outside of my eyes. Love is the greatest pain I have ever had to deal with.

18 JULY

Mount Elizabeth is not a through road, but marks the beginning of the Munja Track. It is pleasantly quiet compared to the Gibb River and Kalumburu roads. The road to the station passes through plantation-like stretches of forest, where, beneath the almost eight metre canopy, the tree limbs are bare and the view is long between the trunks. I imagine that it must feel good to come home through this place.

Pat Lacy is soft-spoken and gentle in manner. In her deep-set eyes, behind thick glasses, there seems to lurk something with which she has made her peace but whose sadness has not gone away. She listens to our story over banana bread and tea. Pat gives us a hand-drawn map and we fill up with water and depart for the Munja Track. We do not get far, however, for just outside the homestead grounds a number of agile wallabies have come into the open to feed. As we wander between the trees, more and more come from their resting places in the shade, until there are wallabies in every direction. I wander slowly through their midst, squatting often on my knees or haunches to see them at their level through the sunlit grass. One has a joey in her pouch and it peers at me with its ears cocked forward.

19 JULY

The sun is slow to reach the ground, where we have camped in tall forest, and in the cool 12°c morning the wallabies are still about. The country we drive through has long, unburnt stretches, and we have a real sense of the land, with the grass a waist-high meadow beneath the trees. All of the Kimberley is forested and the constant presence of trees inclines one to see it as all the same. Today, the presence of wildlife in an unburnt landscape prompts me to look more closely. On our morning's drive, I count twelve distinct forest types. Some are stands of towering pale-trunked karri trees, all 12 metres or more in height; others are mixed scrub and trees on rocky slopes. My personal favourite is the few open grassland areas where thick cabbage gums, with their peach-copper-coloured bark, stand singly against the carpet-like green of the grassland.

Driving until it is almost dark, we make our camp on a small place of flat ground on a hill of sandstone outcrops. The stars fill the sky. As I sit beside a small campfire the kookaburras start calling. Theirs is a strange adaptation: during the day, when they are active, they are silent, and only start calling once they are roosting at night. I have not seen a single kookaburra all day but now they call from every direction. As soon as one begins its screeching, clucking song, another is prompted to join in. Soon, the whole night resounds with their rough, unmusical calls.

21 JULY

At Bachsten Wilderness Camp, Janet Morns and Peter Kelly run a tidy, friendly operation, offering neat, airy rooms and camping with hot showers. We spend the late afternoon hunting the rocky hills behind their small cottage for the rare monjon, a tiny kangaroo that is smaller than a rabbit. Twice we see them, but it is only a fleeting glimpse as they dash through the crevices. In the evening, we work on photographing the northern quolls, bandicoots and golden-backed tree-rats that Peter and Janet tell us frequent the camp at night.

The quoll and the tree-rat, which is the size of a small guinea pig, with an extraordinarily long tail, are obliging. We work until 2 am in the darkness, the leaf litter rustling in the surrounding forest as the night comes alive with small creatures. From time to time there is a great bustling in the undergrowth and a loud chirring, which seems to be the quoll's growl or defiance call. The creatures become so accepting of the camera set in the middle of one of their pathways that they climb up to perch on top of the flash. One curious individual peers into the camera and experimentally licks the lens.

22 JULY

I am busy with the ritual of morning coffee when Beverly returns from an early-morning hunt through the kopjes for the monjon. She is flushed with success; she has seen and photographed this rare marsupial. It has taken patience, a canny intuitive idea of the creature's behaviour and good bush sense for her to succeed. My excitement fuels her happiness, and I get her to tell me twice in detail how she did it. I make a fire under the 44-gallon water drum and we have a hot shower to celebrate. I ask Janet if she can possibly ask the owners, who are coming into camp in a few days, if they could bring a couple of bottles of Chardonnay. It is Beverly's choice of evening drink and we ran out weeks ago. I know it would make a wonderful present.

We are unprepared for Wren Gorge, the small paradise we stumble into just off the track. The gorge is little more than a U-shaped amphitheatre of short cliffs, but in its cup a waterfall drops two steps into a pool, 80 metres across, surrounded by towering cajeput paperbarks and lined on one side with a uniform rank of spiral pandanus palms. The pool is rock-bottomed, deep and crystal clear. In the centre of the pool, two gigantic boulders stand squat and solid above their reflection. One could not invent a more picturesque place. One side of the pool is rock, and behind, where a fig grows wide, its roots wrapped like pythons around the tumble of huge boulders, there is a tall wall of stone, filled with the depictions of life that the ancient peoples made in red, yellow and a purple-brown. Many of the images are literal, men or animals or reptiles, but one seems a map of sorts. It follows a curving twisted line, off which things have been drawn to the side: a pool or a forest. There seems a story attached, for there is the depiction of a man's foot in the same artist's hand off to the side at one of the ends.

Beneath it we strip off and dive into the water. Even though I know that we are well above where any saltwater crocodile has been recorded, I cannot quite relax as I swim out into the huge natural pool. I find myself moving back towards where the waterfall cascades into the shadows, where the water is shallower.

Shortly after the gorge we enter one of the worst sections of the track, and spend forty minutes building ourselves a way through. The sun is hot, the boulders heavy and we are drenched in sweat. Even after our repairs we crawl through in the lowest gear possible, the Land Rover pitching and shuddering as wheels lift free and loose stone slips beneath their grip. Beyond, however, is a lovely country of forested savannas rumpled by tall hills. Several times we are treated to rare glimpses of a view over the country. To the north there is not a single road, track or path between us and the sea more than a hundred kilometres distant. It is the same to the open ocean to the west, and far to the south the Gibb River Road may as well be in another world. Standing on a rocky promontory up one of the hills we have climbed, I feel as if I am on the artery to the heart of our quest, for I am surrounded by wild land.

Today has been exceptional in another regard. It is the first day in our entire journey through the Kimberley that we have not seen cattle, cattle tracks, or dung pats in the sweet-grass valleys. I have read that some years ago thousands of cattle in the Mitchell River National Park were shot in an attempt to keep the country in as natural a state as possible. Perhaps the cull reached to here, for we are close to the southern boundary of the Prince Regent Nature Reserve. If one concentrated effort at eradication can have such broad, long-term results, it gives me hope for the other wild lands of this northwestern corner of Australia, where feral cattle remain a curse and blemish.

We are not beyond the reach of fire here, however, and although we have passed through rare valleys where the grass runs tall from hilltop to hilltop, by far the majority of the country is burnt. The mark of fire in this remote country is irrefutable proof that fires can run for months and far beyond any human-perceived boundaries or imagined limits of time and space. Anne Koeyers had been at pains to explain that pastoral station owners have a sound motivation for setting fires early in the dry season, when a neat mosaic of controlled burns would protect property, encourage pasture, and ensure a cool, minimal burn to the vegetation. But this idea is challenged by the fact that fire has reached here, hundreds of kilometres from any station. Fire runs through this country unthwarted, and it burns all that stands before it. It saddens and maddens me that its regime is perpetuated by people who should know better. Who should be asking themselves: Why?

23 JULY

In the early dawn we find a pair of brolga probing the banks of a small creek. I sit in the long shadow of the tall creek-side trees and they walk past on the far bank, calling low to each other with a guttural croak. The country is fresh in the cool morning, the air trapped still between the hills and the forest. The grass is yellow, brown or bright-green in the black stubble of the fires and there is movement of birds in the high canopy of leaves. At a creek crossing of a broad reach of flat rock there is a squealing din between the dense riverine trees. Looking closer, we find a roost of flying foxes, the individuals squabbling over the prime shade sites. Those in the warming sun are fanning themselves with their tarpaulin wings and scrambling deeper into the canopy in search of shade. Beneath them in the shallow, clear water, schools of fish rush to the plop of their droppings. We wade into the pool over slippery-smooth rocks and photograph their teeth-bared squabbles and quick flights to more shady roosts. We opt to spend the day for the chance to photograph their departure on their evening hunting foray.

Along the banks of the creek, white dragon trees are in bloom. Their pendulous white flowers are shaped specifically to feed the flying foxes nectar in a cup, while the stamens sprinkle pollen on the bats' bristly chins or foreheads. Walking through the close riverine growth to bathe in a pool downstream, we disturb three kookaburras and a pale-white tree frog, which leaps into the stream and swims fast across the bottom to bury itself beneath the leaf litter.

In the afternoon, as I wander upstream with a camera to look in on the flying foxes, I come across a large Mertens' water monitor warming itself on the flat rock beside a shallow pool. Like the birds and the flying foxes, this large lizard seems to have no fear of humans and it lies looking at me as I approach slowly, moving forward in a sitting or lying position, until I am only a metre away.

The flying foxes are still hanging upside down with their wings wrapped around their bodies like shiny black fruit. It is only well after the sun has set that they begin to stir and chatter. While I stand knee-deep in the water below their roost, the mosquitoes feast on my exposed legs. The rocks are covered with fine algae, making them treacherously slippery. I must kick with my bare feet against the rock to remove the algae before I can find purchase, and in the darkness this is sure to attract crocodiles. I scan the water with a flashlight and see none, but am not sure if this is good news or not. These thoughts do little to ease my sense of vulnerability and it is only when the flying foxes start moving that I am distracted and forget my precarious position.

The flying fox in its roosting position is perhaps the size of a kitten, but in flight, with its wings spread a metre or more wide, it takes on a proportion beyond its small size and becomes the mythical creature of our superstitions. Their wings whoosh in the dark, and as the colony comes to life the night is filled with their high-pitched chattering, their kite-like silhouettes moving dark across the stars. It is only after we are done that Beverly tells me that from her position on the opposite bank she could see two crocodiles come close under the cover of the undergrowth to where I stood.

'But they were small,' she says, showing a distance between her thumb and index finger.

24 JULY

Before the first indication of light to the east, the flying foxes have already returned to their roost. In the first sun they seem cold and shivering, and often two cling together, wrapped in the folds of one of their wings. They tolerate us close below them without taking fright, watching us with their bulging Chihuahua-like eyes. As they warm in the sun, they open their wings and preen their wing skin with their long, pointed tongues and their bodies with their well-developed claws. The males have outsized penises that reach from below their belly to their chest. They clean them with their tongues, and the younger, smaller neighbours of the larger mature bats stretch tentatively forward to sniff at their black-sheathed, red-tipped sex. Often, they are noisily rebuked.

The old Munja Supply Post is a mess of collapsed iron sheets and a jumble of metal-framed bunk beds beside a pretty billabong with purple water lilies and white egrets stalking its banks. A stand of boab trees bear the scorch marks of fire, and between there and the Walcott Inlet all the land is drifting with the black ash of recent fire.

Wilderness is a feeling. Remote I like, but wilderness is an elation that rises in my chest. On the high bank of the Walcott Inlet, which reaches far inland, 75 kilometres from Collier Bay, there is not another soul. We are alone. A crocodile launches itself discreetly from the muddy bank upstream, and instead of moving away, as I anticipate, it stalks us. It holds against the flow of the outgoing tide and drifts slowly down towards us. We are high, beyond its reach, and I let it come without moving away. The skin of its head is a lime-cordial green. The eye: yellow and unblinking. It brings its nose right against the bank, directly below us. Its stare is unwavering, its body concealed beneath the mud-stained water, only the eyes and nostrils showing. Finally, it raises its entire body to the surface so that it stands proud of the water and shows its tail. It is more than 3.5 metres long, the pale skin blotched with black. Upstream, we count three more. As the quickly falling tide recedes from the banks, they slide into the current that runs like a river and drift with it fast to the west, towards the distant sea.

A brahminy kite stoops from a tide-side perch, on the steep close bank, to a broad, shallow sandbar starting to show in the middle of the kilometre-wide inlet. It is hunting the schools of baitfish, making a skittering dash across the surface of the water. There are pelicans, spoonbills and egrets feeding in the water, and terns that drift the updraft of the steep bank, stooping suddenly to lift some small prize from the surface. Rainbow lorikeets feed in the red blossoms of the batwing coral tree and little friarbirds sip the nectar from the flowers and chase each other through the leafless branches.

Boab trees are plentiful between the stony steps of the distant low hills of the Charnley River Conservation Park on the opposite bank, and mangroves line the shore at the tide's limit: straight-trunked, tall and dense, banks of green grass at their base. A march of puffy clouds covers the sky, but they dissipate in the dusk, leaving the stars to cast long reflections on the wind-riffled sea. The high, thin call of a dingo carries plaintively across from the far shore. It receives no answer. The night is as quiet as the absolute silence of the stars.

25 JULY

Walcott Inlet becomes a vast, exposed sandbar as the tide drops to its final ebb. The change between the outgoing and incoming tide is sudden and dramatic.

The returning water rises so fast that it churns to a froth of whitewater rapids over the shallow ridges of the sandbar. I measure 100 metres on the shore and throw a branch into the tidal current. It takes twenty-five seconds to cover the 100 metres, which by my calculation means that the tide is running at 14.4 kilometres per hour.

The crocodiles return, drifting on the freight-train current and watching the passing bank for any interest. Over the past few days the flies, too, have made a comeback with the warming weather. Today, they drive me to distraction with their persistence, sitting in the corners of my eyes, on my nostrils and mouth.

On the start of our drive back along the Munja Track, we find a track to the west that leads to an airstrip. But there is no office or staging station, only a broken windsock pole and a sack of wine hanging in a plastic bag on a tree. Beyond the airstrip is a wide wetland, and following the track along its edge we are treated to an unexpected surprise. Hundreds of magpie geese and flocks of brolga take to the air. The wetland is kilometres long, and although its edges are drying, there remains an almost unbroken shallow pan 80 or more metres wide down its entire length. Flocks of birds crowd the open spaces between the rushes. There are white-headed stilts, small flocks of Pacific black duck, pied herons and astounding groups of purple swamphens, but the brolga outnumber all of these, and there are magpie geese in their thousands. Everywhere along the green swathe of the wetland, birds rise into the air above the fringe of the rushes in wheeling flocks. Small terns fly languidly over the shallows and a swamp harrier twists suddenly in midflight to turn on something unseen.

As the day wanes, we walk the edge of the pan, and at the far southern end we find a flock of brolga more than a hundred strong. We photograph them on an open, reedless stretch, where the mud has dried and a short, bright-green plant makes a carpet of green. Heading back to the vehicle I notice seven or eight brown falcons coursing high against the darkening pale-orange sky. They are feeding on something I cannot see, snatching it in the air and eating it quickly on the wing from between their talons. I watch them with binoculars, but they are too high and the prey too small to make out. I choose one that is flying lower and follow its black silhouette against the dusk sky. It climbs a little as if after prey, and then I see it: small and flying heavily towards the night. The falcons are hunting moths.

The honking of the magpie geese and the bugled call of the brolga grow louder as the dark descends. By the time we have made our camp on a low hill overlooking the wetland, the whole valley is a cacophony of sound. It is as loud as a stadium full of people. I sit mesmerised on my camping stool: it is so much life. I hear dingoes calling, adding their thin voices to the din. Within half an hour there is a panicked rush of geese and brolga calling, and the mosquito-like buzz of thousands of wings taking flight and feet splashing up from the water as birds are charged by an unseen predator. They fly low over our heads and I can just make out their dark silhouettes against the stars. Their calling is as loud as bells ringing on a calm night.

29 JULY

As I emerge from the vehicle before sunrise, a butcherbird flies into a nearby tree and, cocking its head, watches me. It flies closer, to a small tree, and then closer still until it is two metres from me. It twists its head this way and that, trying to gauge what I am. I whistle softly, a few notes of the song I have heard them sing. It hops onto a low, vertical branch and repeats them. I whistle them back. It embellishes my poor repertoire with a trill of initial notes and ends with a rolling tremolo vibrato that I cannot hope to mimic. The butcherbird is the songbird of the Kimberley. This individual pauses long enough to hop and flit to the top of the tallest tree and, sitting on a bare branch and ruffling its feathers into a raised puff, sings its heart out.

Rebuilding and negotiating the jump-ups takes most of the morning. Our BFGoodrich tyres, already worn to only a few millimetres of tread, are looking like something the dog chewed. I am amazed they are still holding. On one tyre there is a long cut in the tread that reaches right down to the fibre and steel of the inner lining. On another, the outside edge has been sliced away by sharp rocks in five places. All day I have waited in dread for the hiss of escaping air as a tyre finally gives, but it never comes.

As the afternoon shadows lengthen, antilopine wallaroos rouse themselves and we start to encounter groups along the road. The antilopine wallaroo, a relation of the kangaroo, is a mule-faced creature with a stout tail: the male is an orange-red, the smaller female pale-grey. They appear to prefer open spaces in rocky country. At one particularly large congregation of twenty or so individuals, we turn off the road into the forest. We follow them slowly into the trees, letting them grow accustomed to our presence and photographing them against the light, the bush around them glowing green. I turn back to the road as it grows dark, headed generally west in the featureless run of trees. I am hoping that we will make the road before it becomes too dark to negotiate the many logs, stumps and rocks, when the inevitable happens. There is a sharp, loud hissing and a spewing of dust as a tyre goes flat. A sharp stump has pierced the sidewall, cutting a wide gash. Even worse, it has not ruptured one of the heavily worn tyres on the rear of the vehicle but one of the new ones on the front. Once I have changed the tyre it is too dark to continue and we sleep between the trees. We will find the track in the morning.

30 JULY

At the homestead Pat introduces us to her husband, Peter Lacy, whose family has been at Mount Elizabeth Station for two generations. He is a mild-mannered man with eyes that seem a little bewildered that anyone should be interested in his life. The impression one gets of him and Pat as they sit together is of gentleness, but their history suggests they are made of sterner stuff. Peter made the Munja Track in the 1960s with a tractor and a crowbar: nearly 200 kilometres through virgin forest and rough-rock hills. To drive it today is rough enough; making it then must have required the grimmest determination.

I have noticed that many of the photographs of the horseback mustering of old on the station show Peter working with a crew of Aboriginal jackaroos. I ask Pat about the proposed changes at Munja, the new camp, and about the jackaroos in the photographs. She explains that in earlier times many of the local Aboriginal people lived and worked on the station. A wind of political change then swept the country and almost every single Aboriginal moved to town. The people whose country is the Munja Track area now live in Derby. They have visited the area once in the past two years. On that occasion they shot twelve bustards, known throughout the Kimberley as turkeys or bush turkeys. They told Pat that they managed to cook and eat three before the rest went bad.

Pat regrets the distance now between the station and the Aboriginal people whose country surrounds it. They used to be close with the older generation. Peter grew up with them. Today, they are a distant community, both physically and metaphorically, and growing more so. It is a universal complaint in the Kimberley and especially among the pastoral leaseholders: that relations with Aboriginal communities grow ever more strained. Both Pat and Peter shake their heads. It is not anger but sadness that lies in their silence.

In the evening Peter feeds the horses. They are no longer worked. Helicopters do the mustering, and his back is too painful to ride for pleasure. He keeps the horses now because they are there. He is leaning hard on his quad-bike as I talk to him, the pain from his recent spinal fusion quite apparent, but it cannot extinguish how he feels about the horses, the pleasure they bring to his eyes. Pat gives us four steaks wrapped in tin foil; 'To have with your tea,' she says, for she knows how low our supplies are. It is not the magnitude of the giving but the way that it is given that leaves us overwhelmed by the kindness of the Lacys.

2 AUGUST

The land is serene, a quilt of yellow sunlight and dark shadow beneath the broken cloud. Finches fly up from beside the road. I see cat tracks in the dust of the path. Big tracks for a cat, about the size of an Australian dollar coin. The Australian Wildlife Conservancy's research estimates that there are about five hundred feral cats on the 360 000 hectares of Mornington Station, each with a home range of three to six square kilometres. I have noticed dingo tracks almost every day, but very seldom have we seen cat tracks except when exploring river-beds. Again, the research done by the AWC shows that because dingoes have a marked preference for using roads to patrol their territories, and dingoes catch cats where they can, the cats tend to avoid the roads and use the creek beds.

The Kimberley is a land of trees. Human eradication of cats is ultimately limited by the terrain, and the vastness of the remote country renders all attempts futile. Like the cane toad, on which millions of dollars have been spent in unsuccessful eradication attempts, it seems that the feral cat is now a part of the Australian ecology. If this seems of little significance, then it is sobering to digest the research figures on how much feral cats consume in the wilds of the Kimberley. Taken not from observation, but by counting the stomach contents of feral cats in the Mornington Wilderness, the average feral cat kills and eats up to ten small animals a day. Most Australian mammals are nocturnal, and the smaller ones almost exclusively so; the cats walked out of the house into a trove of riches. Small cameras attached to large tomcats on Mornington show that the cats kill animals the size of quolls, and also reveal another trait: they kill because they can. Retrieved video footage shows the cats leaving an animal they had caught by chance, when not hungry. Research has shown that in just one day

Australia's feral and free-roaming domestic cats kill approximately 1.3 million birds, 1.8 million reptiles and 3.1 million mammals.

In the hot afternoon, we spend two hours with Laura Smelter discussing fire. We tell her we have issue with Kimberley fires on two counts. First, fires seem to be started out of habit, or ingrained custom, rather than for any plausible reason. Second, in a landscape with few effective boundaries, any fire purposefully begun is done with absolutely no idea of where it might end. This leads to the burning of the country to an extreme degree. EcoFire, Laura tells me, works on the premise that the Kimberley is fire-resistant country, brought about by thousands of years of the use of fire, to varying degrees, by the different Aboriginal communities. This evidence comes from verbal histories, and is also borne out by core samples, I was told separately by a geologist, who added that there was a period in the relatively recent geological history of Australia when there was no fire in the Kimberley. He was using the absence of fire as an argument to question the claimed period of occupation by Aboriginal people, but for me it held interest in that if what he said was true, then the Kimberley did not *have* to burn, as the assumption seems to be. EcoFire works on the hypothesis of fire as necessary to a healthy environment. I acknowledge that no fire in the Kimberley is a moot point, given occupation of the land by people, and with the fuel load of combustible vegetation that develops in the dry season fire is inevitable, but it would be interesting to know how this vast landscape of subtropical forest would evolve without burning.

10 AUGUST

Climbing to 2 900 metres, we look down on the country where fires are the outstanding feature. Smoke billows thicker than big industry, carried west by a stiff breeze. Chris, our pilot, tells me that it is clearer today than most. Often, he says, you can only see the ground directly below; the horizons obscured by smoke haze. We fly north, crossing land, then ocean, then land again, as the scalloped coast runs east then west again. The deep inlets and bays are separated by broad peninsulas, with small islands lying in random disarray in the shallow sea. The names of the places below sound like pirate country: Kings Sound, Collier Bay, The Funnel, Secure Bay, Raft Point, Prince Regent and Prince Frederick Harbour.

We land on the beach on Naturalist Island at the mouth of Hunter Creek.

Odyssey, our catamaran for the next ten days, is lying close offshore. When the helicopter clatters off the beach we are left with silence. I have looked towards this coast from far inland, prevented from getting to it by a terrain too rugged for roads or tracks. The land is creviced hills of almost pure stone and dense with forest. It is as wild and remote and ruggedly beautiful as the Kimberley can get. The full moon is rising behind cliffs that are fire-coal red in the last light. The sea is calm and the moon's reflection lies yellow on the mauve water.

12 AUGUST

We motor on the flood of the tide, as the crocodiles do, into the tidal valley of the Prince Regent River. The river gorge runs roughly west to east, carrying the sea an incredible 48 kilometres into the rocky hill country. We ride the tide's flood between sheer sides, into some of the wildest land we have seen. Not a single road. Not one hut. No abandoned endeavour. The shore, covered by the rise of the tide, is lined with mangroves. The amount of water rushing inland is staggering. The river is 740 metres across, and for 48 kilometres between dead low and peak high tide today it will rise over nine metres. The high tide will reach the final inland terminus of the Prince Regent Estuary three hours after the tide has begun to fall at the narrows of its entrance, in Saint George Basin to the west. Debris and crocodiles float on the tidal river.

13 AUGUST

We are on the sea early. The low tide is at dawn, and as it pushes up we plan to follow it in a small fishing skiff as it floods into the mangrove creeks. The barramundi too will follow its front, lying at the edges of its advance to ambush the mullet and other baitfish. At one promising creek, where a tangle of dead trees forms the kind of hiding place that barramundi favour, I am facing the stern watching the drift of my bait when Pete Warihana, our guide, shouts, 'Watch out!' Spinning around, I see a three-metre crocodile coming fast at me from the bank. It slows as I face it, raising my arms and rod, but it is only when Pete does the same and hisses loudly that it stops and dives away. I have just relaxed from the incident, and Pete is facing the opposite bank, throwing a lure between the snags, when the crocodile surfaces again out of the murk two metres off the stern, its eyes locked on Pete's back. At my cry, he whirls around and whips his lure out of the water and slams it down on the crocodile's head. It dives in an angry swirl.

14 AUGUST

On the bridge, only the red glow of the instrument panel shows in the darkness as we motor into the first crease of dawn. There are skull-and-crossbones markings on the navigation charts to indicate where unsurveyed reefs lie close to the surface. We glide through a maze of channels and islands, past tidal creeks, rocky shores and occasional beaches. Our course is never straight; even in the sea there are shoals and whirlpool currents caused by deep holes and the extreme tides. There is not a single light anywhere. It is a remote wild sea and not one to navigate lightly in a boat. I find I have sympathy for the owner of a yacht we find abandoned at a mooring in Kuri Bay.

Kuri Bay is a place of ghosts. On the shore, a small settlement stands silent, abandoned. Frangipani trees bloom between the ordered, sprucely painted bungalows, and the steep concrete road is still whole, not yet overtaken by the weeds. This was formerly the site of a Paspaley pearl farm. A recent fire has already claimed some of the buildings on the shore, but otherwise it seems like a village whose population has gone over the hill for a wedding celebration and will soon return.

Camden Sound, in which Kuri Bay lies, echoes with failed human endeavour. In 1864, three vessels arrived here, coming to what was supposed to be the new Eden of frontier farming. They brought with them one hundred and twenty farmers, ten constables, a magistrate and four thousand five hundred sheep to settle the new land. The entire scheme had been hatched based on the report of two government surveyors, who had written glowing accounts of a lush country of green savanna and pasture and clear-running springs. When the three ships arrived on this hard shore of stone-strewn hills it soon became apparent that the surveyors had never stepped ashore. More than half of those who arrived took one look at the prospects and remained aboard to return home to Victoria. Those who stayed did not fare well. There was no water. The rocky hills were thin-soiled, and the grass was coarse and unsuitable for sheep. There was no savanna, but there were men with spears who found the sheep a pleasant and easy meal. Their spears also found Constable Walter Gee, killing him on 17 September 1865, shortly before the entire settlement was abandoned, nine months after it was begun.

Perhaps the most curious death of the whole disastrous escapade was that of Antonio, the horsemaster for the expedition. Antonio was not a trusting man, and having sold all his worldly goods to invest in the new land, he carried around his belly a money belt filled with gold sovereigns. He fell overboard in the bight of the southern coast of Australia and before the lifeboats could be launched the weight of the gold had carried him to the bottom.

Today, there is nobody here. Even the spear-throwing peoples are gone. Some were killed by later settlers. Others were forcibly removed during the Second World War out of fear that they might aid an enemy landing here. To those very few who returned or remained, there came the policy of educating and modernising, and the last Aboriginal people were taken from the land. The 'stolen generation' never returned.

15 AUGUST

Montgomery Reef is a phenomenon unlike any other: it lifts the sea above the sea. The reef is gigantic, 400 square kilometres in extent, with several islands near its centre. What makes it unique is that it is almost completely flat. As the giant tides in the region recede, the water on top of the reef remains. The centre empties over the rounded edges in a shallow cascade of white water. It gathers in indentations in the reef into small rivers that, as the tide drops, grow faster and faster until they are torrents. We motor in the tender into a 15-metre-deep fissure that runs into the heart of the reef. It is low spring tide and the top of the reef is many metres above our heads. The edge is a gently rounded slope that rises like low hills from the sea. Water runs constantly over the sides, and in places forms a torrent so strong that it pushes our twenty-person tender sharp and hard to the side.

There are turtles in the channel, hawksbills and loggerheads. They drift to the surface, breathe briefly, and dive again. Egrets and herons patrol the reef, while an osprey dives deep into the channel and lifts an emperor fish into the air. It perches on a termite-like mound of brain coral, the fish flapping desperately in its talons. It stares at us for a moment, then focuses its attention on its catch, killing it by tearing out the gills.

The entire reef is coral. Corals growing on corals, and their diversity can only really be appreciated with close inspection. There are browns and yellows and pinks and greens, as well as occasional lacings of red. There are clams with purple lips, brain-like mounds and small pointed clusters of stag's horn. A small lace-like fan coral vibrates in the wind. Clustered shoals of tiny fish patrol the edge, and a large batfish hovers on the edge of a torrent that pours off the reef in a white cascade.

On the *Odyssey*, I climb onto the cabin roof where I can see over the top of the reef. The shallows glitter in the bright sun for kilometres all around. In the far distance the lagoon they form rises to a thick forest of mangroves and finally the familiar red cliffs of an island. The vast shallow sea of the reef top stands eight metres above the ocean. This singular place is now to be protected by inclusion into the proposed Camden Sound Marine Park, which will protect not only Montgomery Reef but also all the coastal waters as far north as Brunswick Bay at the western entrance to the Prince Regent estuarine system. There are, too, murmurs of extending this protected zone south to include the Horizontal Waterfalls and the network of bays that form the convoluted coast of the northern shore of the Yampi Peninsula. I would hope that those who are steering this ambitious project will consider including under its protection some of the coastal land as well. To the north, much of the coast is already protected by the Prince Regent Nature Reserve. To extend this south to protect all the coastal regions of the Camden Sound Marine Park and the northern Yampi Peninsula would be an unprecedented gift, not only to all future Australians but to the world as well. This is some of the wildest, most unspoiled country we have seen. I cannot imagine the importance it will come to have as a pristine place of both land and sea to generations in fifty, five hundred and a thousand years' time.

This place is, however, not without its threats, for as unblemished as it seems, for the little reach of the steep shores that one can see, it has already been probed and sounded and sifted for what the mining companies might extract of value from the ground. Already Koolan and Cockatoo islands, off the northern shore of the Yampi Peninsula, have been decapitated by iron-ore mines. We see their lights in the night sky, a city-bright intrusion into the shimmer of the wild night constellations. The islands are Aboriginal Lands, and the temptation of the easy money of royalties from mining operations makes little land safe from exploitation.

At Raft Point we climb a steep saddle to lie on our backs and contemplate the mouthless, feather-headdressed Wandjina painted on the roof of a wide rock overhang. The painting has recently been refreshed, which renews and maintains the Wandjina's power. The power of a creator. Many of those who visit here will see it only as rock art, but I am an African and I have been raised among peoples with powerful cultural beliefs. It has taught me that no matter what deity stands on your pedestal, the most important ingredient is belief. When people believe without reserve or doubt, things happen that defy rational explanation. I wonder, as I sit in the cool shade of the rock, about the contemporary power of the Wandjina, and about how the Aboriginal creator feels about its people destroying islands for money.

18 AUGUST

We arrive at the jetty in Broome in the mid-morning. Ashore, I learn that yesterday my father had two motor-vehicle accidents in a row. The tone of my brother's email detailing the incident is sombre. Surprisingly, my father is fine. He is ninety-one years old and lives alone, and although we have all known that this time is coming, the event is sudden and shocking. We must wait until evening to speak with him on Skype. When we do, and our picture appears on his computer screen, he claps his hands with delight. 'There you are. Hello, Bev.' His face is filled with a grin that wrinkles his eyes and he is rocking forward on his chair.

'Pete, I just don't know what happened,' he says, his arms folded across his chest. My brother's doctor suggests a series of small strokes. Perhaps, but even before we left he was prone to falling asleep anywhere at any time, sometimes in mid-conversation. I hope against hope that it is only this.

19 AUGUST

We are camped on the shore of Roebuck Bay, the lights of Broome across the water. We have a second Skype call with my father, during which he agrees to see a heart specialist. When I say goodbye and the screen goes blank, I feel very far away.

Dinner is almost ready when a vehicle drives slowly down the track to our camp. It stops behind us and its lights are turned off. After a time, the driver switches off the engine and I walk out to confront whatever it may be that is coming. The driver is shirtless, and I notice a few lateral scars like traditional body scarification on his shoulder as he flicks a lighter to his smoke.

'Where are you from? What ya doing here?' he asks.

'I'm from around, how about you?'

'Come to talk to my grandfather. He was born here.'

'I've just been talking to my father too. He's in Africa.'

'Africa! Yooh! That where yuse from?'

'Ja, South Africa. My old man just had an accident. He's ninety-one. Seems like this is a good place to be talking to the old people.'

'What's your name, bro?'

'Peter.'

'Greg.' We shake hands.

'Came to talk to the old man?'

'This was all my grandfather's place,' he says with a sweep of his arm, his cigarette carving an arc of red in the dark. 'This is Yawuru country.'

'I've been all over the country. It's good country.'

We talk for a while of the Kimberley I have seen, of the coast, of the crocodiles.

'One day my grandfather call me and say: come and look here. He got a net in the sea and there's a croc sitting there chewing on all his fish.'

'I can't believe all these people who walk into the water in the rivers and the sea up north. There's serious crocs up there.'

'That fella fuckin dumb.'

'Oh! The guy that just got taken in the Adelaide River?'

'Fucker got his line snagged, so walks into the river to get it free. Can't see shit in the river water, croc came right out of it.'

Greg reaches under his seat and comes up with a beer, which he cracks, and takes a long draught.

'You got kids, Greg?'

'Grankids! Two. The smallest one is so big.' He puts his hand half way up his thigh. 'He's already talking. I shaved his head the other day. Now it's all bristles, like a brush.'

'And the other kids?'

'One other. So-big.' He puts his beer across his chest. 'Got blue eyes. His father from down south.'

We talk some more and Greg finishes his beer. He grinds out his smoke in the dirt.

'You okay if I stay here?'

He slaps me on the back. 'This my grandfather's place. You stay here as long as you like, Pete.'

22 AUGUST

After a long day's waiting, my father's diagnosis, when I Skype him in the evening, is good. I can feel the tension that has been building in me release like a balloon in my chest. I feel light, and he persuades us to continue with our plan to cross the Tanami and Great Sandy Desert via the Canning Stock Route, after which we will return home.

23–24 AUGUST

Some 618 kilometres to the east, we arrive outside Halls Creek in the late evening. Our crossing of the desert will be more than a thousand kilometres of track; in the morning, we fill every fuel tank and container in preparation. Water too will be precious.

After a twisting track through the bush, the blue shimmer of Lake Gregory in the dry desert savanna seems more mirage than beauty. There are flocks of brolga everywhere, and along the shore the trunks of trees killed by a year of high water look like spears of an army thrown at the shore. There is a raft of black swans drifting close by the edge, and on the track near our campsite we find an emu with chicks. The adult is almost delirious with panic. Rushing away from our advance, it hesitates, rushes back to a dawdling chick, and then rushes away again, its feathers blowing in the wind like long fur. The single dingo we come across 50 metres off the track behaves similarly, running as if the gates of hell were opening behind it. Theirs is the fear of the hunted.

In Billiluna, I had stopped to ask directions. A woman was decorating her yard fence with balloons and brightly coloured crêpe for her four-year-old granddaughter's birthday. 'We're having bush turkey,' she told me. The wind is blowing as I sit absolutely alone in the night by the lake, pinned to the earth by the stars, and I wonder if that is what became of the emu's mate.

25 AUGUST

Before we reach the first well of the Canning track, the corrugations shear off the three bolts holding in place the double Old Man Emu shock-absorber system on the Land Rover's rear axle. Spinifex, with its sharp pointed stems, is unpleasant to walk through, but lying in its needled cushion on the hot desert sand to make the repair is a singular torture. Each prick through the skin stings, and every movement finds sharp points all over my body, even though

I am lying on a heavy tarpaulin.

The flat monotony of the country is arrested in the afternoon by low rocky hills to the east. We walk into the hills to Breaden Pool and Godfrey's Tank, two small catchments of surface water hidden in the small gorges that lie in the shadows of the valleys. Although we have seen camels in the vicinity, I am interested that none has come to the pools to drink. There are only the tracks of birds. The camels are dropping their calves, and we see several young camels, gawky on their legs, following their mother's slow amble through the trees. The camels are yet another domesticated animal that, abandoned to the wilds of the outback, have thrived. Although there was a government mass-culling project between 2009 and 2013, which reduced the camel numbers from around a million to three hundred thousand, they are to be found throughout the desert, and camel tracks remain fresh on the sand as we travel slowly to the southwest.

26 AUGUST

A purple dawn all too soon becomes a day of hot buffeting wind that runs through the yellow undulations of spinifex like a horse's mane. At mid-morning, the shock absorber bracket on the opposite side of the Land Rover gives way. I am short on bolts, and when I crawl under the vehicle I see that the damage is worse than I thought. The entire mount has torn free of its weld to the axle. We limp to the next well, a few kilometres south, where I attempt a makeshift bracket with chain.

While I am working, a convoy of seven vehicles arrives, travelling north. The occupants bundle out and demand our permits. There is no introduction or inquiry about our predicament, which, tools spread around the vehicle and myself deep under the chassis, is quite clear. Their attitude is officious, bullying, but their contact is never quite direct, as if they see me not as an individual but rather as a representative of some broader entity. Like many groups of Aboriginal men whose company I have come into, they seem to be possessed by a generic anger and resentment of non-Aboriginals. When Beverly produces our permit, it becomes apparent that nobody can read it and they lose interest and drift away. Their mechanic is as reluctant as the rest to help us, and the only assistance we are given is advice: 'Turn back. You'll die out here.'

Inwardly, we fight the decision to return, but it is the only sensible option. At the moment, we can still drive out. If we continue into the dune country towards distant unknown small settlements, we could well break down and our predicament could be far more serious. It is with heavy hearts and slumped shoulders that we start back on the almost 400 kilometres that we have covered. As we drive through the hot afternoon and into the evening, both of us are silent with disappointment.

27 AUGUST

The desert is lovely in the predawn, soft and tranquil. However, heading north, the corrugations soon begin in earnest. I drive with my teeth gritted, my shoulders a knot of tension. In one particularly bad section I stop, unable to stand the hammering shudders a moment more. I walk along the track and measure the hardened ruts. They are half a metre apart and 12 to 15 centimetres high. Climbing under the vehicle to check whether any more damage has been done, I notice that an oil leak has begun on the transfer case. It shifts another step away the possibility of our returning to the desert.

On the road to Halls Creek I come to a decision. It percolates slowly into my consciousness. Every cell in me wants to return. I don't give up; it's just who I am. But my mind has another argument, one that weighs the situation. It would be foolish to return, and out there foolish can have dire consequences. We are alive and fending for ourselves. It is lucky enough that we have managed the hundreds of kilometres coming out on our own. We are not going back. This is the end.

On the road, I slow down for a snake that winds its way across the open dirt. It is a king brown, the biggest we have seen, around three metres long. Its body is a pale khaki-brown and as thick as my wrist. It raises its head and neck off the ground as we stop, watching through the open window. Its head is square and heavy, the eye black, and its fearless appraisal of our presence fills the warm, still morning with a sense of trespassing on treacherous ground. It lowers its body to the earth and moves soundlessly into the bush. For a minute we sit without speaking, watching where it has gone, the hot air frangible with tension.

I recall the Aboriginal paintings of snakes. They hold something more than threat, or an aesthetic appreciation of their form; regard perhaps, as if imbued with a sense of awareness from the snake's perspective.

It is how I feel as we move on into the shimmering distance; as if this land, like the snake, is more than the sum of its parts. Something powerful that draws one in until one is changed and, instead of looking at the country, one is within it and looking out.

Each of the Aboriginal paintings I am carrying home has a story. Tommy May's pale-yellow landscape within a surround of multiple hues of blue in which a neon-orange kangaroo and a dingo flee, which he haltingly explained as a story warning against the ache of loneliness we will suffer if we drive out from the land those with whom we share it.

Another is of a turquoise ocean crammed tight with a shoal of barramundi, between which red-rimmed, pure-white eggs float freely. These are the souls of unborn children. A pregnant woman must go and sit in the water to capture a soul for her child before it is born. The last is of a symbolic narrowing of hills through which the thousand experiences of life, a jumble of multi-sized dots, must pass. The place into which they emerge is calmer, a place of symmetry and constance.

I must return to Africa now, to my father, for he is where it began. His seeing and awareness and love of the outdoors were his gifts to me, which I must now carry back to him. I will begin with the paintings. I will unroll them slowly and tell their stories, and of the places where we found them, for they are songlines, a map of a journey carried in our heads.

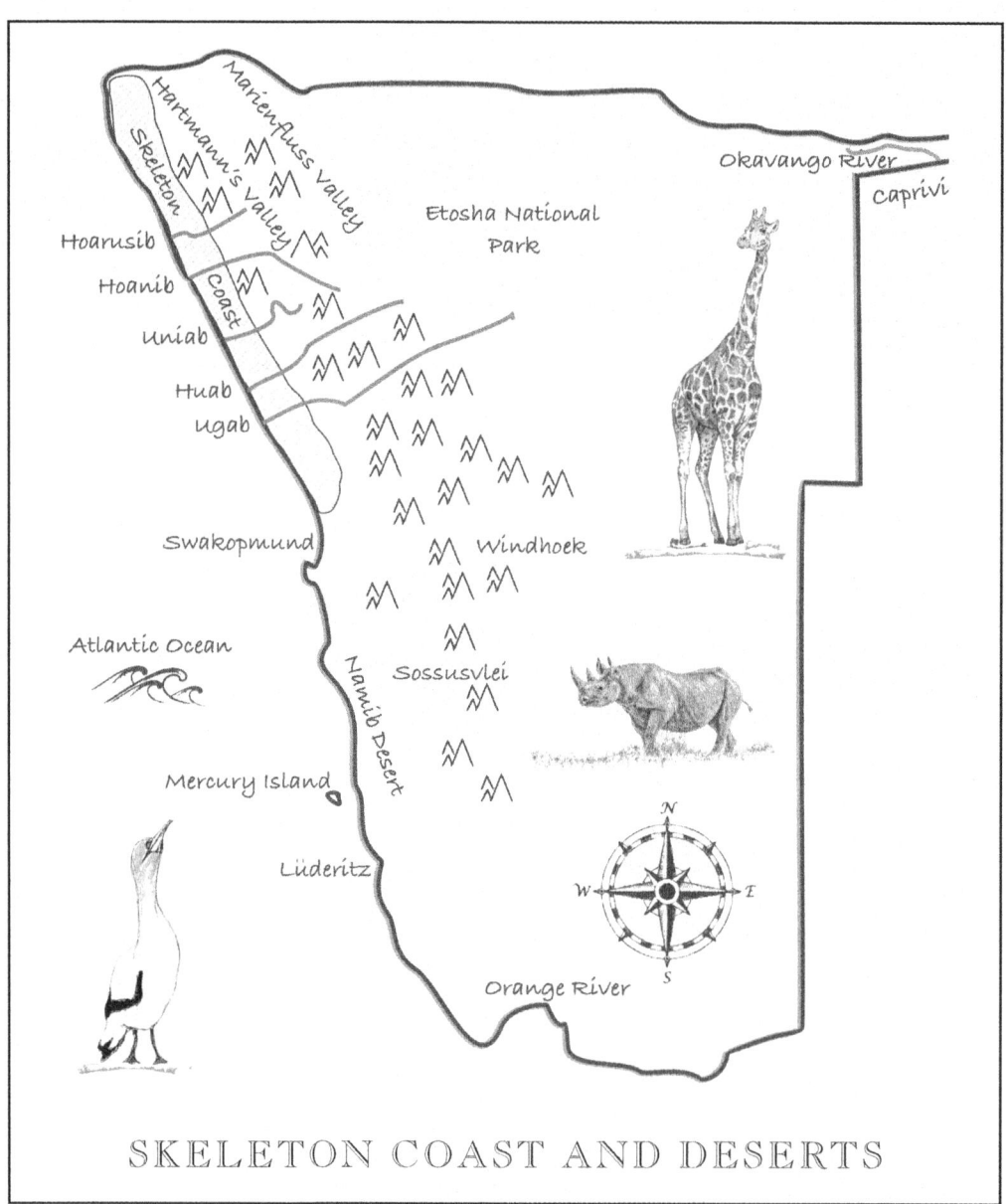

Marienfluss Valley
Hartmann's Valley
Skeleton
Coast
Hoarusib
Hoanib
Uniab
Huab
Ugab
Etosha National Park
Okavango River
Caprivi
Swakopmund
Windhoek
Atlantic Ocean
Namib Desert
Sossusvlei
Mercury Island
Lüderitz
N
W   E
S
Orange River

SKELETON COAST AND DESERTS

VII

AFRICA
The Skeleton Coast, Caprivi and Deserts of Namibia

Out there, where the flat distances dance in a blue mirage, where stone is broken by the heat, one comes to stand before these wild denizens with regard. To see a rhinoceros here is to see it fully: a fantastic figment of evolution.

4 APRIL 2015

What is it about leaving that is such a bitter sweetness? I feel it now as I close the door, turn the key and, climbing into the Land Rover, turn north. My whole life I have done this, turned my back on certain loves to face the promise of others. The urn with Blaid's ashes still on the shelf where I can talk to her, my father's falsely cheerful smile as he waved, standing lonely on his overgrown verge, in exchange for the taste of dust in my nostrils and a chance to visit other friends, friends I have missed. I drive through the arid, granite-domed landscape of western South Africa with elephants on my mind.

We cross the Orange River into Namibia on a single-vehicle pontoon. The river is the colour of milky coffee and is whipped to whitecaps by furious gusts of wind. Tall reeds bend and twist like a movement repeated down a chorus line. The Namibian immigration officer is a petite woman with striking aquiline features and almond eyes. She is coy when I question her about her obvious pregnancy. 'This month.' I ask if it is her first, and she smiles shyly, turns her gaze into her shoulder, and holds up two fingers.

The land outside is barren and stony. The high hills all around are naked black or dark-orange rock. As we drive north into Namibia, the land rises steadily from the river valley. At Rosh Pinah, at an array of informal shops beside the road, four women are having their hair braided into long plaits. They laugh lightly at my query as to where I might buy a SIM card for my mobile phone. It is Easter Saturday and everything is closed. Inside the tiny, blue-painted, tin-walled salon, pictures cut from magazines are pasted on the walls. Outside, on a white plastic chair with one broken leg, a woman breastfeeds her baby. Their English is fluent and I am reminded that it is almost twenty years since we last worked for any protracted period in Namibia, and independence and the passage of time have brought exciting changes.

The road runs straight into the vast vistas that define my memory of this country. There is nothing growing tall enough or in sufficient profusion to interrupt the eye. Flat-topped mesas and breast-shaped peaks rise out of the plains that stretch 50 kilometres wide. The wind raises dust off the ground, casting the distance in an opaque blue haze.

Where the road climbs through a rocky valley towards the height of the mesas there are hillsides covered in enigmatic quiver trees: a pale, flaking-barked aloe whose branches radiate in round clusters from the single trunk to terminate in finger-like, curving yellow blades. We surprise two klipspringer, quick and agile across the rocks. Further in, there are four kudu standing in a gorge, their large pink-centred ears cocked forward. They watch us with the frozen stillness that pre-empts flight.

Where the road issues out onto the high plateau, three mountain zebra turn to watch us from the high sides of a close hill. I stop the Land Rover and climb out to glass the surrounding country. My anticipation is quick in my chest. It is this that I remember, the reason we have chosen this country for our African wilderness: the vast, sparsely populated country where wild animals still roam in a place long settled, and it quickens my pulse to find it still the same.

5 APRIL

The moon is setting between a pair of pointed hills as the sun rises. The first shafts of light gild the grassy plain to a luminescence. Birds chatter in the tree beneath which I am sitting in the chill shadow: sociable weavers and a striking crimson-breasted shrike. A barking gecko close behind me gives one last loud chirruping call before retreating into its burrow for the day. This is my fastness, the place to which I am tied. A love so deep-seated, it seems to rise from every pore. I came to it in my youth and, responding to the voice it stirred in my soul, I never left. It has given my life passion and purpose. I lay down my pen and breathe it in.

At a cistern in the Namib-Naukluft National Park that wild horses frequent, a flock of ostrich is sitting in a clustered group in the dust. Animals begin to appear from far out over the plains, black dots in single file, like the spokes of a wheel drawn inward to the hub. Two horses trudge tiredly across the bleached stony ground to come and drink. The ostriches rise as a group. They wait, blinking, for the horses to slake their thirst, and then crowd the water's edge, scooping the water into their open bills with thrusts of their long necks. The wind is hairdryer hot. I set up a remote camera on the edge of the man-made cistern. The moon will rise almost full tonight and I am hoping to catch horses and oryx in the moonlight.

The moon crests the distant mountains as I complete my set-up, and I am about to stand up and leave when I hear the drum of hooves approaching. I am caught in indecision. Behind me, the beat of the hooves breaks into a trot. I dare not turn my head to look around. And then they are beside me, snorting the air at my presence. They pause for a moment on the far side of the water, heads held high. They are two dark shapes that stand in silhouette against the sky. One stamps its hoof at my low, hunched form. They step forward, snort loudly once, then walk right up to the water and drink noisily. The moon climbs yellow into the sky and shines gold on the rippling water.

6 APRIL

Beneath low cloud and drifting fog, the harbour town of Lüderitz is hung with a melancholy that seeps beneath my skin. The waterfront is industrial, the town centre losing ground to the elements. Those few buildings that owners have sought to restore seem only to accentuate the decay of their neighbours. Cement bubbles to reveal rusting steel, and paint peels away. In a pink house with broken, purple-framed windows, a man stands in the opening of a window sagging loose on one hinge, talking on his cellphone and gesticulating with his free hand.

On the rises overlooking the sea, the middle-class houses seem cut from the same cookie dough of square, faded face-brick. There is nothing green, almost nothing growing, and the earth and the world feel empty. Even the clustered housing of the labourers and fisherfolk, painted in bright primary colours, in the low valleys back from the shore, fails to raise the mood of sombre functionality garnished with decay. Only the old town, rising up the steep sides of the tall central hill, has the historical character and charm to lift the spirits, but it is largely hidden behind slab-walled structures.

8 APRIL

At the Fisheries office in town, Kolette Grobler is an amiable, middle-aged woman who walks barefoot through the historic building, keeping her efficient staff happy without seeming to exert any bureaucratic authority. She is wary of our request to visit Halifax Island, but after a time agrees to allow us to accompany the monthly census of penguins on the island, on the condition that we are accompanied by her staff and adhere to their instructions.

Her hesitation to allow photographers into a sensitive area is something we are encountering more and more, as digital photography and the capabilities of modern cameras open the doors to many more would-be professionals. The head of the Namib-Naukluft National Park told me of one 'professional' nature photographer who, after persistent requests, was granted special permission to stay overnight at a place within the park where skeletal remains of century-old dead trees stand in stark contrast to the high dunes that surround them. Once he had finished his work he took an axe to these ancient trees, so singular that they are one of Namibia's noted attractions, and built a bonfire for the night. Every subsequent photographer to work in Namibia has reaped the consequences of his actions.

I assure Kolette that we will not abuse her trust, and half an hour later we are aboard the Fisheries Research Vessel *!Anichab* (pronounced 'Incanigub'). The skipper, Sakkie Jason, is a soft-spoken man with a gentle disposition. As we motor slowly out of the natural cove that forms Lüderitz Harbour, he is animated as he tells us of the pods of orcas they saw yesterday and the noisy flapping of schools of hundreds of dolphins on which they were closing. His disappointment is tangible as he explains that darkness and fog prevented them from knowing more.

Fog greets us as we round Diaz Point and we land on Halifax Island in a world of opaque silver-grey. African penguins waddle stiffly aside as we climb the short shelly beach. The ghostly drift of the fog, the abandoned building with glassless windows, collapsing beams and penguins nesting beneath old shelves, where glass bottles and old jars stand covered in moss and shed feathers, are all emblematic of this coast.

Returning to the harbour, there is lightning in the sky and three fat raindrops hit the windscreen, but nothing more. It has not rained here for four years.

9 APRIL

In Aus, a lone young woman sits on the shaded, tiled veranda of the Bahnhof Hotel eating a huge portion of chocolate cake layered with cream and chocolate icing. Across the road an old woman with dusty, worn shoes and a faded, waist-less dress collects an armful of dry sticks for firewood from the trees lining the street. We drive into the hills north of town, with clouds gathering over their exposed granite peaks. The clouds turn dark and close out the sky. We stop on a high saddle of a small dirt road and wait. Lightning flashes silently between the shadowed stone hills and the air becomes chill as the clouds grow dense, flush-ing a dark purple – and then it is raining. The rain comes in waves, its tattoo soft, then loud, against the vehicle body. A horse-drawn cart approaches along the track: a man and two boys returning from collecting firewood. They are wet through and smiling.

'Good to have rain?'

'Very, very good to have rain,' the man nods animatedly beneath the hood of his bright-orange seaman's sou'wester. The boys' smiles are so wide that they seem to ache on their faces.

'I wish you a wet journey home.'

The man laughs. 'Yes,' he says. 'Thank you very much.'

10 APRIL

We climb the granite domes in the pre-dawn, photographing the shallow pools of rainwater gathered in the depressions and the far reach of the desert beyond. There are oryx, ostrich and cattle. Continuing north, the land falls in an unbro-ken sweep to a valley 20 kilometres across. On its far side, red dunes rise like a wild sea, cast into sharp crests and steep troughs. Storm clouds are gathering above the dunes and the distance is invisible beneath their deep shadow.

Our isolation is interrupted by the arrival of a motorbike. Jakob Engelbrecht is circumspect at first about our presence, as we are unwittingly trespassing on his unsignposted property. He has had problems with people shooting his cattle, and others rudely dismissive of his authority. Finding us to be neither, he quickly drops his reserve and becomes animated and friendly. He tells us that Eureka is his family farm, with seventeen of the immediate family members living in the farmhouse at the base of a domed hill that he points to in the west. Last night, they slaughtered two cattle and he carried the offal into the veld for the hyenas and the jackals.

'They've got to live too,' he says.

He nods sadly when I ask whether they do not disturb his stock.

'We corral the sheep and goats every night,' he explains. 'But two months ago, we had to shoot a cheetah that kept on killing our calves. We didn't want to, but it wouldn't stop. We've had lions here too, but the neighbour shot those years ago.'

When I ask him how many children he has, he is proud of six of his own.

'I don't want to admit it in front of her,' he says, indicating Beverly with a dip of his head, 'but my wife and I aren't married.'

'She won't mind. Things have changed a lot in the world these days.'

'Yes,' he says, and then adds almost as an afterthought, 'but maybe not in the eyes of the Lord.'

Lightning flashes below the clouds gathering over the hills. Their shadow drifts over us and gives instant relief from the sun's bite.

'Rain fell last night close to Aus. Looks like you're going to get some of it.'

He grins. 'I've been talking to the Lord about that. I asked Him really nicely,' he chuckles infectiously. 'Maybe this time. Maybe.'

11 APRIL

Driving the western edge of the Tiras Mountains, out on the bare ground we come across a pale-yellow Cape cobra, sliding lithe and quick over the sand. The contrast of the red earth and blonde snake is striking and I climb out to make some images, keeping a healthy distance. The snake stops, turning its head towards me. I back towards the car. It follows. I move to the side photographing its swift, silent glide across the open ground. Beverly is photographing it out of the window. Suddenly, I realise the snake's intention: it is between Beverly and me.

'Move the Landy,' I shout loud and urgently. 'And close the windows.'

Beverly lowers the camera from her eye and realises in that instant that the snake is headed directly for her. It is moving quickly. She must scramble into the driver's seat, but her apprehension gives her the jitters.

'Drive! Drive!'

She is fumbling for the keys when the cobra reaches the rear wheel. It lifts its head into the chassis, sliding up into the dark. Beverly starts the engine. If she drives forward she will crush the part of its body still in front of the tyre.

'Close the windows. Now!'

Through the glass her face is fearful. The Cape cobra is one of Africa's deadliest snakes. Her lips are moving, but I cannot hear what she is saying. I inch circumspectly closer.

I get down on my knees a distance from the Land Rover, but the sun is bright and the chassis is in deep shadow. I can see nothing. I work my way cautiously closer, but there is no sign of the snake and I decide that perhaps to drive will be the best thing to scare it out. The road is corrugated and stony.

'I'm coming in.'

Approaching the Land Rover from the front, I climb in quickly. We drive for fifteen minutes over the rough road, but when I stop and climb out gingerly with a torch, the snake is still there. It is coiled on top of the fuel tank, between the twin struts of the chassis. In the hope that it is a hot and uncomfortable place, I pull into the shade of a large camel thorn tree. I need to write, and in the hour or two that it will take I am hoping that the cobra will overcome its fear and make its way down, tempted by the scent of the tree and the twitter of the birds. While I write, Beverly does her research and keeps watch, but when I am done the torch reveals that the cobra has not moved. We drive on.

At midday, it is over 40°c and we stop beneath a camel thorn where sociable weavers pour from their domed nest like water out of a shower head. Once more we leave the snake to a long period of quiet, while we up-end bowls of water from a reservoir over ourselves to bathe. It is still there when we return. It feels strange to climb into the Land Rover and know that there is a deadly venomous snake beneath your feet.

As we drive I realise that the cobra could be with us for quite some time. It is clearly reluctant to move, and until it does it is coming with us. The thought crosses my mind: if that is going to be the case, I might give it a name. Lionel, perhaps? That sounds like a fitting name for the kind of distant relative or acquaintance who does not get along with anyone, who pitches up unannounced, and stays for a week longer than they should.

After two hours of driving the hot corrugated road, we stop at the farm Spes Bona, which straddles the crossroads, and bother Esther Vermeulen with our story and the strange request to use her hosepipe. She seems quite undaunted, and by the time I have driven around to the road in front of their garden, Hennie Vermeulen is waiting there with the hosepipe. He stands in his socks in the dirt, holding a long metal pipe. The blast of water gets the cobra moving, but it still does not come down. Hennie suggests I try moving it with a steel fence dropper, but I cannot get a clear view of it and am afraid of damaging my fuel pipes.

'The water won't chase it out,' he says. 'These snakes swim.'

Then he tells me that a cobra is not really much to worry about. It is black mambas that bother him.

'Mambas? I didn't know you got mambas here.'

'Ja, old Jannie Theron shot one up in the mountains near here. Twenty-four feet three inches long, it is a world record.'

'That's a monster.'

'Ja, I had one kill four of my dogs once. They cornered it up on that kopje over there and before I could shoot it, it had bitten all four. The dog that got the furthest was only seven metres from the snake before it died. All the others fell right there. That was a small one though, only about seven feet.'

'So, what do you suggest I do to get rid of the cobra?'

'Petrol.'

'Petrol?'

'Ja, snakes don't like petrol. Spray a bit of petrol in there, that'll bring him out. You can get petrol at Betta.'

At Betta, we repeat the saga of our day to the farmer, who is welding something that looks like a massive steelwork sculpture for the garden.

'No. Fuck them! Tell them they must go,' says a neighbour standing by, who backs away as we tell the story and picks up a wooden fence pole, waving it in our direction. The farmer laughs and waves me to the petrol station that stands on the other side of the farm buildings.

The petrol works. The snake reacts immediately, moving from its hiding place and sliding fast along the struts of the chassis, but there are now five people watching, squatted at a safe distance around the Land Rover, and it does not come out. After a time, it grows still again and I approach gingerly with another squirt of petrol out of the short pipe. The cobra does not react. I can see its coils clearly between the steel beams. After five minutes of its staying dead still, I prod it with a long length of wire. It does not move. I bend the end of the wire into a deep hook, snag it around the snake's body, and pull. The head drops free and five people leap backward. But the cobra is dead, its head drooping lifeless, its tongue protruding. I pull the rest of its body free. It is longer than I am tall, about six feet eight inches. It is a pale straw-yellow, the head small, the eyes pure black beads. It makes us sad and heartsore to have taken its life. We had no idea the petrol would kill it.

The farmer's son wants it. 'Noppies, Noppies,' he calls to a labourer across the road. 'Have you got a knife?' He wants to skin it and hang the skin on his wall. He is a chip off the old block. In his father's cavernous garage, attached to the house, there are stuffed kudu heads hanging on the wall and two cheetah, mounted life-size, snarling and fierce. As I drive off, the boy drags the snake's head in front of my tyres.

'It's dead.'

'I want to be sure.'

The body quivers slightly as the vehicle's weight crushes its brain.

'Heiya!' The boy drops the tail and runs.

13 APRIL

Sossusvlei is a unique phenomenon, where the flat-floored river valley cuts into the unbroken dune sea of the Namib-Naukluft Park. The valley floor is sand-free and the red dunes tower along its fringe as it pushes 60 kilometres into the desert sands before it finally concedes defeat. 'Vlei' is Afrikaans for wetland and Sossusvlei is the river's terminal, where the very occasional flow of flash-flood water is dammed and sinks forever into the sand. It is a remarkable landscape, without peer in the world. Red dunes, with their sharply defined ridges, draw curving lines of light and dark shadow as they descend from their 300-metre heights to the valley floor.

A windstorm blows up and raises a pall of thick dust across the valley. The sun shines through it from behind, burnishing the dust to silver, and it etches the stark outline of tall dead trees into monochrome relief. A lone oryx stands still in the sudden tempest, its tail whipped up by the wind.

17 APRIL

The staff are late for our scheduled 8 am departure from Lüderitz on the RV *!Anichab*, this time to visit Mercury and Ichaboe, the offshore bird islands. We load our equipment aboard the *!Anichab* and wait. Three hours later Abraham, one of the staff of Ichaboe, finally appears. He is surly and without apology or explanation as he loads the plastic crates of his belongings for his months on the islands. Skipper Sakkie Jason shrugs his shoulders, but his usually gentle demeanour has a grim set. We are very late and he already has his doubts as to whether we will arrive at Mercury Island with enough light to work.

The *!Anichab* is a tubby, top-heavy, fibreglass vessel that rolls heavily in the small sea that runs sideways to our track north. One cannot walk without clinging to something, and a piece of loose equipment on a lower deck slams periodically into the side. We are sharing the female scientists' cabin with Olivia, who manages the stores on Ichaboe Island. The narrow bunks are tiered to the ceiling along the outside wall. Olivia crawls into the lowest bunk and draws the curtain. She will lie there for the entire duration of the two-day return voyage, emerging only to vomit and, once we are anchored, to perform her duties as stores manager. The telecommunications technician wedges his large person into a booth in the dining area opposite, and watches action movies and soap operas at full volume without pause for the whole of the eight-hour voyage to Mercury Island. Even during meals, he only glances down briefly to load his fork before his unwavering stare returns to the screen.

I sit on the foredeck talking to Ipula, the youngest and newest member of the crew. He is filled with a bright enthusiasm, his mind alive with plans. He tells me he likes to work, and it is not a boast; I had already noticed him among the crew. He has studied, and his position as second engineer places him above the deckhands in authority. It is clear that some of the older crew resent his position.

The day is calm, with a warm sun reflecting off the smooth sheen of the blue-green sea. Yellow-nosed and black-browed albatross drift stiff-winged along the face of the swells. White-chinned petrels patter up from where they have been sitting on the water. For a time, small pods of dusky dolphins surf the bow wave.

I stand right on the blunt prow, photographing them directly below. They are playful and quick. We are making eight knots, and they accelerate away in their chases with a few vigorous flaps of their tails, then leap high and clear of the surface ahead. Their bodies are black above, white below, with a stripe of grey in between; the lines of division are a curving sweep from head to tail, reminiscent of the curving Buddhist symbol for peace.

At Ichaboe we anchor in the lee of the flat profile of the island. On shore, a man emerges from the cluster of yellow buildings on the northern tip. The black mass of birds stampedes away. The man moves slowly, leaving behind him a swathe of clear ground in the seething birdlife. Walking out onto the jetty, he lowers a hook on a pulley. The transfer of people and equipment to the island is hasty. We leave them standing on the jetty like castaways, the equipment and stores piled around them in great white sacks.

We arrive off Mercury Island in the dark. The fog is so thick that we can see nothing of the steep profile of the buttress of rock that rises sheer from the sea. I know it is there only from the reek of the guano and the braying of the penguins. We anchor in its lee for the night. Even in the calm, the *!Anichab* rolls to and fro with a heel of ten or fifteen degrees. Sleep comes a blessing to my discomfort.

18 APRIL

In the pre-dawn the fog has thinned sufficiently to reveal the shape of the island. It is a steep-sided rise of pure rock, its surface serrated and jagged, its height a sharp dragon's-back ridge, perhaps a hundred metres above the sea. Every inch is covered in birds. On shore, Beverly and I meet the jovial Rian Jones, who reminds me of Happy in the story of *Snow White and the Seven Dwarfs*. From beneath a beanie, pulled low over his ears, long hair escapes in a spring of untamed curls, and his eyes twinkle with a lively glint. He stands barefoot on the chill, wet concrete of the jetty, wet guano squidging up between his toes.

'Watch out, it's slippery,' he warns as he gives us a hand up out of the *!Anichab's* inflatable tender. The ground is a slush of wet guano, green and glutinous in the shade of the jetty's cover. The ammonia reek is overpowering. Penguins crane their necks to peer at us from their nests. The island is a condominium of breeding birds.

'About 85% of the total world population of bank cormorants breed right here,' Rian tells us as we make our way up to his quarters, an old guano-collecting building now painted a cheerful pale-blue. The rare bird breeds on the steps where we walk, on the plank that supports a water pipe over a span between structures and on the concrete top of the water cistern. They are pure-black with a yellow eye. In the sheen of the fog-filtered sunlight, the covert feathers on their wings are jade-green.

'It's going to be slippery. Unfortunately, you've got a north wind that brings the fog. Barefoot is best.'

But my feet are not hardened to rough walking and the idea of guano clinging to my feet persuades me to keep my boots on. Rian takes us on a route directly up to the highest ridge. At first the climb is easy, but at the top the rock is treacherously slippery, and my boots slip like ice skates on the fog-wet guano that covers the smooth, weathered stone. My heavy camera pack unbalances me and, gingerly, I put my hands down for extra support, dusting the smear of guano on my trousers with distaste. It becomes worse as we crest the ridge. It is like trying to climb on smooth ice. Only in a crevice, or on a sharp rock edge, do my boots give purchase. I must use my hands or fall. The ridge is narrow, and the sides plunge sharply to the sea below. As Rian moves forward, the ridge narrows further. Even standing on a stepladder makes me giddy with vertigo. Here, I abandon all dignity and distaste for the cloying guano and edge along on all fours, negotiating the descent on my backside.

The birdlife is like bees. The skies around the rock swirl with black dots. As we edge forward over the treacherous ground, Cape cormorants rush from our presence like waves sucking back from the shore. The penguins and gannets are quite the opposite: when we stop to make images of the scene, those closest to us often move forward to peck tentatively at our boots or camera lenses.

Rian is gentle with the birds, waiting for them in places where our presence might frighten them onto a treacherous cliff, or shooing them forward when they block our path with their curiosity.

'After nineteen years here I speak pretty good penguinese,' he says, encouraging a few reluctant penguins along a narrow ledge above a place of flat ground, crammed edge to edge with nests. Once every two weeks, he and his assistant count all the moulting penguins on the island. It takes two weeks for an adult penguin to complete a moult, so Rian and his colleagues are able to establish that there is a permanent population of just over sixteen thousand penguins on the island. Added to this are approximately two thousand seven hundred juveniles that return each year. The Cape gannet breeding season is over, and only a few hundred, with mostly fledged chicks, are still using the island. During the summer season, that figure swells to two thousand two hundred breeding pairs. Mercury Island is also a critical haven for the bank cormorant. During the last summer there were around two thousand five hundred and five breeding pairs, the largest colony on earth.

The fog is lifting and we can see across the narrow strait to the shore, where the rusted skeleton of a wreck stands high on a beach between rocky promontories. The guano dries quickly in the lifting fog, and our climb back to the research station is easier but still treacherous along the ridge. Rian accompanies us onto the *!Anichab*. After four months on the island, he is heading back for a few weeks' leave with his wife and daughter in Lüderitz. I look back at the island as we wallow out to sea. It is a smudge of land, steep and hostile, but it could not be more important in the chain of life that inhabits this cold and desolate coast.

At Ichaboe we anchor close off the jetty. A line is drawn ashore to pull the hoses from the vessel that will discharge 15 tonnes of fresh water into the concrete cisterns of the research station. Burly Tony Delport has a neat clipped beard and dark, brushed hair. Ichaboe has been his province for the past twelve years, and he has names for every small feature of this flat rise of land that stands just clear of the sea. The island is six hectares in extent, its entire perimeter protected by a two-metre-high sea wall.

The ground is black with Cape cormorants. Underfoot, the ground is spongy and soft with accumulated guano. In the mid-19th Century Ichaboe was far more renowned than it is today. Men fought and died here over what was then the stinking 'white gold' of guano. Thousands of men stood jealous guard over claims they had staked over the island, and in three years they cut 10 metres off the surface of the island to be used for fertiliser in distant European fields. The island changed shape from a low, domed hill to a flat plain, with a rocky rise slightly off centre. Today, guano is still harvested here every other year, but Tony restricts the harvest to areas that will not disturb the nesting penguins or gannets.

23 APRIL

The gravel road from Windhoek west to the coast at Swakopmund is one of the loveliest drives we have made anywhere. There is a tar road to the north that draws most traffic, leaving our road peaceful; we pass only three other vehicles the whole day. The road winds through a crumpled landscape of steep shale hills, thick with low acacia trees. The country is green after rain. At the Boshua Pass the road drops suddenly out of the hill country into a low savanna punctuated by granite inselbergs whose knobbled ridges peter into single outcrops on the fringe of the Tinkas Flats of the Namib Desert. A lone giraffe stares at us. It stands motionless, a single kopje behind it to interrupt the flatness of the land to the west.

As we move toward the sea the sparse vegetation thins further until the ground is entirely bare, a vast plain of sand and rock without hills. This is the northern extremity of the Namib-Naukluft National Park, which runs unbroken from Lüderitz, far in the south. It is one of the largest national parks in Africa, covering almost 50 000 square kilometres. Out in this desolate place we come across a small group of Hartmann's mountain zebra. Several of them stand on the top of small rocky outcrops, perched there as if the sand that surrounds them burns their hooves.

Swakopmund is a flat town of low buildings and wide streets. Palms have been imported in an attempt to break the monotony of the featureless land. On the beachfront boulevard, a high fog keeps the sun at bay. The sea is a dirty brown, with cold, small breakers rolling in to the shore. A woman, in maroon tights and matching sweater, climbs from the small taxi parked close beside us, licking her fingers and wiping them on the wrapping of her fast-food lunch. She walks down to the sea, pulls down her tights and urinates in the sand.

In the afternoon, we meet Jockel Grüttemeyer and his wife, Monica. Jockel is a whirlwind of energy, bright-eyed with enthusiasm. We spend the whole afternoon talking, poring over maps, writing notes at places where no annotations have been made on the chart. Our project is about wilderness, and Jockel's passion for his country, his intimate knowledge of the land, his first-name familiarity with the people of the wild places, breathes life into our work. In the vast country that surrounds us, Jockel is opening our eyes to an encyclopaedia of opportunity.

We camp tonight far up the dry Swakop River Valley, where steep hills of bare rock preside over arid canyons. While I am showering, the Southern Cross emerges out of the darkening sky. It is an old friend, a constant I have known since I first walked out into the wilderness.

25 APRIL

The road north from Swakopmund follows the coast. It is a salt road, in which salt has been mixed with gravel in the top layer, wetted, compacted and layered again. It is smoother than tar. The landscape is flat and featureless: tight stunted scrub dots the barren sand wasteland, like pepper sprinkled on the ground. It is made ugly by the crisscross of 4x4 tracks. In this place of no rain, the surface crust of the desert is a fragile layer that is soft underfoot. A vehicle driven over the crust breaks through, leaving deep ruts in the sand. With no element to repair these cuts – no rain, no vegetation growth, no loose surface sand – what little windblown sand there is takes a long, long time to mask these scars. Not a week, nor a month, nor a year, but usually a lifetime. A few moments' thrill will remain as an epitaph long after the perpetrator is dead. No tombstone inscription could more poignantly express humanity's selfish and callous disregard for the earth.

Cape Cross is a rocky headland where more than a hundred thousand Cape fur seals come ashore. It is named after the cross erected here by Diogo Cão in 1486. Cão, a celebrated Portuguese navigator, was looking for a sea route to the East and was the first to sail this far south along Africa's western shore. There are two aspects of Cão's cross that give measure to how remarkable his feat was. The first is that on the cross is the inscription that it was planted 6 685 years after the beginning of creation, and the second is that it was more than three centuries before it was sighted again, by a Captain Messum in the late 1840s. Messum was hunting for guano sources and the stench of the seals probably lured him ashore.

The stench today is the same and is complemented by the din of the calling of seal mothers and pups, which blocks out even the sound of the surf crashing against the hard shore. There are pups everywhere, saucer-eyed, roly-poly sausages of pure-black that suckle at their mother's teats or lie in tight huddles, dozing in the cold wind.

In a few months' time men will come here armed with pick handles and club them to death for their fur. Today, Namibia permits an annual harvest of eighty thousand pups for their fur and six thousand bulls for their penises, which are used as an aphrodisiac in Asia. The culling takes place in the colonies at Cape Cross, Wolf Bay and Atlas Bay south of Lüderitz. The Namibian government argues that the seals consume about 700 tonnes of fish a year and must be killed both for the survival of the fishing industry and to create jobs. Research has shown that the ending of seal culling does not impact negatively on fish populations but rather improves them. The effect on job creation is minimal, with just one hundred and sixty part-time workers currently employed in the culling of seals. The method used, crushing a young seal's head with a blunt wooden club, is never openly discussed. I am told in defence of the practice that it is an emotional issue, but to my mind it is barbaric. Seal fur is a luxury, the most unjustified reason for anything to die.

26 APRIL

The 40-kilometre-wide coastal plain is stony desert that in the east rises to the ridges of the Messum Crater, formed by the rims of two long-expired volcanoes lying side by side. It is 20 kilometres across, and from the peak of the hills in its centre the landscape of naked earth sits in an ochre-hued amphitheatre. Across its cold, dry sands, welwitschia plants grow in a scattered profusion.

The welwitschia is one of the desert's strangely adapted plants. It grows close to the ground and sports a few broad, flat green leaves that look like lizards' tongues laid out to dry. Its flowers are a spangle of cone-shaped orange tonsils grown on short, multibranched stems at its centre. In association with the plant are the brightly hued welwitschia bugs, *Probergrothius angolensis*, which spend their whole lives in the small world of the centre of the welwitschia. The plant grows so slowly that its stem is as woody and gnarled as that of ancient trees. A big one stands under half a metre high, with leaves as wide as it is high, lying nearly two metres out across the desert ground.

On the first plains that descend through rough shale ridges to the Ugab River Valley, it has rained and a short flush of grass gilds the desert in a gossamer green that shimmers like water in the strong wind. There are oryx and springbok on the plains punctuated by fields of domed granite boulders. The animals are skittish, however, and break into a run when we approach closer than 500 metres. Only four ostrich, which we come upon suddenly around the bend of a hill, watch us without fear.

27 APRIL

A strong wind rocks the Land Rover until late into the night, but in the predawn the world is still again and there is a sweet, honeysuckle fragrance thick on the air. The sun shines in shafts through broken cloud. We find herds of zebra and springbok spread out across the plains. All of them bolt at our approach, fleeing in a panicked run, their compact groups raising a trail of dust in their flight. Only one thing makes wild animals this wary of humankind: hunting.

Namibia is a world leader in community-based conservation. The population is small, just 2.2 million people in the whole country, and rural communities are tiny, while their lands are vast. With the exception of productive agricultural areas and veterinary cordons, the land is not fenced, and rural communities have a long history of protecting and conserving their wildlife. Pastor Gustaph Tjiundukamba, chairman of the Kunene Conservancy Association, wrote in a press release: 'We are the true custodians of wildlife in our country, Namibia. We have inherited the spirit of our ancestors who have always lived with wildlife.'

The conservation of the wildlife that roams the unfenced wilderness of the vast community-controlled areas has brought real benefit to rural communities. Tourists come to see the game. Lodges offering accommodation abound, and every possible wilderness experience is available to each niche of the market, whether trophy hunters, photo hunters or adventurers. During a spate of good rains spanning a decade, game grew plentiful and another utilisation was adopted: shoot to sell. I have read of the slaughter, of freezer trucks parked on the plains while pick-ups race across the open savanna, shooting everything. Cows, calves and whole herds of wild game. The meat is sold into butcheries across the country, and the conservancy communities receive the benefits. This utilisation breeds animals that run, that are terrified of vehicles. I see now that it is the Land Rover that is causing panic, and so we leave it standing on the track and walk out into the kopjes and low hills. The animals watch us approach. A few trot away, two young springbok pronking in the stiff, four-legged bounce that gives them their name, with their backs arched and white spinal ruffs raised. But there is none of the panicked flight that the vehicle elicits.

We find a raised vantage point on a low shale hill and sit still between the rocks. Below us two valleys come together in a V-shaped plain. Shadow and sunlight lie scattered across the scene like leaves adrift on a smooth stream. The vista is vast, 40 kilometres at its shortest, more than a hundred kilometres to the north where ridge upon ridge of hills run to blue. The earth is ours. Ours alone. When I scan the distance all round with binoculars I do not see a single human artefact anywhere, nothing but open landscape and its wild game. It is more than five hundred years since Diogo Cão set foot on these shores. It is remarkable that this place has remained so pristine. I breathe it in, deep into my chest, this precious gift of a morning that is a rarer find than any gold or glittering stone.

28 APRIL

In the cold dark of the predawn the stars are as bright as frost overhead and the fragrance of some sweet scent lies like a quilt on the land. I walk to the top of a low hill, keeping my silhouette to the jumble of round boulders so that I do not stand out from the land below. Game is all around: zebra on a far plain, oryx in small groups in valleys and on hillsides, and springbok everywhere, a group of fifty grazing, 150 metres from where we are camped. On the second 360-degree scan of my surroundings I see something I had missed on the first pass, a distant animal standing still beneath a tree. I watch until it moves. There are two, and they are cheetah. They walk purposefully towards where the game is trickling into the hills, their shoulders hunched, their tails held in an upward curve, just the tip twitching. We drive to a point closer to the mixed herd of springbok and oryx that are nearest to the cheetah and moving in their direction. We set out on foot, keeping to the low ground behind the profile of the hills so that we cannot be seen. We work our way closer, and I peer over a saddle. We are 200 metres away. We move quietly around the hill, position ourselves between the shale outcrops and wait. The springbok drift into view, grazing. The herd is scattered, a few moving along the hillside, but most on the valley floor. They are feeding, and it takes an hour for the whole group to move past. I have scanned the opposite hills repeatedly but have seen no sign of the cheetah. Perhaps they chose another direction, perhaps they were full and not hungry, just moving through, but cheetah are remarkably sharp-eyed creatures and I suspect that they saw us without us seeing them.

The sun is high and warm and we leave the plains and descend the road into the Ugab Valley. A two-wheel track wends between scattered loose rock, following the course of arid streambeds. Our going slows to a crawl, and the valley narrows to a high-sided gorge, the rock black and fissured high above our heads. It issues us suddenly into the Ugab Valley bottom.

There are stands of 15-metre-tall acacia trees, dwarfed by walls of stone that lean in striated layers. It is a vivid cross-section through the formation of the earth, when the crust was so hot and the forces so huge that rock could be bent and folded like waves in the water. We find fresh rhinoceros tracks and then those of a whole pride of lion, including small cubs. But the river has become closed in by reeds and holds no promise of photographs. Of the road north, my map warns: 'Hazardous track, not recommended. Extremely sharp rocks. Lions. Caution'. It sounds like the kind of place where few will venture. My kind of place.

29 APRIL

I climb a pointed kopje that stands alone and wait for the sunrise. To the west the desert is a vast fog-shrouded plain with a few isolated hills. The air is cool. To the southeast the Brandberg dominates the horizon in black silhouette. To the south the Goboboseberge rise above lower ranges whose ridges form the pleated divide of the Ugab River Valley. From the foot of my lone kopje, the shale ridges run north–south, sharply defining the alternative lines of rock and yellow grass like striped serpents of some prehistoric world. The land is bare. There is not a single tree to be seen anywhere and what little brittle dry grass there is cannot hide the earth. The rock is black. The soil runs from iron-oxide red through ochre to pale yellows, all of it bleached by the sun to faded pastel hues. A lone oryx is all the life I see. It is dwarfed by the landscape, a tiny speck in a powder-dry world.

In the afternoon, we take a track that runs smoothly down to the bed of the Huab River. There is fresh lion spoor in the dust of the track and a single oryx walks away from our arrival into the surrounding stony hills. The valley is not as dramatically cut through rock as that of the Ugab, but is surrounded by tall mountains, most more than a thousand metres high, framing every view in a grand amphitheatre. In the riverbed, water issues from a reach of verdant reeds 200 metres across and several kilometres long. The river has flooded recently, and in the cracking surface of drying mud, grass is pushing forth in a blaze of green. I am surprised that there is little game taking advantage of the luxury. Perhaps the grass is not palatable. I pick a stem and chew it. It is luxurious on the tongue, soft as spinach leaves and juicy, but the aftertaste is bitter.

30 APRIL

I sit on my haunches and take in the surround of mountains as the sky pales with the coming light. All mountains create drama in a landscape, but here they cut short every horizon with their bearing. It comes to me that in a temperate climate this would be a place busy with people. The peaks would all be named, trails marked to their summits. There would be water sports in the river and a golf course in the lower hills. The land would be divided and fenced and fought over and each would fashion their fragment of it to their own idyll. But the desert offers no excess. Humans must be frugal to live here, and it seems that there are but a very few who are. We take of the earth's plenty, convinced of our right to it. So entrenched is this conviction that we will defend it if we perceive that our fair portion is slighted. But it is not ours – we just take it. The desert, where there is nothing but the magnificence of the earth, bare, without comfort, teaches us that.

As we crest the saddle that makes the height of the Huab Valley, the entire country to the west lies under a quilt of fog. It runs in tapering streamers into the valleys. Mountaintops rise through it, like islands severed from the ground below. At Bergsigpos, goats are drinking at the trough of a windmill. A small black dog of confused heritage rushes, barking, at the Land Rover. These dogs are a common feature throughout the herding communities that coexist with wildlife. They are used to deter predation of the small stock that wanders untended into hills populated with predators. The dogs are raised from puppy-hood with the young lambs and kids, sharing their pen and sleeping with them. It creates a remarkable bond between the animals. Once they are weaned and released each morning to wander in search of grazing and browsing, the dog goes along as an integral part of the herd. In the presence of predators, though, where the stock runs, the dog turns barking to confront the aggressor, success-fully giving the defenceless sheep and goats time to escape. It has become a highly successful strategy and is encouraged throughout Namibia's community conservancies, since it saves farmers from losing stock and conservators from having to eradicate predators that kill domestic animals. The dog ignores us but bristles at the now-stilled Land Rover. I am taken by its fearlessness, for it is no bigger than a lamb itself.

In a dry streambed there is an elephant path that runs straight towards a dis-tant windmill. There is fresh dung along the path. I have not seen an elephant for more than four years and the promise of elephants is like an invitation to a party with old friends, friends I have missed. High on a flat top of the closest hill a lone bull stands silhouetted against the sky. I frame him in the view of a long lens. He flaps his ears and throws his trunk in the manner of one entirely content with life. The waxing moon stands behind and above him. Of all the creatures on earth, the elephant is my totem, my fastness, my coming home.

I MAY

It seems that we have in the last day crossed some invisible line with regard to vegetation. There are trees in all the valley bottoms, and a few isolated, hardy individuals on the hillsides. There has been rain, and grass flushes between the stones. From the height of a steep, rock-strewn descent the hills appear to flow with green as the wind stirs the tasselled heads of the ankle-high grass. It has brought the game, and the hills are alive with the movement of distant life. We descend into the warm valley, passing small herds of mountain zebra that trot with the high-stepping pace of quarter horses through the stone strewn around. There are oryx, and on the valley floors herds of springbok, several hundred strong, flow like quicksilver over a tabletop as they run from our advance and then stop, bunched tight, to turn and watch us pass.

Our way is abruptly interrupted by the double veterinary cordon fences that run east–west through the landscape. The twin fences divide the foot-and-mouth areas of the north from the disease-free south. The fences run right across Namibia, and are a condition of the European Union to safeguard against diseased beef contaminating European stock. The first fence is as high as my chest. A second, to the north, is double the height of the first. In several places elephants have broken through the double barrier to reach the grass and green trees on the other side. We watch herds of antelope and zebra flow through the breach. A spotted hyena scrambles beneath the broken lower strands, dragging a carcass half as big as itself in its jaws. It stumbles on the tatters of trailing skin as it runs.

Poachers Camp is a spring that seeps into the land in the middle of a valley, cupped in a semicircle of mountains and tall hills. The valley is so green that it seems an imposter in the midst of the arid country. Giraffe step slowly, coming to drink as we pause beneath a tall mopane tree for respite from the sun. In the afternoon, we retrace our track. The elephant bull has come down from the mountaintop to drink. His grey hide is rust-coloured from the earth beside the cistern where he has dust-bathed. As the windmill rattles in a stiff wind the elephant places his trunk over the opening of the pipe that delivers the water into the reservoir. He holds it there, collecting the fresh unsullied water and then lifts his trunk carefully to his mouth, tipping the contents in, not spilling a drop. He repeats the process over and over, with no sense of haste. When the windmill stops turning, he waits. The absolute lack of urgency, the calm, is infectious. It is one of the reasons I am so fond of elephants: they put time in its place. In their presence, time is not a line; it seems more like a pool, a history of ages held still.

2 MAY

'It's easy to criticise, looking from the outside in. Doing is much harder.'

We are talking to Garth Owen-Smith about the shoot-and-sell policy of the conservancies.

'The intention is to be perfect, to do it right, but there are always mistakes. It's how we go forward. Learning from the mistakes.'

Garth is a pioneer, if not *the* pioneer, of community-based conservation in Namibia, and Namibia is now well versed in this practice. He is a veteran, now five years retired, who seems as busy as ever. He is in his early seventies and his reputation is held in high regard, from the lone herder out in the barren hills to the hallowed halls of European governments. His name, as an introduction, is a credible reference in every wild corner of this wide land. He is as I remember him twenty-five years ago, when we lived for a year among the Himba people of Kaokoland. He is barefoot, his trousers frayed away around the ankles, dirty at the knee. His shirt has been washed too many times, the colour bleached out and the collar showing threadbare. On a New York or London street he might be tossed a coin, but out here it is another badge, that of a man too distracted by passion to be bothered by self-image. Or perhaps a mind whose passion demands that humility is a prerequisite of self-esteem. His and Margie Jacobsohn's dedication to the idea that communities can and should conserve their historical lands to their own benefit, incorporating traditional rural farming with wildlife conservation, has built an empire that I think not even they and those who stood with them dared dream of.

The first conservancies were registered in 1998, four of them. The conservancy audit of 2013, only fifteen years later, is almost too incredible to believe. Seventy-nine registered conservancies, covering nearly 165 000 square kilometres, with 175 000 residents, represent more than 50% of all communal land and nearly 20% of all land in Namibia. It is almost exactly the same amount of land again as is conserved in Namibia's national parks. The conservation of their traditional communal lands by rural Namibian communities brought in N$72.2 million in returns for the communities in 2013, and since 1991 has contributed N$3.92 billion to Namibia's net national income. But it is not just the people who have benefited from community conservancies. The elephant population grew from seven thousand five hundred to nearly twenty thousand. Lions in the Etosha region were able to expand their ranges westward to the sea and northward to the Kunene River, their numbers having increased by 600%.

It is interesting to analyse where the conservation of land and its flora and fauna leads to benefit for its people. The 2013 audit revealed that thirty-nine communities had entered into joint-venture tourism enterprises, generating nearly seven hundred jobs, while forty-four concessions had been granted trophy hunting rights, giving more than two hundred and sixty people work, and generating the greatest single direct revenue earner for the communities. Hunting fees brought in N$20 882 315 in 2013, whereas tourism generated N$9 568 742. However, it is important to note that most of the N$23 982 130 paid in salaries and wages to community members came from the tourism sector. I have occasionally, in our time in Namibia's wilds, seen women gathering plants or gum from trees, and so it is interesting to know that indigenous-plant and grass harvest brought communities N$5 186 265, and had 2 762 people out reaping the natural harvest of the land. Community members also received 542 280 kilograms of meat from game hunted on their lands. Of all the activities by which rural communities benefit from the conservation of their land, the shoot-to-sell policy of hunting animals en masse to sell as meat represents only 1% of their total income. It, however, is a complex and emotive issue that fires perhaps some of the hottest debates regarding sustainability and ethics.

In the face of two years of serious drought, in 2012 and 2013, the shoot-and-sell quotas set in the good-rain year of 2011 became unrealistic. Garth explains. 'It was a mistake. Many of the conservancies are in arid areas where the climate is fickle. We have learnt that now, and we are having a meeting here,' Garth taps the table, 'on the eleventh, to remedy it.'

In our material society, with esteem based on price tags, and success gauged on how much one can collect, I have found greed, envy, egotism and condescension, but I do not recall having found greatness. My esteem of value runs to some other worth, a worth whose measure shines from within, even through clothes washed a good few times too many.

3 MAY

In the red dawn sunlight, a black rhinoceros emerges from a field of red basalt stones. Reclusive and wary by nature, black rhinoceros take fright easily, so we keep our distance. This one has been dehorned – the horns cut off to make it worthless to poachers. Perhaps for some this degrades the impressiveness of the rhinoceros, but for me this fascinating relic of a beast retains all its rhinoceros-ness. A wild artefact, making bizarre and fabulous the arid stone hills of its chosen realm.

We leave it be and follow the rough track over an escarpment, descending into the heat. It is our lucky day. We find a second rhinoceros, a female. We are on a collision course on the track and I stop and wait, my breath loud with adrenaline in my ears, for the black rhinoceros, as much as it is retiring, has little hesitation in charging when confronted. She stops, stares at us with her short-sighted eyes, her ears turning like radar antennae. The animal is stocky, heavy muscle bulging over the foreleg, the neck thick and rounded holding up the heavy head. The muzzle is deeply wrinkled all the way to the tip of the pointed upper lip. Snot runs in small rivulets from the inside corners of the wide nostrils, which are flared as she tests the air. She shifts on her feet, her step springy like an athlete anticipating the start of a race. Time holds still. The rhinoceros backs a few paces, swinging from side to side, then moves slowly off into the bush of the wash in which we are stopped. Fifty metres from us she backs into the shade of a dense mopane tree. She stands still for perhaps fifteen minutes and then lies down to sleep. It feels like a compliment.

As the sun sets a ball of red in a purple sky, I look out over the country that runs like an undulating calm sea to the north. Beverly comes up beside me. We are utterly alone. I take her hand, kiss her, and throw my arm out toward the scene.

'Welcome to my world.'

She smiles up into my face, comes to stand right against me.

'I've always been here,' she says softly.

4 MAY

At a tiny settlement of houses, between mature mopane trees, we meet Emil Roman, one of the three brothers who live there tending small herds of cattle, goats and sheep. The elephants come every second night, right between the three houses, the gardens protected by flimsy fences, to drink at the reservoir that feeds their water trough.

'If we think the water is not enough we release some from our storage tank,' says Emil, pointing at the water tank that stands beside a slowly turning windmill. 'They don't bother us. They like to come when it has gone still, when everyone's asleep. The bulls come first. Once they find it okay, then the cows come with the calves. They keep the calves surrounded, protected. They come fast. The other night one wanted to eat my fruit,' he says, indicating with a tip of his battered hat a pomegranate tree less than a trunk's reach from the fence. 'I just told it to shooo! And it went away. One night, last week, some people fired warning shots on the road.'

'Shots?'

'Ja, the elephants were in the road. I had to go up there to see what was happening. I told them, don't be afraid, be careful. We stand right here in the vegetable patch and watch them, me and my wife and children.'

'In the dark?'

'Ja, in the dark.'

Emil Roman is a soft-spoken, self-effacing man. The elephants trust him. I feel I could do the same.

5 MAY

The town of Kamanjab is little more than a crossroads. In the dimly lit trading store, Himba women, in traditional dress of a calfskin skirt and naked torso, squeeze past me in the narrow passage between the shelves, leaving on my clothes smears of the red ochre that covers their skin. Their headdress and decoration are fancier than in the 1980s when we made our book, *Himba: Nomads of Namibia*. It has been embellished with personal flourishes, fashioned in brass wire, buttons and shells that are a lavish fashion statement. Their long red-ochred braids now end in a burst of free-springing hair, like roses at the tips of long stems. They smile when I greet them in my smattering of long-forgotten Otjihimba: 'Wapenduka.'

'Wapenduka nawa.'

'Eh, nawa, nawa.'

There are Herero women too, in their traditional Victorian skirts bolstered from beneath into a wide flaring train. One is in a matching outfit of almost neon orange; her bonnet, which flares to points on either side of her face, is fastened with a brooch of silver and lime green, matching the brooch on her breast and belt at her waist. An old Herero woman wears gigantic sunglasses, stained with dust and rimmed by broad frames of plastic tortoiseshell, in the dim interior of the store, too driven by appearance to be bothered by the fact that she can hardly see the shelves. Himba men, it seems, have abandoned their tradition here. Most dress in Western style. We see only two who wear the traditional cotton skirt of the men, their traditional hair plaits or scarves abandoned for the common shaven head. Apart from the two, I cannot tell Himba or Herero or Damara apart.

6 MAY

A jackal wakes us in the early hours of the dawn. We are camped a short way off the track in a stand of mopane trees, between bouldered kopjes. The jackal is close; I can hear it grunt as it builds up to its distinctive call. It is high-pitched and eerily plaintive, like that of the coyote and the dingo. It is both exhilarating and comforting, a signature of the African night.

In the morning, there are mopane bees, a plague of them. These tiny stingless bees, a little bigger than a fruit fly, buzz in swarms around our heads, seeking the water in the corner of our eyes and the salt of the perspiration around our faces. There is no respite, except to climb inside the Land Rover. I have never heard of their honey, but they make a tacky wax that withstands heat. The wax is black and prized in keeping worked skins supple. Almost every traditional drum I have seen in Africa has a wad of mopane beeswax at its centre that the drummer warms over the flames of a fire and spreads across the skin before playing.

The Palmwag Concession, a massive area of 5 500 square kilometres, sandwiched between the Skeleton Coast National Park to the west, Kaokoland to the north and Damaraland to the south, is a brittle-dry fragment of its full self. When we last visited, there were elephants in the reed-filled riverbed and eddying herds of game across plains, blonde with tall grass. Today, there is only heat-beaten stone. There is no life. Nothing for any life to eat. Tufts of dry yellow grass, as brittle as uncooked spaghetti, snap and disintegrate beneath our feet.

The barren earth makes me think of the community conservancy policy of shoot to sell. Here, eight years ago, the stipulated take of a maximum of 5% of the estimated population would have run into hundreds of animals. Today, it would not even be one. But the ground is not littered with carcasses; the game has not died en masse of thirst and starvation. In this land without fences, the game has moved. That is its tradition, its instinct, its survival. To my mind it therefore becomes critical, when considering any harvest of wildlife, whether for trophy or for meat, not only to do an annual count but also to take into account the animal densities in neighbouring conservancies. Not to do so is to take an unrealistic view, one that does not bring migration into account. Invariably, the animals will congregate wherever food is most plentiful. To take 5% of those herds that are in abundance will, in reality, translate to 5% of those same herds being taken every year, because each year the animals will move and tend to centre in whichever area has the most food. The current model, then, only considers abundance in terms of relatively local terms, and does not weigh perhaps the most important consideration of abundance in terms of all the land.

In the evening, we camp on the fringe of a wide, dry river valley. A single black-backed jackal comes brazenly close as I cook ribs over an open fire. It steps right into the firelight, a metre from my grid, and although it skips away if I move quickly, it is unafraid. We put our shoes inside the Land Rover in case it takes a fancy to old leather.

Later, I lie on the ground in the darkness with a camera, imagining an image of the bold cunning of this long-persecuted, long-surviving predator, which will eat the berry fruit of the trees to survive out here. I purse my lips and call to it as one would a puppy. I see it on the fringe of our light. It does not come. I call again. I can see only the vague outline of the trees, the pale silver of open ground, shadows. I call again, longer this time, searching for its shadow moving on the vaguely lighter ground. I hear it to my right, moving closer through the dry grass. I purse my lips, calling, luring it in to my photograph: from low down up towards a star-filled sky. It keeps coming. It seems to be making too much noise for so small a creature. I peer into the murk. I see its shadow moving, but it is too big for a jackal. It is big and it is close. It keeps coming even though I have stopped calling. Its footsteps crunch in the brittle grass. It is a rhinoceros, coming straight at me. I bolt for the Land Rover. Only 12 metres separate us. But the rhinoceros is more startled than aggressive. It stops and peers at the dark, misplaced apparition of our vehicle. It watches for a few moments, then turns and walks on up the low hill of the riverbank into the night, as my heartbeat thuds loud in my ears.

7 MAY

I glass the country from a high ridge. By my reckoning, the closest other human is 60 kilometres away as the crow flies, to the southeast. To the north, it is double that. We see a lone rhinoceros in a wide area of small undulations and walk out to it as the sun crests the eastern hills. The wind is blowing strongly and we keep ourselves where the rhinoceros cannot pick up our scent. It stops every so often and browses on a variety of bushes. At the milk bushes, a candelabra-like euphorbia with clustered, thin, green upturned stems, it plucks the stick-like growths with its prehensile upper lip, biting them off. I have tried the same with my hands; the stems are remarkably tough, fibrous, and immediately ooze milk-like sap that dries tacky and cloying on one's hands. The rhinoceros suffers none of my struggle, nor my discomfort.

The track runs north over a naked earth. Any life out here is incredible, but a rhinoceros is a miracle. It is as much a phenomenon as a polar bear out on the barren waste of the ice sheet. How does so large an animal survive in a land so deprived, so hot? Out there, where the flat distances dance in a blue mirage, where stone is broken by the heat, one comes to stand before these wild denizens with regard. To see a rhinoceros here is to see it fully: a fantastic figment of evolution. It shapes one's understanding and sense of privilege at being able to stand aware of our remarkable earth. It is unbelievable that someone could kill it for its horn. It is a barometer of how far humankind has drifted from any connection with the earth. It burdens me with shame: that our most singular gift, to be able to think outside of instinct and to reason, should have led to a path so far from humility, from consideration, from love. It is in our cleverness that we have moved away from the desert rhinoceros, to the point where we have made idols of ourselves, a god above all things. A god that can demand the killing of a life so special, so rare, that in Asia its horn can be ground down into powder and sold as an aphrodisiac. The desert is a poor place for tears. It soaks them up and leaves only heat and stone.

8 MAY

The rhinoceros come in the darkness before the moon is up, the light crunch of the stones beneath their feet betraying their presence. At around midnight, when the moon slides into the night sky, they are standing there, at the spring, drinking without haste. One is particularly interested in the camera we have set up, which is surrounded by rocks, low on the ground. It approaches the camera from various points of the compass, stepping slowly, bringing its feet gently to the ground, its head held low, nostrils flared. As it gets nearer, it stops, pauses, takes one step, pauses again. Coming forward, drawn by curiosity, but untrusting, tense and ready to spring away. Like a dog sniffing at a still, coiled snake. When the camera shutter clicks, it springs in fright and then drops its head and charges three quick steps with a deep snort. I worry that it may flatten the camera but it stops just short and turns away, trots 10 metres, then turns back, head held high, ready for treachery.

In the morning, we find rhinoceros again, but they are far warier in the daylight than at night, and run at the sound of the Land Rover until they are gone over the hills. In the Hoanib River Valley there are lions too. One of the reasons we have come to this isolated snaking oasis of trees that cuts its valley through the desolate desert is to talk to Dr Philip 'Flip' Stander, a recluse renowned for his singular dedication to the desert lion, the subject of his research for more than twenty-five years. He does not have a home, and instead lives his life out of his research vehicle. He avoids people as though they were a pox. He lives, eats and sleeps in the desert with the lions. He wears only shorts and scribbles all his notes on the skin of his arms and legs. Recently, however, there has been love in his life. Emsie Verwey is doing brown hyena research and assisting with the administration of Wilderness Safaris' Hoanib Camp. She is tougher than she seems, with a penchant for the challenges of a life in the remote corners of this harsh land. Every time we see her it is at a place where the road ends, with only frontier beyond and all around. She takes us to meet Flip, whom she says she has habituated to visitors.

He stands, shirtless in the sun, scratching at the calluses and scabs of old wounds. He is animated about the lions, past and present. We talk of our work, of the years in between. When we last worked in Namibia, a pride – one of the first to take up a territory beside the sea – had been poisoned by cattle herders when they ventured inland.

'I try not to get bitter, their life is so hard already. But there are always new lions, new life. We move on.'

Flip has told us that a group of five young males, 'The Five Musketeers', is down somewhere in the riverbed below the camp. The wind blows out of the dry hills as hot as the exhaust fan of a blast furnace. It is 35°c when we lie down on the sheets and try to sleep. Somewhere in the dark timelessness of night the wind stills and then switches to the west. There is a shiver of cool in its first stirring, and when it grows earnest and blows hard it brings with it the chill of the sea. The temperature plummets in minutes to 12°c and we must crawl beneath the duvet to sleep.

9 MAY

There is mist in the riverbed. It creeps like a tide into the hills. We climb an isolated rise that stands alone on a gravel plain. From its top, the country lies wide below us, the fog drifting inland from the west. Its tendrils creep into valleys of rock and sand, casting the riverine anna trees into relief. It is the desert at its softest. The sun rises and lights the furthest fingers of the fog with an orange-gold, as it spreads between the confines of the hills. A giraffe plods heavily across a drift of low sand dunes, a dark silhouette in the blue-grey tide of land-bound cloud. Red-billed francolin shout their call from the branches of dead trees, loud in the silence of the muted desert morning.

In the afternoon, we track lions south into the valleys that rise away from the riverbed. We find them in a place of steep-sided kopjes of tumbled red rocks. There are five young males peering at us from between the giant stones. All of them are collared for research. Ugly, fist-sized protrusions that hold the batteries and satellite communication units stick out from the base and side of each of their necks. The lions are soporific in the heat. Sleeping from time to time, one raises its head, blinks at the surroundings and then flops back down again. We stay with them until dark, when they move out of the rocks and come to lie in the open of the plain. As it grows dark I use the headlights of the Land Rover from time to time to focus on them, making wide images of the lions in the surround of the jagged rock hills and barren plains. When they finally move off, we camp right where we stand, not wanting to disturb the plain where they have gone to hunt. We have just sat down to supper when Flip's spotlight bounces through the window as he drives up the valley towards us. I climb out to greet him, averting my head from the bright light until he draws up beside me. I turn smiling to greet him and invite him to join us for a dinner of calamari and couscous, but I am cut short. Flip is terse and angry. He blusters that we cannot use any kind of light around the lions. He has been trying to get these lions habituated, and this kind of thing may do irreparable damage.

We are stunned into silence. It is common practice in wildlife photography to use flashes, and we have not used spotlights to search for animals or in night-time photography for decades. I begin to apologise profusely, but Flip backs away, calling out to meet him at the lodge the following day.

10 MAY

We try to make our morning cheerful, to imbibe the hope of a new day's cold dawn, where a thin rope of fog lies between the high-cut sides of the river-bed, but for two photographers whose primary concern is for their subject, last night's encounter with Flip has left us upset.

Finally, a giraffe walking the sharp bank edge, the mist cotton-woolling behind it in shades of silver and purple-grey, helps us to focus on our work. We find fresh elephant tracks in the riverbed, a lone bull, and we track it upriver to where it is standing waist-deep in the green abundance of a woolly caper bush. It is browsing slowly upriver and we move ahead of it, waiting for it to catch up, making images as it comes. It is without haste in all it does. It seems to savour each moment, slowly chewing the seedpods of the anna tree that it plucks one by one from the branches, or picks up from the ground, shaking the sand from the roots of a short shrub that it has pulled free with the curved tip of its trunk, throwing it over its head so that the sand lands cool on its neck and trickles down its shoulders. It calms me to be in its presence, like having tea with an old friend whose thoughts are sound with wisdom and bring balance once again.

At the lodge, I seek out Flip. He is at Emsie's tent, making a fire in the sun. He is as angry and unassailable as he was last night, and although he hears me out as I explain that we do not have a spotlight and did not follow the lions with lights, and used the vehicle headlights only to be able to focus as they lay around our vehicle, he remains unreconciled and walks away.

'Emsie, what must I do? I can't fix this if he keeps walking away.'

'It's the lights. Even he doesn't take pictures at night.'

'Why doesn't he just say so? Of course, we won't do it if he doesn't want it. He's barking at the wrong dog. We play it completely by the rules, but we've got to know what those rules are. I was talking to him yesterday; he knows that we want to photograph the desert lions, but he said nothing about lights or flashes or photography guidelines.'

When Flip returns, he is calmer. Now that I know the crux of his issue, I apologise, and we are able to resolve the matter without further fuss. He offers me a glass of wine.

'I've still got to work today, but instead I would like it if you offered me your hand.'

When I walk away, ten minutes later, he stands.

'Thanks for coming to talk it out.'

11 MAY

As we drive into the riverbed the cold through the open windscreen stings our faces. I glance at the thermometer: it is 10°c. We have only just turned upriver when I see lion tracks in the dunes. They are fresh, still sharp-etched, the sand held crisp by the cold, which the heat will later cause to crumble. We find The Five Musketeers lying on a low dune amid the fallen limbs of a dead tree.

One of the lions grows suddenly taut with the tension that possesses them when they sight potential prey. It holds its head low and moves quickly forward, with its body dropped into a slight hunch. The others watch. When a second lion from the group follows the first, I abandon our position photographing them against the remarkable backdrop of the dunes and circle slowly around towards the riverbed where the two have gone.

On the edge of the riverbed I see a lion, low in a crouch behind dense bush. Up and to the right, a mother and baby giraffe are browsing a low acacia. We position ourselves wide, with a clear view of the open riverside and switch off the engine. The giraffe steps out from behind the acacia, and a lion we had not seen erupts up the bank from behind. It charges, throwing spurts of sand as it digs deep for purchase in the soft sand. The giraffe is at full gallop, its calf hugging its side. The lion runs, gaining a little. The calf veers right, away from its mother. The lion charges toward the gap, forcing the divide. But the distance is not closing, their speed the same. The lion sprints on, the ground growing hard. The baby giraffe has fallen slightly behind its mother, which has turned back towards the riverbed. The other lions come bursting out of the bush 50 metres away on the giraffe's flank. The female turns abruptly into a gap in the bush. The calf is galloping behind, its tiny legs high in panic. As it swerves after its mother, the first lion lunges at its flying hind leg with an outstretched paw. As the tiny giraffe breaks clear through into the riverbed, the lion that we first saw crouched low behind the bush, and which has all the while remained parallel with the chase, launches itself at the neck of the baby in a tackle that erupts in dust. A giraffe calf is not a large meal when shared between five male lions, and within an hour only one is left gnawing the large bones, the others retired panting to the shade. Their bellies are distended, their heads thrust into the low branches to avoid the attentions of the mopane flies.

At midday, we load our gear into a Cessna 210 headed for the coast. Gert Tsaoseb is there to meet us. African wilderness is his passion, and we have a remarkable afternoon together driving through the country. We talk in Afrikaans, his second tongue after Nama-Damara, of wilderness, of wild things, about our work and guiding, and of the changes in the land. Gert is more earnest than genial and has a shy smile that comes lightly to his face. On foot, we follow a lone elephant into the coastal dunes and attempt in our images to express the incongruity of finding the largest land mammal on earth in this wasteland of sand.

When I tell him on our return that he should know that if I die on the way home that I have died happy, his smile creases his face. 'These pictures will be good for Namibia, I think,' he says.

14 MAY

In the hours before dawn a Cape fox is scouting the bush around our camping place. It is a petite creature, about the size of a small dog, but longer in the body. It runs lightly, its paces short like a tripling horse, so that its body is still above the carriage of the quick movement of the legs. I squat motionless to watch it and am surprised that it would come so close. Perhaps two metres distant, it stops to regard me, raising and cocking its head. It has one of the most beautiful coats in all of the animal kingdom. A dense fur of silver-grey, like moonlight on pale sand. Its muzzle is a mustard tan and the tip of its dense bushy tail tapers to a dark grey. It seems too delicate a creature for this harsh place.

On our way downriver the female giraffe that lost her foal is still in the same place. The fog drifts past her, the sun a luminous silver-gold coin shining through the translucence. She is standing in the riverbed, like someone too early or too late for a train and confused by its absence. She stands without eating, staring downriver, then turns in a half-circle to look in the other direction. Her head turns to every movement. Her ears swivel to the nuance carried on the first cold zephyrs of the wind.

Once, years ago, I saw an owl do the same. Its companion killed by a speeding vehicle, the owl returned night after night to where the carcass lay. It called from the branches of a tree, flew closer, and called again. The silence that was its response was perhaps the most potent epitaph of death that I have yet experienced.

15 MAY

To find elephant and lion in so hostile, so barren, so arid an environment is to wonder at the possibilities of nature. Life is the singular phenomenon of our planet, and to see it here is to recognise how fierce a drive life is. It does not give up – it fights. As we sit in the path of the slowly browsing elephant herd I look about the riverbed. It is hemmed on each side by buttresses of sheer rock and stony hills. It is a place flogged by heat where no more than a few drops of rain may fall every other year. And yet in the sand of the riverbed there are deep-rooted trees and copses of bush, even a coarse grass growing in isolated tufts. To this narrow ribbon of wealth, a flurry of life is drawn.

The sand is pockmarked with the indentations of passing feet. From the platter-sized discs of elephant tracks to the tiny pad prints of genet and the scurrying paths of mice. In the trees, every crown is a flutter of movement. Eagles pant in the shade and tiny sunbirds hover over some small delight. There is no silence, there is life. Birdsong is constant, its overture, its contentment, in the fullness of living. An oryx snorts a warning at our still presence; an elephant rumbles its long, deep contact growl; a baboon barks. To see life fully is to be overwhelmed by its force, and much more so in this place, so hostile that it defeats rock.

The elephant head upriver, feeding exclusively on the fallen seedpods of the anna trees. They amble from tree to tree, stop in its shade, and sweep the ground with the tips of their trunks, smelling the pods. They pop them into their maw one by one, as one would eat popcorn. They are so singular in their eating only the pods that I decide to try one. It crunches at first like popcorn, but quickly reduces to a toffee-like gooeyness that is sweet. There are flat, brown seeds, the size of a bean, too hard to chew, that remain. I spit them out. The elephant swallow them like watermelon pips, reseeding the river as they go. Each time the elephant move past us, I loop ahead until they have grown so familiar with our presence that they pay us no heed as they walk past only a few metres away. After several hours, a young cow grows curious, and as she passes in front of the long-stationary Land Rover, she pauses and stretches out her trunk towards the bullbar. The tip hovers briefly a few centimetres away before she coils up her trunk against her cheek, a gesture of uncertainty. She reaches out again and this time touches the metal. The tip of her trunk is like a hand held in a V. She moves it over the rounded bar, then retracts the tip to put it in her mouth. She shows no reaction but to blink her eye, which at this short distance is amber, brown and large, set all around by long lashes. She moves away. On the black metal, a wet smear remains.

16 MAY

I write for the morning in the shade of a pair of anna trees. The acacia anna is the stalwart of the riverbeds. These two are typical. Their boles are contorted, perhaps eight metres around, and give rise to a slanting trunk that soon splits and divides. Their tops are domed, the understorey trimmed in a flat line at the maximum reach of giraffe and elephant.

In the afternoon, as the sun leaves the valley early behind the mountains, lions make their way out of the dense bush to lie on the open sand. There are four, two males and two females. All are young, the oldest male only just starting to show signs of a mane. None of them are collared. They watch us fixedly, as if our vehicle were a new phenomenon to them. After a time, the younger male gets up and, with his head lowered, stalks a few steps towards us, his eyes intent. When we make no reaction he abandons his pose, looks around and lies down again. One of the females follows his lead, closing in, but her head is cocked to the side on a turned neck, more wary than brave. After an hour, they have all come to within eight metres of the unmoving Land Rover, inquisitive, as they peer beneath the truck or lift their heads to see better our movement in the cab. As the sun leaves the mountaintops they yawn and stretch and walk, with that lazy roll of their paws so typical of confident lions, down to the water to drink. Once they are satiated, they return to their curious inspection of the Land Rover. They are bolder in the darkening twilight, and although they circle wide around the front they steal slowly close in behind, very close. I can see only the haunches and tail of one male in the rear-view mirror. I am afraid that it might try to claw or bite the spare wheel at the back, and I pump the brakes to make the lights flash, but the lion does not react. I turn the motor over, just for a second, enough for it to cough. It does the trick, and the lion jumps away.

They lie close around us, watching, the tips of their tails twitching. As we drive slowly away they come to their feet and bound after us, happy at last that we are moving, that they can make chase. It is a boisterous scene in our rear-view mirror, the lions tackling each other, rolling head over heels and then bounding up alongside the Land Rover again. My last view of them in the mirror is of two sitting in the track, with that look a puppy has when you take away the shoe or sock it has been playing with.

17 MAY

At the height of the land, the valley opens onto the wide Giribes Plains. The land is dry, a fall of bleached yellow between black hills. The distant mirage wavers like a watery reflection in the heat. The last time we were here, rain had fallen. The plain was a Serengeti of game, herds drifting across the wealth of grass. We stayed for three days, alone in an Eden of the pristine earth. I have read that the Giribes was the site of one of the recent shoot-to-sell events where wildlife was shot en masse for the benefit of the community. Glassing the whole slope of the Giribes, I see only three zebra.

We turn west towards Fearless Pass, climbing into the stony hills and finding two giraffe in a place so barren that they seem a figment of the heat. In the middle of these arid plains of stone, so wide that their horizon crops distant mountains just below their tops, there is a lone pool of water. It is two metres across by three metres wide, situated in a slight depression, with the ground all around trammelled by animals. The pool of Ganias Spring does not overflow; it is just a pool, and for 30 kilometres in any direction there is no other surface water at all. I walk a circuit around the tiny artesian spring, reading the ground as to what has walked out of this barren horizon. Ostrich and oryx are the most common, with a few brown hyena following the game paths in, but on the far side there are elephant tracks. Two of them, a few days old. I look up at the land, at the far horizon. I turn a little and look again until I have covered all points of the compass. It is featureless, vast. There is no beacon at the spring. How did the elephant come here? How did a poor-sighted pachyderm that struggles to see a human at a hundred metres navigate to this tiny point in all this hostile space? By memory, I am told, but what memory is that? Smell, the feel of the ground? What map do they carry in their heads? By what sense is their place in the land known?

The Aboriginal people of Australia sing a map of the country through which they move – the songlines. Perhaps elephant do the same. I like the image of that, an elephant song, sung and heard only by them. But we do not know. It is a mystery fitting to this place. Perhaps, for once, we should leave it so. For what excitement is there, what adventure, if every stone has been turned over, every secret known?

19 MAY

The hot berg wind that was a breeze yesterday is a gale today. We pass the small settlements of herders, tiny houses built of dung and clay, windowless, with tin roofs. Most are no more than three metres by two – enough space for a bed. Living is done outside in the shadowed lee of the walls, or under a tree. Three women, two Herero and one Himba, sit on old paint tins and crates beside a water point, where the ground is grazed so bare that only white stones remain. A cow with curved horns, all sunken skin and hipbones, splatters liquid dung on the ground. Two donkeys watch, unmoved, batting their ears at the flies.

In Sesfontein the stock has eaten away every morsel of vegetation, including the roots. It leaves loose earth that the wind lifts in sheets so dense that visibility is reduced to 10 metres or less. We drive with our lights on through the dark storm, grit in our mouths.

At a rough shack offering tyre repairs, five men are sitting idle, staring through the swirling dust. 'This is the only rain we get here,' one of the men tells me. He is wearing a bright floral Hawaiian shirt and sunglasses so coated with dust that he must rub little circles to see me. His hair and skin are the pallor of the grey clay dust.

22 MAY

In Etosha National Park, we sit after dinner at a waterhole where six rhinoceros have come down to drink simultaneously. It is a spectacle in the secret nocturnal life of the black rhinoceros, and it happens 20 metres from where we sit. A cow and her almost-adult calf are in a standoff with a bull. All three stand motionless. The calf huffs a rhinoceros warning. The bull takes a half step backwards, and stops. The tension crackles between them in the stillness. The cow charges, her calf a half-step behind, right at her side, their flanks touching. The bull drops his head towards the charge, running backwards. The cow huffs loudly and repeatedly. The bull retreats 12 metres, then stands his ground. The cow breaks her charge 1.5 metres from the bull. Their horns almost touch. Almost. The bull swings his head ever so slowly to one side then the other. The cow has her head up. The bull steps backwards. The tension eases. The bull backs up, turning to the side. The dust of their scuffle billows around them in the halo of the camp spotlight. The cow turns to the water, her calf touching her side with its shoulder as she walks. She stops at the water's edge to drink. The calf walks into the water. Right in, until only half its head and the top of its back show above the water. The cow follows, wading to the centre of the waterhole where the two of them stand side by side in that stillness that humans find impossible to emulate.

A stillness that waits without waiting. Humans are aware of doing it and need a word to describe it as a conscious action. We call it patience. But wildlife knows a stillness beyond patience, which is just stillness. In that stillness, they hear and see all around them. It keeps them alive, for in that stillness any infraction is betrayal, a reason to flee. The quarter moon is setting behind the two prehistoric beasts up to their shoulders in the water. Their ears turn to every sound.

25 MAY

A cheetah walks across a plain. The air is cold. The coats of a herd of black-faced impala are all raised to keep them warm. They graze on as the cheetah walks directly towards them out of the sun. But the cheetah does not stalk, it walks. It stops once to look back over its shoulder. It raises its head to see better and then turns and walks on. Even with the glasses I cannot see what it is walking away from, but if a cheetah fears it may lose its meal to another predator, a hyena perhaps, then it will not expend energy on a hunt. The impala see it only when it is dangerously close. Their alarm snorts fire through the still air like shots. They stand with their necks erect, their ears cocked forward, stamping the ground. The cheetah walks on without a sideways glance. A blacksmith lapwing keens as it dive-bombs the cheetah, just as it is swallowed by the trees.

We see the domed backs of elephant through obscuring green and stop in an open area, perhaps 70 metres across, hoping they will cross where we can see them. It is a small breeding herd and they come directly towards where we wait, stepping into the open in a line. At a large, spreading mopane tree in the middle of a clearing, the lead cow stops. The others join her, huddled into the shade.

There are four cows, a young bull and seven young elephant, from tiny calves to six year olds. For a while they shuffle a little, touching their flanks, scraping a trunkful of dust to throw on their foreheads. Gradually, they begin to doze. A cow closes her eyes, rocks slowly on her feet and rests her forehead against a tree. A calf lies down. Another follows. They lie stretched out on their sides, trunks extended. After forty-five minutes, all seven are lying asleep on the ground. They lie all higgledy-piggledy between the legs of the cows, heads on their feet, rumps resting against their legs. They sleep with their eyes closed. I had never noticed this before, but I have never seen this happen before. We are only 20 metres from them, and they came to us, lay down, and went to sleep. I have been invited many places but seldom have I felt so privileged.

One of the tiny calves grows restless. It steps on another as it moves to its mother to suckle. All four feet of the one asleep rise off the ground, but it just raises its head, flaps an ear, and lies back down. For a long while it is quiet. An ear flaps. One lets go some air. Then there is a new sound. An elephant is snoring. The air wheezes out and saws back in. It is the cow closest to us. The last section of her trunk is flat on the ground and the sharp angle of the bend seems to be making the snoring sound. She starts awake, like a child nodding off in class. The snoring stops, but she falls asleep again quickly and snores again.

26 MAY

The twitching tip of a tail gives a lion away as it sleeps in the grass. We wait, and the tip of a tail becomes a battle-scarred male lion, who walks to the site of its night kill and sniffs the ground. There is nothing there but flattened, blood-smeared grass. As he ambles back towards us we hear a familiar sound: the meowing of cubs. The lion flops down in the shade of a copse of sedge grass. We hear the cubs again but cannot see anything. About 20 minutes later, a female sits up right under our noses. The cubs tumble around her. She moves away from their tail-biting ebullience and attempts to suckle, to lie off on her own under a copse of sedge close to the male. Another female sits up. She too is lactating. There are six cubs between the two females, no more than a month old, and they gambol over the prone adults and bite the twitching tail. A military helicopter on anti-poaching patrols flies low, close behind. The females take fright and leave the small but effective cover and walk out towards the edge of the pan. Their cubs, no higher than the short savanna grasses, follow behind. They walk in a strung-out line, the two females ahead, the cubs in between, with the male bringing up the rear. They are walking towards a lone, umbrella-topped acacia tree set in relief against the flat barrenness of the pan. It is postcard Africa.

27 MAY

It is dark when we set out, the stars still in the sky. I am not an early riser by habit, but I do love the dawn. It comes gently. A tide of light, crisp with cold and full of promise. The black sky morphs into an opal paleness, all pink and mauve along its edge. Hemingway wrote that the world is 'true at first light'. Perhaps in the short hour of calm before the world wakes, we can be true to ourselves, and in truth there is hope and possibility.

I drive, not wanting to be anywhere except moving through the country with a sense of the earth as company. It is moving that helps me stand still in my soul. Dawn is the time when contentment comes to me unbidden. When I can savour the richness of being alive without purpose or distraction. I would make it longer if I could, but the sun lifts above the horizon and the light it casts sets the world in motion once again.

Doves and sandgrouse come raining out of the sky to drink at a circular waterhole in a palette of flat, open ground. As the sun grows warm they arrive in flocks of hundreds, dark clouds that shift in shape, expanding and drawing out, only to compress again, catching the light as they turn. A young lanner falcon hunts them along the water's edge. Time and again, it charges into the middle of a flock that it takes by surprise, but it seems unable to single out an individual prey within the melee. Eventually, by sheer weight of numbers and dogged persistence, it is successful, snatching its prey mid-air in a chase that is a blur to the eye. Its flight is just centimetres above the ground, like a fighter jet, except here the speed is silent and the air instead is filled with the call of sandgrouse. Not raucous or excited, but liquid and fragile, like petals that are thrown out to flutter to the ground.

Due to strict regulations set in place in an attempt to curb the recent spate of rhinoceros poaching, we are escorted to a hide in the non-tourist area by the sector ranger and three policemen in brown camouflage fatigues. They stand in the back of an open Land Cruiser with automatic rifles slung over their shoulders, the curved magazine flat across their chests. The waterhole is situated in a natural shallow depression of bare ground, and one can see the game approach from hundreds of metres all around. Zebra and wildebeest in long lines, groups of springbok that rush out of the bush in a mad gallop and set all the other game into a panic. Flocks of ostrich, which whirl with fanned-out wings at the pandemonium. Oryx that stand and stare and wait for another animal to drink first. The elephant, plodding heavily, emerge from the bush in the full heat of the day.

Only the elephant drink at the cistern, finding the outlet pipe and closing it with the tip of their trunks, slowly drawing trunkfuls of clear water. The elephant are all bulls and among them there is a distinct hierarchy. The smallest last, the biggest first. All of them are massive, and it is sometimes difficult to determine which are the larger ones, but as they approach the cistern it becomes clear. If the approaching elephant is larger, the smaller elephant at the pipe gives way.

As the afternoon wanes, the pan grows quiet. A small group of giraffe follows one of the game paths to the water. The male with them sees another male approaching out of the trees. The male abandons his harem and heads directly towards the second male. Without preamble, they break into a fight, standing flank to flank, and swinging their heads in wide arcs to slam them into their opponent. They are in earnest and several times they bodily lift the legs of their rival with the force of their blows. The leader of the harem seems to be winning when he swings his neck fully two hundred and seventy degrees, dropping his head as he does so to strike an upward blow with his thick, blunt, bony horns directly into the chest of his adversary. The dull thud of the contact carries clear over the open ground. The opponent staggers backward, but the giraffe that delivered the blow staggers too. For a moment, it seeks its balance on the rubbery legs of a punch-drunk boxer. It stumbles to the side. The other giraffe recovers and follows. The first giraffe, clearly unsteady, backs away. It has nearly knocked itself out and the fight has gone out of it. The victor, by default, sniffs at the females of the other's harem, but none are in heat.

The sun drops to the horizon. Of all the hundreds of animals here during the day, not one remains. The day pauses on the cusp of change. A gust of wind carries the first hint of cold. A jackal trots down to the water, opening the stage for the night to begin.

29 MAY

In the afternoon, we travel to perhaps our favourite waterhole in all of Africa. We have worked here twice in the past, and because it is well removed from the tourist areas, the wildlife is not wary of humans and our encounters with game have been exceptional. As we come out through the trees, its familiar vistas open before us. My heart quickens. We are to spend three days here. We have chosen the time of the waxing moon so that we can work into the night.

The plains around it flow with the ebb of animals. I feel like a stone in the centre of a stream, as they part around us to close again behind. It is not yet dark when the first rhinoceros walks up to the stationary form of the vehicle. It comes close, stops, and raises its head, ears cocked forward as it snorts the black-rhinoceros huff, which is like the first puff of a steam train leaving a station. When we do not react, it drops its head and shuffles backwards and to the side, all the while keeping its head towards us. It is patient with its interest in our unexpected presence, several times intimating a charge to provoke us to move. Finally, satisfied, it turns away and walks off to drink.

Behind us a leopard has come out into the open. It keeps low to the ground, sinking to its belly to drink where the overflow seeps into the ground. When its thirst is slaked, it sits up and stares at us for a long while, and then, with a flick of its tail, stands and melts into the deepening shadow of the trees.

30 MAY

The sandgrouse arrive on the cusp of dark. One or two pairs land in the dust close by our open doors. They utter a few calls, the signal for a deluge. The hush of evening is filled suddenly with the flutter of wings. Dark shapes race across the last orange-red of the sunset sky, to drop to the ground. They rain from the dark like a waterfall, until there are so many that their calling drowns out all else and leaves me feeling as if I were sitting neck-deep in a gurgling brook.

31 MAY

The moon is almost full, lions are calling, and the dark forms of animals detach themselves from the shadow of the trees to grow distinct against the pale relief of the plain. A breeding herd of elephant are so thirsty that they break into a run, their grey hides almost blue by the light of the moon.

They are gone when the lions finally reveal themselves. A lioness and her six-month-old cub. The cub is curious and approaches a rhinoceros cow and calf, which stand with their heads lowered to drink. The lioness warns against it, calling her low grunt to the cub and moving away. But the cub ignores her and walks right up behind the rhinoceros, until it is only a few paces away. Through the lens I can pick up on its indecision. The cub is so close that the attack would require no more than a bound or two, but the rhinoceros is massive. It turns its head to check with the lioness, and just then the rhinoceros cow smells it, whirls around, and without hesitation charges. The cub is slow off the mark and stretches to a sprint, with the rhinoceros scything to within a metre of its tail. There is the thunder of the rhinoceros' heavy charge, the calf coming with its mother, but the cow does not carry the charge long and comes to a huffing halt in a billow of dust, the lion sprinting away. When the lion cub realises that it is no longer being chased, it stops and turns to face the now-distant rhinoceros. It looks around and sits down nonchalantly, suddenly brave. When its mother comes walking up, it crouches low and pounces on her, holding onto her rump. The lioness stops and lies down. The rhinoceros have turned away.

I JUNE

We drive north to a waterhole at the centre of a half-kilometre-wide sandy waste. The ground all around is littered with elephant dung. A three-storey observation tower dominates the open ground. A strong wind is blowing, raising a pall of dust. A few kudu sprint into the wind.

It was here, in this crude steel structure, enclosed by waist-high shade cloth, that one of the most incredible discoveries in animal science was made. In the mid-1980s, in Amboseli in Kenya, Joyce Poole and Katharine Payne discovered that elephants communicate with subsonic rumblings, sounds inaudible to the human ear. Twenty years later, Caitlin O'Connell made an even more remarkable discovery: elephants listen through their feet. Using seismic detection equipment, usually used for earthquakes, O'Connell discovered that when elephants stand still and listen, something anyone who has spent time in the African wilderness will have witnessed, they do so not through their ears, but rather through their feet. Listening to, or, more accurately, feeling the seismic tremors from as far as 3.6 kilometres away and far beyond their ability to see, an elephant can not only tell of the approach of other elephants, but is also able to distinguish the individual elephant that they are listening to.

In the evening the wind dies, and as the sun sinks low, breeding herds of elephant come marching out of the bush in single file. They disperse around the water and for several hours we sit deep in their midst as they drink and socialise, spread out in groups over this featureless, flat sandy place. It has just grown dark, the almost-full moon a hand-width above the horizon, when two elephant bulls erupt into one of the rare occasions that I have witnessed real aggression. The dispute is over quickly, but the dominant bull is so incensed that it chases the loser, bellowing and screaming, at a full run, across the open ground and into the trees. All night, bulls come singly, in pairs, or occasionally three together, to the clear water of the spring. In the light of the high moon we see a full-maned lion cross into the open. It stands statuesque on the plain and roars, filling the whole space. The elephant bulls stand quite motionless, their trunks turned upward, listening.

4 JUNE

At midmorning, the wind fades away. An hour later, the game start to drift out of the trees, relaxed in the warming day. Among them is a dehorned rhinoceros bull that plods wearily through the soft sand. It comes to drink. As it lowers its head to the water I see a deep gash with a trace of blood in the centre of its forehead. It looks like a bullet hole. I use the binoculars to examine it more closely. It is a crucifix-like wound where two deep gashes from fighting have crossed each other in the centre of the bull's high forehead. The raw tears are centimetres deep in the hard hide.

In the last decade, 7 245 rhinoceros have been poached in southern Africa, their horns hacked off, their bodies left to rot. Removing the horns of rhinoceros is an obvious solution. Although it detracts from the look of the animal, I personally see no contest between seeing a hornless rhinoceros and seeing no rhinoceros at all. This morning's rhinoceros, with its battle scars, however, brings another aspect to the debate. If one dehorns a bull rhinoceros without doing so to all the other bulls in the vicinity, then humankind's well-intended intervention has ramifications. A dehorned rhinoceros is at a distinct disadvantage. For females, they will be inadequate to the task of defending their calves. For the males, the gene pool of the strongest, biggest, largest-horned rhinoceros is thrown out of balance. The natural selection process becomes subject to those animals that humans have missed, so that a rhinoceros with its horn is king. The rhinoceros we are looking at clearly thought itself up to the fight, but its horns are stubs. Its opponent lacerated its face and twice scored telling blows. The rhinoceros defecates and urinates on the large midden, 10 metres back from the water, scraping the midden like a dog with stiff back legs after it is done. It lies down right there and falls asleep in the sun.

5 JUNE

At Namutoni, we stop to offer our thanks to Kapofi Immanuel, the warden for the eastern sector of Etosha National Park. Kapofi is a strong-faced man with a gentle handshake and soft-spoken demeanour. There are children's crayon drawings on the wall. I imagine he would make a generous and caring father. Our talk soon turns to rhinoceros poaching. The east has suffered the brunt of the recent spate of rhinoceros killings.

'I am sleepless,' Kapofi tells me in earnest, seated at his desk. I notice that his hands are clasped tight together, one fist interlaced over the knuckles of the other. He is subconsciously squeezing his fingers, squeezing hard.

'You just need one person to talk.'

He nods. Eventually someone will. We both know that. Guilt is a strange brew. Truth invariably percolates to the top. Until then, Kapofi Immanuel must remain vigilant, and that means sleepless. I look again at the drawings on the wall. There is the outline of a 4x4 with a person, big and strong, behind the wheel. There is little doubt that this quiet, articulate man is a hero in someone's eyes.

But he is another kind of hero too: the unsung kind. There are many, their names unknown, who fight against the tragedy of wildlife extinction. History will not remember them. Their walk is not in the limelight. It is hard and in the dark. Their fight is against the very nature of humankind: greed. Greed erodes the soul of man. Rhinoceros will not be the first on the list that enumerates the extent of our disregard for the earth over our own desire. But if we take that train all the way to its final station, then it is plain that, in our inability to address our selfishness, our greed, we will eventually bring about our own undoing. Kapofi Immanuel takes the persecution of the rhinoceros seriously. All humankind should, for we too are standing somewhere in the shadow of the last rhinoceros.

11 JUNE

This morning there are two young lanner falcons hunting the flocks of doves and sandgrouse that come to the waterhole to drink. Their individual techniques are markedly different. The more dominant of the two is a speedster. It comes low and fast, bursting into the flock, where it selects an individual and, using its speed, knocks it hard like a punch. If it misses taking hold of the sandgrouse immediately, it still has the opportunity on a fast banking turn to overtake the stunned bird.

The more reticent of the two falcons uses stealth. As the number of animals around the water increases, it comes in low from the surround of trees. It flies without haste and chooses its course between the densest concentrations of animals beside the water. As it clears the fringe of animals at the water's edge it climbs steeply a metre or two into the air. Its sudden appearance puts the sandgrouse into panicked flight all around. The falcon chooses its target and snatches it mid-air as it passes. The stealth method uses far less energy and its success rate is higher than that of the speedster.

The number of animals around the waterhole increases steadily throughout the morning, until we sit in the heart of a phenomenon. More than a thousand animals surround us. Springbok lie down and sleep less that two metres away, all around the Land Rover. Wildebeest mill in groups, where the cows and calves lie or stand and sleep, while the bulls rub their foreheads in the dust and mud. When they stand again, the hair between their eyes is coated with white dust or black mud, making them look like clowns. Zebra wade right into the water, while the oryx toss their horns to open a space to drink at the concrete cistern. A breeding herd of eighteen elephant arrive and, chasing all the other animals, they annex the cistern for themselves. The game is now so dense that we literally cannot see the water. When a young elephant bull turns and charges a wildebeest that has come too close, the sudden movement ripples through the animals like a rug waved in the wind, the rush of hooves a sigh against the ground. It is only during the wildebeest migration of the Maasai Mara, of Kenya, that I can recall ever sitting in the midst of so many wild animals, gathered so tightly together. It is as if they are herded in a corral.

14 JUNE

I make a fire from gnarled pieces of hardwood that we have collected in the veld. We seldom make a campfire. It is a luxury that requires a surplus of wood, which is rare in the desert regions, and time, which is rarer still when one works until dark. I cook the meat over the hot coals, and just before it is done I throw twigs into the fire and sear steaks on the flame.

After we have eaten, I stand beside the fire and stare into the flames, fixated by the tongues of yellow that lick around the coals. The world around me slips away. My mind grows still, no thought, just the mesmerising eddy of the red heat that glows bright, then dull, as the breeze fans the fire. I feel its radiation through my hands and on my face, and it courses from there to some inner place, like oil on water, until all I know is calm. And, when I look away, there are the stars. Although they have names, they remain mysterious, and I feel insignificant and happy at the same time.

16 JUNE

At a cattle watering point, in northern Kaokoland, I find Tonga Mbinge sitting beside his white mare. The horse is in a lather from the long ride from Ombombo. The pump in the well is broken, he tells me. It will take two days to fix. Cattle are milling everywhere, their lowing and bellows filling the air. Their hooves raise a cloud of fine dust that lies over the place like thin mist to the height of their backs. There are horses too. They lick the walls of the low drinking trough, but there is no moisture there. I realise, looking at their number, that this is horse country. There are no roads into the hills. Every stockman must use horses for his work. The horses are stocky, short in leg and body, but they are spirited and strong-boned. In our pause at the watering place several other men on horseback have drifted in through the trees, curious about the presence of outsiders in a place without interest to tourists. They do not dismount. They lean forward or back in the saddle, resting their elbows on the horse's neck or rump. Some carry short whips, which they flick at the stock to move them away. There are two reservoirs serving two troughs, spaced a hundred metres apart. The reservoirs are both about half-full, but the stockmen do not run the water into the troughs.

'Do the elephant come here to drink?' I ask Tonga.

'Yes.'

'Many?'

'Many, many.'

'How many?'

'Twenty, fifty, seventy,' he mumbles. 'A hundred,' he says finally, with certainty.

'When do they come?'

He points to the west, about two hand-widths above the horizon. At the top reservoir, the ground is littered with elephant dung. That convinces us to wait.

We have yet to see the interaction between wild game and people and their stock. We retreat into the shade of some distant thorn trees, away from the dust of the milling, thirsty animals. There is an elephant path there that comes through the trees. It is a metre wide and full of fresh tracks. At 4 pm we rejoin Tonga and his fellow herders at the empty water troughs. There are more cattle than before, several hundred standing in the shade of the thorn trees. The air is loud with their bellows.

'Ondjou.'

I look up. Two young bull elephants are coming out of the bush. They hesitate only briefly and show no alarm at the shouts of a group of teenage youths, who hang nonchalantly close to where we are parked in the shade. The elephants advance on the reservoir, waving the cattle away with their trunks. When they have drunk their fill, by reaching over the high rim of the reservoir with their trunks, they move off into the trees and browse a few branches. They are clearly waiting, and are unperturbed by the arrival of vehicles and the voices of men come to discuss the repair of the pump not more than a hundred metres away. Cattle and horses mill around them.

'Ondjou!'

This time it is a breeding herd. They are on the fringe of the trees, but they are waiting there. For twenty minutes they wait, until the vehicles drive away and there are only the voices of the men around us and the mournful lowing of the cattle. When the matriarch elephant moves forward, the young ones surge ahead too, charging the cattle with their ears flared, and kicking dust. The herd encircles the reservoir, lifting their trunks over its high edge as they drink. The calves cannot reach the water. The adults are still drinking when the next breeding herd comes out of the trees in single file, walking fast. The youngsters, who cannot reach over the rim, turn and chase the cattle from the trough. But there is no liquid there, only wet mud. Elephant crowd around the reservoir and more are coming. A group of six comes slowly out of the trees, and almost immediately behind them another group appears down the far view of their path. Elephant issue from the trees, walking in single file, until there are more than a hundred in the area around us. It is more elephant than we have seen at one time anywhere. Anywhere! In the Chobe, of Botswana, perhaps we have witnessed similar scenes, but there the elephant were spread out along the length of the riverbank. Here, they are in one place, almost shoulder to shoulder.

The bolder young bulls approach the second reservoir closer to the road, where the horsemen give way. From where we sit the elephant are backlit against the gloaming of gold in the west. They kick up a tail of golden dust as they hurry to the water. The walls of the second reservoir are much higher than the first, higher than the elephants' heads. I watch the gold-rimmed silhouette of the first bull to arrive there. It raises its head, until its tusks rest on the rim. It then levers itself bodily off its front feet, walking forward a pace on its back legs. Its front feet are more than a metre off the ground and its head is over the rim. It drops its trunk to the water, sucks a trunkful, and then lowers itself back slowly to its feet while it drinks.

The horsemen have retreated. It is getting dark and the elephant are soon crowded around the second reservoir. The matriarchs lead their calves to it, but the wall is also too high and its trough entirely dry. The calves squeal when they are jostled in the melee. They trumpet wildly, chasing cattle away, but it brings them no water. The mothers' desperation for their young is palpable. They march fast and determined between the reservoirs with their young in tow. They flare their ears at us where we sit in between as they pass. I see a female drop a trunkful of water onto the ground, beside the concrete wall. Twice she does it, but it is trampled to mud before two small elephants have had barely a mouthful.

17 JUNE

Opuwo is a sprawl of shanties on bare ground, rising up a hillside. The roads and the few trees lie under a thick coating of grey dust. Around the main intersection a few new buildings rise freshly painted and smartly signed, while vendors hawk cigarettes and plastic goods and taxis weave slowly between the potholes. A man drives a flood of sheep down the tarmac, and from the top of an overloaded pick-up, with a shattered windscreen and broken lights, a trussed-up goat blinks at me. At the fuel station an articulated truck is offloading sacks of stock feed, while a pregnant Tchimba woman tries to sell us her goods. She is naked to the waist, her bulging belly wrapped around with a fire-engine-red sarong. She sells bracelets made from PVC pipe, and cloth dolls, out of a large blue plastic tub.

The young people of Opuwo flaunt their cellphones as their badge of modernity. A girl in faux-leather boots and tight-fitting minidress has earphones plugged into her ears beneath a hairdo of tight, blonde braids. She holds her phone before her as she walks, scrolling up and down. A young man in fashion-ably torn jeans, a Billabong T-shirt and sunglasses is wandering the fuel station and taxi area, talking incessantly on his phone. I suspect he is faking it. He is watching people to see that they see him.

In character, the town is the same as it was nearly thirty years ago, only bigger. The noticeable change is the lack of tradition. Some Himba women are to be seen, with their calfskin skirts, plaited braids and ochred bodies, but the men are indistinguishable from the rest in their Western dress. I can pick a few out, in the loitering groups, who adopt the typical Himba stance of leaning one-legged on a stick, the other leg locked against their ankle or calf. Some still wear the skirt of pleated cloth, but they wear T-shirts too, and beanies emblazoned with 'Minnie Mouse' or 'Go Bokke!'. The Herero women are unmistakeable in their colourful, bell-like trains and matching headdresses. Herero men, however, are Western in their attire, with a fondness for Pilbara- or fedora-style hats.

The bottle store is doing a brisk trade. Inside the massive new Agricultural Union, the racks are stacked high with goods, and the staff unpack bottles of All Gold tomato sauce, Chinese plastic washing tubs and saddlery accessories. Aside from a bent old Himba woman who buys half a kilogram of maize meal with an ochre-smeared twenty-dollar note, I am the only customer and they do not have what I want. The single cashier is bored and does not meet my eye.

The newly completed road to the west winds into a country of stony, tall hills, where giant baobabs stand leafless in the sun. Their trunks and limbs are smooth-barked, a pale grey that changes to dark maroon on the sides away from the sun. Many are giants, with huge girths. They resemble sumo wrestlers squatting in repose, and those on the ridges stand as bold buttresses against the invasion of the sky.

19 JUNE

In the hope of encountering desert elephant, we leave the road and head over-land down the Hoarusib River gorge. We make our own path, pushing forward through heavy sand and picking our way, jolting in low gear over rounded stones. The world is left behind, the mountains rising sheer to shut out all else. It is just us and the riverbed, between buttresses of rock hundreds of metres high. We are tiny in its grasp, the sky a pennant of blue directly above, the rugged drama of naked rock rising vertically all around.

There are frequent seeps of water against the foot of the cliffs. The unseen river below rises to the surface, forced up by the rock to appear shallow above the sand for 20, or 50, or 100 metres, before vanishing again to leave the surface bone dry. Lapwings fret and dip along its edge, with algae thick and green growing mat-like where it lies in still pools. We see no life, bar the lapwings and flocks of rosy-faced lovebirds, and in the wet sand along the water's edge the signs of stock are very old. The country is too steep to be anywhere but in the riverbed, and that makes this a transient place for all life: no food for stock and a place of entrapment for wildlife, with the exception of baboons, whose tracks are the only common presence in the sand.

Where the gorge begins to widen and the sides become hills again, we bathe in a shallow pool where schools of translucent tadpoles lie torpid in the cool water. Half an hour later, while I am writing, three girls appear on the far side. They hang back in the trees, peering at us around ledges of rock. Gradually, they make their presence known by monosyllabic utterances, thrown out like stones into a still pool to see what moves. It takes them forty minutes to finally come and sit in the sand beside us, raising their hands in shy greeting. The eldest, in her early twenties, is carrying a baby. She wears a plain, dark-blue smock that hangs baggy on her petite frame. The next in line is in her late teens, a shower cap on her head. She is painfully shy and cannot bring herself to look directly our way, except when our heads are down or our backs turned. The youngest is possibly eleven or twelve, and stares at us with candid astonishment from behind the elder's dress. When they finally speak, it is to ask for food. They eat the oranges, crackers and cheese and drink the cordial that we give them with the atmosphere of a Sunday picnic, opening and closing the packaging with an air of ceremony each time a portion is doled out.

Downriver we pass their onganda, or traditional dwelling, two domed struc-tures of bent poles, to the height of a man's chest, perhaps 2.5 metres across, covered with dried dung. A few articles of clothing and bolts of cloth are thrown over the top. In the riverbed, the sand around a strong seep is busy with the imprints and droppings of their stock.

An hour on we come across a young boy, no more than twelve years old, walk-ing alone upriver. He stands aside to watch us pass, and when we stop to greet him and ask if he is okay, he too asks for food. He is carrying only a small day-pack but it is completely flat. In the rear-view mirror, I watch him walk upriver, without looking back, slowly peeling the mandarin we have given him.

20 JUNE

Yesterday it took us the whole day to drive almost 70 kilometres down the Hoarusib riverbed. What sign we saw of elephant was very old. This morning, as the valley begins to open out from the final confines of the gorge, we cut our first recent sign of elephant activity. As we come around a sharp bend in the tall hills, the valley broadens suddenly before us, like a pipe emptying into a dam. The hills fall away into long rises of red stone to the west, dimpled beehive-domed hills of pale ochre to the east. The western sky is cut short by the rise of the Etendeka Mountains.

The riverbed widens to more than a hundred metres, a meandering sand snake, shaded by dense stands of anna trees, camel thorns and leadwoods. Giraffe stare down at us from within their sea of green and ostrich trot out of our way into the stony surround. Away from the riverbed the earth is bare. We climb a promontory to look around. The desert is wide around us, a barren red ground that falls sharply off peaks nearly a thousand metres high. Along a mountainside, ostrich walk in a line, a necklace of black dots around the narrow rise of a ridge. There is a peppering of giraffe between the trees, and there are springbok too, but no elephant.

In the riverbed, I zigzag across the sand as we head downstream looking for signs. There are tracks now, and dung, but all are too old to bother with, until suddenly the fresh spoors of two elephant cut diagonally across the track. I turn along their tracks to the west and find the elephant a few hundred metres on in a close stand of trees. There are two cows, two young bulls and an older bull. One cow is old and dust-baths herself almost continuously. The day is hot, 34°c, and the elephants use the loose sand lying in the deepest shade to cool themselves, throwing it onto their backs, behind their ears and along their bellies.

I have been given figures of six desert elephants left in the Hoarusib River below Puros, and nine in the Upper Hoarusib. When we last photographed them on a trans-Africa journey in 2000, there were at least twice as many. I have not read any published hypothesis as to their decline, but suspect one of the major causes to be increasing human presence, both tourists and pastoral-ists, along the river course to which they are bound as a food source. Their recent history is tragic. The elephants wander up and down the riverbed and its fringes, according to the seasons, moving more than a hundred kilometres along its length from inside the boundaries of the Skeleton Coast National

Park in the west to beyond the settlements of Otjui in the east. Summer rain, with its season of grass and surface water, allows them to move away from total reliance on riverine vegetation. Winter draws them back. In the last few years, a researcher wanting to establish their ranges botched the tranquillising process while sedating an elephant to fit a radio collar and killed it. Another elephant was shot after it trampled a tourist to death at Puros. I have little doubt that the elephant was not at fault.

Puros is a hub of Himba culture, popular with tourists, and is community land outside of the rigorous rules and regulations of national parks. The elephant coexist with the nomadic pastoralist Himba in a relationship of mutual regard and avoidance. The sight of an elephant in the vicinity of people or settlement seems to make tourists behave in extraordinarily dangerous ways. Because there are cattle around, people walk up to elephants as if they too were domesticated, ignoring the ample warning signs elephants give when they feel uncomfortable or upset. It takes persistence to get an elephant to charge you. Like almost all wildlife, their preference is to walk away.

A third elephant was killed by a farmer in the east: he shot the elephant for raiding his crops. But, not wanting to be blamed for killing an elephant he knew to have value to community conservancies, or to be prosecuted for shooting it without the requisite authority, he shot it in the gut. The elephant walked for several days, 50 kilometres into the quiet heart of the sanctuary of the Hoarusib River gorge, before it died.

21 JUNE

As we cross the Hoarusib northward onto the narrow 4x4 route into the stony hills, a Himba woman and her children come to the road ahead of us. The children clap their hands and stamp their feet, singing a traditional Himba song. Their mother has a small stand made of branches beside the road where a few bracelets and carved wooden oryx are for sale. As soon as we stop she puts out her hands.

'Omakaya?' (Tobacco).

She is pregnant. Around her are seven children, their song dying away as they stare at us. Her hair is cut short, her body un-ochred and she wears a shabby light-blue dress. Somehow the move away from the traditions of culture has robbed this woman of something, a certain dignity, and left her depleted by a world she appears to want to embrace and yet is clearly not within her understanding.

Himba tradition provides a quantifiable hierarchy and social structure. For the Himba to step out of it is as complex as my trying to step into it. This woman cannot possibly see herself as I do. Abandoning tradition is not uncommon, but it is an unconsidered view. Tradition is a cultural heritage, and to abandon it to appear modern in Western terms, as many do, is to fail to recognise that modern is a state of mind, not of dress or social practice. I have encountered very few who have retained their culture while adapting to the ways of the contemporary Western world. Without exception, they have been individuals whom I admire. What they have achieved takes wit, fortitude and courage. They are the few among the many.

22 JUNE

In the last dark, a pearl-spotted owlet calls and then there is a stillness. Absolute silence. I can hear the breeze sigh quietly through the bush, giving the false hope of a nearby stream. We have come north and east of the Khumib into a country of high hills. Flat-topped mesas and domed kopjes fall sheer and then steeply to the valleys. They are wooded with drought-stunted trees. In the silent sunrise, I search their flanks for life, but find none.

We drive north and in the heart of the valley we come across a church with a corrugated-iron roof on galvanised steel girders. The walls are gabions of stone packed between steel mesh. In the centre of each wall, white rock has been used between the red stone to create a cross. A steel cross, painted silver, shines bright in the sun at the apex of the roof. The door is locked, but through bolt holes we can see that the interior is disused. A concrete pulpit is the only adornment, and 'Immanuela' is inlaid crudely in a mosaic of red and blue. Bird droppings coat the floor.

Missionaries did not, historically, have great success with the Himba. A sedentary populace is necessary for religious teaching, and the Himba are nomads. Today it is not politically correct to use bangles, baubles and mirrors to attract converts, despite the Himba tradition of adornment, and so this church, donated by the Dutch Reformed Church of Namibia, remains empty, a well-meaning gesture with no depth of sensitivity to the place.

The Himba are for the most part animist. There are places on our map that denote 'holy mountain', or 'holy plain', implying a spiritual worship, a credo of belief and the practice of ceremonies. But never in our year living with the Himba, in 1989 and 1990, did we ever see any religious ceremony or encounter any place reserved for ceremony. And in 1989 Himba culture was prevalent and steadfast, not fractured as it is today. The Himba believe that time flows from behind us, overtaking us, that we are looking forward into the past. Ancestors are therefore important figures, and they are revered and consulted, especially around the central fire in the onganda, but it would be wrong to say they are worshipped. No shrine is built to them and no ceremony conducted aloud to their name. The Himba do have ceremonies and rites, but these are cultural and without religious overtones. There are no shamans, not even traditional healers, as there are in other southern African cultures. We have seen spirits being exorcised from an ailing Himba woman: it was done by the women of her community, who performed a series of rituals and danced themselves into a trance state. Whatever the core of Himba spirituality may be, it is practised quietly and alone.

Beyond the church we turn north, on an even less distinct track, to visit a place special to us. Otjihaa, as we remember it, was a spring of good, clear water that pooled into a natural basin of calcrete rock. It was a place that the Himba came to when all other water had dried up, traditionally at this time of year. Otjihaa was striking for its classic perfection of how a spring looks in one's imagination: a clear pool in a surround of white stone with a fringe of sedge and short grass. We spent a month there, among a people far removed from the modern world and yet entirely satisfied and content. It was the place where I fully realised how well the Himba knew and lived in the desert. To get there we had departed Etanga at dawn and driven the whole day over a torturous road to arrive in the early evening. Wapenga, a Himba friend of ours, had departed Etanga after us on foot. In the evening, when we arrived, Wapenga was already at Otjihaa waiting for us. He and his younger brother had crossed the mountains on foot, traversing the rocky landscape wearing sandals cut from rawhide.

It was my first introduction to what it means to walk like a Himba. When you ask a Himba where a place is, the answer, as he points, is 'Ngwi!' which shows only the direction. It is the emphasis given to the hand that is pointing that shows how far. A little tip of the wrist indicates it is close, anything from a few hours to half a day away by vehicle. A straight arm and hand, given a short, sharp shake, means it is far, at least a day's journey. A little click of the fingers during the short shake means it is very far, definitely more than a day's travel and possibly more than two.

While we were working on our book on the Himba, Wapenga proposed to Kata. To do so, he and his father, Kathetaura, a man then already in his mid-sixties, walked for two days through the desert without food and water. They arrived at our camp in the evening and for half an hour exchanged civilities before asking for something to drink. The Himba are desert people and walking is their way across the land.

The road into Otjihaa is as I recall it: a slow, narrow track through rocky hills, the ground loose stones of white quartz. But the settlement of scattered ongandas is different and we cannot find the spring. We stop at a large onganda of several huts and a big stockade on top of a hill to ask directions. Walking in on the family having their morning meal around the outside fire, little has changed since our time here twenty-five years ago. Some receptacles are plastic, not gourds, and the two elder sons wear partial Western dress, but everything else is traditional Himba. The women are all naked to the waist, their bodies and braided hair smeared with red ochre mixed with butterfat. An old man wears a threadbare jacket unbuttoned over his shoulders for the cold, but his waist is girdled with the loincloth of pleated blue cotton, and on his chest is a neckpiece of a pair of steenbok horns fashioned into a pendant between wire beaten into angular silver beads. The young unmarried women have kohled their eyebrows and around their eyes.

The eldest son directs us to the water. The spring is gone, covered over by a slab of concrete. Above the calcrete ridge a borehole has been sunk into its core, a diesel engine thumping in a steel cage of oil-soiled ground to pump the water into a polycarbon tank on a steel stand. From there it is piped to a concrete trough in the centre of a pitch of bare earth. All the ground around is trampled to dust. The tap below the tank is broken. The situation is the epitome of how the efforts of well-intentioned development NGOs can go awry.

Throughout Kaokoland, we found that springs and traditional wells have been replaced with boreholes. The intention, Garth Owen-Smith tells me, is to provide a reliable, uncontaminated source of water for rural people, keeping the water they drink separate from that contaminated by stock. When we were last at Otjihaa, the stock were kept from the spring by a barricade of thorn branches, the water scooped by hand into a trough hewn out of a tree trunk. Today, I watch women fill their receptacles from the mouth of the pipe that delivers water to the trough. Dogs and goats lick the water from the same opening when they are done. The Lister diesel engine, with its heavy flywheel, is shuddering violently on its mounting, and when I look more closely I find it is held down by rocks. I open the cage and take a look. Bolts have sheared off, been stripped, and the boltholes have been worn to gaping tears in the mounting blocks. Men gather around. I ask them to shut down the motor and for an hour and a half I repair it with what bolts, steel plates and washers I carry as spares. Before I start it again, I shake the 220-litre fuel drum that stands to the side of the cage. The men shrug and turn their palms upwards. 'Opuwo.' It is finished. They swing the heavy crank, close the compression and the engine fires quiet and steady. Each of them in turn grips its top, shoves it to confirm it is fast and then turns to thank me with a nod and a smile.

But my repair will last only a few months and then it will break again. In the meantime, the men must sell stock to buy diesel to pump water that was once free. And when the pump finally wears out one day, or the engine topples off its worn-through base and breaks irreparably, and no one can be found or brought to this remote place to fix it, these people will chop through the concrete that covers the spring and open it once again. The water will still be there, but what was once a spring, beautiful as one could imagine, will just be a place of water surrounded by debris. It is a reminder that good intentions are not sufficient. If one is to interfere in the lives of other people, then what is required is a lifetime commitment, not a once-off visit.

23 JUNE

Van Zyl's Pass is one of the roughest roads in Namibia. It is a narrow descent, through steep-sided hills, and then sheer mountains, into the valley of the Marienfluss. The only alternative route is a detour, hundreds of kilometres to the south and west, and yet is the route most people take. The going is rough and tricky. Several times we must build a way, using stones to fill out deep hollows, where the track descends over steep faces, or where rock cuts through wide crevices. The angle of descent is in places so steep that to use the brakes is to slide, and I am forced to let the Land Rover edge downward in low-range first gear, the wheels turning to keep traction and give me steerage.

The track is long, with little respite from the concentration required to avoid slicing the tyres on sharp rock, wrecking the sump or suspension on boulders, or plunging over a sheer edge. At midday, we reach the final mountains that drop away to give a wide view of the Marienfluss Valley to the west. We stand on a cliff edge looking out. The track we must take can be seen clearly between the trees hundreds of metres below. It makes me apprehensive for the final descent. My trepidation is not unwarranted. At the first incline, we cross a short saddle, and then the track drops so steeply that I cannot see it beyond the bonnet of the Land Rover.

'No,' Beverly says. 'No way … There must be another way around.' But there is not. The hill is sheer all around. Beverly walks in front of the vehicle to guide me as I drive. I pitch the Land Rover forward into the narrow gap of rock, nurturing it inch by inch with low-range diff-lock and my foot light on the brakes. I can feel the wheels come free to hang in the air, as they climb over, into the void. The vehicle is slanted at more than 50 degrees, and the rear wheel is above my head. If I lose traction and slide, then there is just a short width of perhaps three metres of gravel before the precipitous edge. My feet are perspiring. The suspension clangs and rings as each wheel drops to full extension, and is then pushed up again to full compression, against the deeply uneven rock face. The Land Rover sways in slow motion, first hard to one side and then the next. I am grateful we are not top-heavy. The rear wheels come almost level, in a short sharp curve of gravel, and then there is a second pitch as steep and as long as the first. This one, however, is more shale than rock, with grit of loose stone. I take it an edge faster to keep the wheels turning, so that we do not begin a slide that cannot be stopped. The descent is still steep enough that Beverly slides on the loose stone as she walks ahead, but with the gearing I am able to keep control of how we go.

There are three or four other rough and strenuous climbs, down over jagged rock, but none with as dire consequences for failure as the first. On the final rough section, with one rear wheel off the ground, the other's grip breaks and we slide for a moment towards the edge, before I hit the brakes and swing hard to stop. It is only a second, and the danger more distant, but there is dust in a cloud and our hearts are racing. At the bottom, we celebrate with a cup of tea in the shade of the mopane trees.

The Marienfluss Valley is wide and flat, ending abruptly to the east and west in a margin of sheer mountains that rise more than a thousand metres to narrow the sky. We drive north over gentle red dunes that run into the pale bleached ochre of the sand.

At the Kunene River, the northern border of Namibia with Angola, we stay in the Himba community campsite, enjoying the privilege of a steaming hot shower, a roofless flush toilet whose one side is a dense mustard bush, busy with birds, and a shaded campsite right on the river's edge. We have the place entirely to ourselves. Not a single person approaches us or appears anywhere, not even the person designated to collect the fees.

24 JUNE

Sara, the Himba manager of the community campsite, comes to greet us. She wears Western dress, speaks good English and totes two young children. She assists me with my Herero: 'Ohau-uto ye-ura' (the car is full), 'Omakaya aand' (my tobacco is finished) and 'Ombura' (rain). She tells me the last time it rained here was in 2011.

Beverly is discussing with her the book we made on the Himba, telling her of working with Margie Jacobsohn and Garth Owen-Smith. She knows Garth, and has heard of Margie. I mention Chris Eyre, who was the Ministry of Environment Conservator for Kaokoland at the time.

'Chris?'

'Yes, Matwi.' I mention his Himba nickname.

'He's coming here.'

I look at her quizzically. 'He's very sick.' We have had reports of the deterioration of his health, the last indicating 'not more than a week'.

'He's dead.'

'He's dead, but is coming here? You mean, they are going to bury him here?' She looks puzzled.

'Put him in the ground here?'

'Yes, at Obivango, on Friday. He is coming in a helicopter.'

Beverly and I exchange looks, and instantly a decision is made. We will stay. Chris was a friend. Obivango is only 30 kilometres away.

25 JUNE

Three vehicles arrive with about twelve passengers. They have come to coordinate arrangements for Chris Eyre's funeral: one to collect firewood, one to carry water, one to shoot a springbok and an oryx, all of them to collect people from remote encampments. The coffin will only arrive tomorrow: the funeral will be on Saturday. Trucks have been dispatched from throughout Kaokoland to bring the people here. They are arriving already.

Chris Eyre was a recluse. I imagine him watching the preparations, waving his pipe, stem outward, and saying in his loud voice, 'Nah, nah man, no need for all this fuss.' His was a genuine self-effacing humility, part of what made him so popular. Chris was the black sheep of a society family of thoroughbred stud breeders in South Africa. So complete was his move away from that world that I only once ever encountered him indoors, attending to paperwork in the Ministry of Environment offices in Windhoek. Of the times we stayed with him in Opuwo, at his ministry-supplied house, it was difficult to think of it as his house. He slept on a sofa on the veranda with his Jack Russells. The rest of the house was occupied by Himba, come for some or other purpose to town. He gave us the luxury of our own room, where we slept on our camping mattress on the floor. The rest of the house was communal. On one occasion, when he was forced to take leave, he spent the time at Etanga talking to Vetamuna, the chief of the Himba, about matters on which he sought guidance. Vetamuna would later grant Chris the right to make his home on Himba land, an act that made him an honorary Himba. He was the first white man to be honoured in this way by the Himba people.

Chris had a fierce conviction and wisdom that stood far outside bureaucratic rules and way above the policies of discrimination of that time. As a conservation officer, he practised his own discretionary law that had at its core two ingredients: fairness and right. It drew people to him. He would give his days to the discussion of a situation in some remote valley, sleeping out on the ground until it was done. The people of Kaokoland came to value his counsel and sought him out on wider and wider issues.

Chris was also renowned for his bluntness. Several times it nearly cost him his job. In public meetings, in his dusty uniform, he would stand up and voice without any ambiguity an opinion in direct opposition to government policy. His bias in these matters was invariably in favour of the Himba, and it made him famous – too famous to fire. You could never reach him by telephone, but he could be found by asking at almost any onganda, no matter how remote, across all of Kaokoland, a region the size of the Netherlands. He was a man for whom a gathering of three or four was sufficient company, and yet he had given enough of his time to be held in regard by thousands. The last time we saw Chris was in 2008. He was retired, living under a tarpaulin with his two dogs, repairing buffalo fencing in the east to keep himself occupied. We will never forget our time in his company, and tomorrow his three-day funeral will begin.

A wind starts blowing out of the east in the late evening. By darkness it has become a gale. It lifts the sand from the ground, until the air is filled with it. It dims the stars and invades everything. When I go to bed there is sand on the sheet and in my mouth, gritty between my teeth. I can hear the wind in the darkness, tearing at the three tall palms between the river and our camp. It feels as if the world is mad at something.

26 JUNE

By morning the wind has died and we drive with Sara to the site of the funeral. She has brought a small travel bag, her six-month-old baby boy Believe, and five litres of water. In Africa, it is common for people to have a traditional name and a Western name. Sara is Uasesenga. I ask her whether, in her youth, she dressed as a traditional Himba. When she says yes, I ask about her hair.

'Mmmm, in two plaits,' she motions with her hands the traditional plaits of young girls that start at the back of the head, coming forward over the forehead. They are stiffened and protrude out above the eyes like tusks.

'When did you decide to wear Western clothes?'

'At school.'

'Did you decide, or did you have to ask your parents?'

'I decided and told my parents I wanted to change to Western clothes. They said it was fine.'

'Why did you want to?'

'The ochre gets on everything: your clothes, your desk and you cannot write because it goes all over the paper and so you wash and cut your hair and put on clothes.'

'All the children at school?'

'Yes. All of them.'

'So, there is no ceremony you have to make to change from traditional Himba to Western?'

'No, you just do it.'

'Can you change back?'

'What?'

'Yourself, to traditional Himba now, if you want to?'

'Yes.'

She smiles and shakes her head, plucking at her floral blouse. She has 'Himba' crudely tattooed on the outside of one bicep.

On a plain to the east of the small scattering of ongandas, at Obivango, vehicles are gathered beneath the trees. The Marienfluss Valley is divided here by a sawtooth ridge of mountains. The western divide is narrow, hemmed in by the high rock, the valley floor of red sand softened by a sparse setting of trees. People have set up their camps in a radius around the confines of the hills. Smoke drifts across the scene. A myriad tents are arranged in the shade. There are hundreds of people. They have come by helicopter, truck, 4x4, car, motorcycle, donkey, and on foot. A small, white-roofed marquee stands at the centre. There is a PA system, and generators buzz out in the sun.

At 5 pm Chris Eyre's coffin arrives by helicopter. There are too many hands that want to lift it. It is carried on the shoulders of the men to the marquee, a troupe of women in full traditional regalia following behind, ululating their high stuttering keen. Groups of men advance on the coffin, their thin yellow sticks held high. Their feet are a stomping shuffle that lifts trains of dust into the still air. Their leader advances, the men a tight cohort right at his back.

'Rrrrt!' the leader calls.

'Hah!' the men shout, short and hard, with a stamp of their feet.

'Rrrrt!'

'Hah!'

'Rrrrt!'

'Hah!'

'Rrrrt, Oh!'

'Hah! Hah!'

It is the traditional greeting of the dead, and its echo comes magnified off the hills. It takes several minutes for the procession to reach the marquee, where the coffin is laid on a white tarpaulin on the ground. The group of women form a semicircle and start a rapid, staccato clapping. They are bent slightly forward, stomping their feet in a sliding backward step. All of them wear a loincloth of turquoise blue. Their skin and braided hair glisten with the sheen of red ochre in the last afternoon sun. The greeting of the dead continues into the dark, the groups of men wandering far off into the trees, their shouts growing distant, then coming back.

As the traditional greetings for Chris's soul die down, the speeches of remembrance begin. Anyone may join in, and the PA system broadcasts it so that the whole valley may hear. The speeches of remembrance will go on until midnight, and until then no one will eat or drink. Our campsite overlooks the proceedings. The multiple campfires glow like fireflies in the night. Around the marquee people sit for a time and then drift away, to return in a while. Three young children in loincloths and polka-dot vests play tag and dance around the coffin. We stand on the periphery of the gathering, looking at the faces and listening to the people who take their place to talk of our friend.

A Himba man, older than myself, one of the few in full traditional dress, wearing over it a twill jacket against the cold, stands to speak. 'Matwi was like a father to me. And I say, like a father, because he treated me in every way like you would expect a father to treat their son. You might say this cannot be, because he was white, but he was, and now my father is dead.'

A young man, tall and gangly, in fashion sweat top and shiny black exercise pants, is next. 'Matwi was like a doorway. Through him you could go to where you wanted, be how you needed to be. A doorway to life, to make you believe. He is gone now, but the door is still open. He opened it in every one of us.'

Vehicles continue to arrive all through the night.

27 JUNE

We are woken at first light as the men once again begin their greeting of the dead. Women's voices respond, ululating from the surrounding hills. The sun is slow in cresting the tall amphitheatre of rock, but as its warmth reaches the marquee the speeches begin. The first is a tribute to Chris's actions, sung by the cohort of men. There are tales of his bravery in front of lion, of saving people from angry elephant. The leader sings a line, then the cohort shouts, 'Hah!' The leader sings the next line, then, 'Hah!', until the whole tale is told. At the end, a man fires two shots into the air over the coffin.

It takes until 2 pm for all the speeches to be read. A Herero chief, in a faded black pinstripe blazer, puts it succinctly: 'The organisers have asked us to be brief, but we are Herero and we will not be brief. We will pay this man the tribute he deserves.' Some are the heartfelt eulogies of friends, some a litany of Chris's work, some propose a full memorial ground here, some preach conservation, and others use the platform for political hectoring. But all of them acknowledge that an era has changed with the passing of Chris Eyre.

The final procession is vocal. No voice is still and the stamping, running feet raise the dust to the wind. When the Herero priest prays, the men hold their hats over their faces. I'm sure that some are as grateful as me for the disguise, for it is hard to slow your tears.

29 JUNE

I stand silent before the view. From my feet orange sand falls steeply, between high ridges of eroded black basalt, to the Kunene River, unseen beneath a kilometre-wide tongue of silver fog. To the north, the mountains of Angola are etched black in silhouette, ragged ridge upon ragged ridge, like a land of giant dragons lying asleep. Shafts of sunlight run in bright, sharp spears, divided by the high barriers of rock.

There is no drinking water in the Hartmann's Valley, so we have rationed ourselves to use the water that we carry only for drinking. We used it carelessly at the outset, certain we would find a well or a bore, but we found none, and I reckon we have enough water remaining for five to seven days. My estimate is conservative, but experience has taught me to err on the side of caution, and to be happy if I am wrong.

Without water out here, the land takes on an entirely different tenor. Thirst is something very few people know. When it comes to our principal needs, thirst is hunger's twin. It is an urgent force of life. You can die of thirst, and out in the desert that possibility comes closer than anywhere else. While we were making our book on the Himba, when the country was more remote than today, we found a prospector who had lost his way. He was delirious, close to death, from thirst. When you pinched the skin on his arm it remained standing up straight. There was no moisture left in him.

30 JUNE

If you walk into a cathedral alone and hear a single chord on the organ pipes, struck loud and sustained, then you would have an impression of the force of the view that greets us. It is one of the most singular we have seen in all the world.

We are hemmed close, north and south, by granite inselbergs, gigantic domes of blue-black rock. To the east, looking back the way we have climbed, the valley floor lies hundreds of metres below, a gently cupped palette of orange and pink sand. Towards its edge, hills of black stone begin the sharp unseen descent to the Kunene River Valley. The distance, east and north, is layered ranges of mountains separated by a fire-smoke blue haze. To the south, the sand valley grows broader, isolated hills of rock standing like ships at anchor in its pale-orange sea. To the west, the land falls gently from the heights in a slope of sand, hued like the colours of a bleached rainbow.

Nothing grows on the sand, not a single tree. Oryx, in groups of two or three, wander the vast barren range, nibbling at things unseen. Far to the west the land becomes dunes, the final distance invisible beneath the thin blue line of coastal fog.

We turn southward, stopping frequently. The scope of the earth here is immense, so expansive that it weighs upon you, as if you were pressed down beneath some giant palm. We pass through vast swathes of red land, where garnet gravel lies fine on the surface. In other places a slope is a run of purple ground that surrounds isolated round boulders the size of a house. Ochre and orange run with the sand from the edge of the mountains of broken granite, like dyes come out in the wash. At night, the Milky Way arches over the silhouette of mountains like the centre pole of a tent holding the unfettered sky aloft.

I JULY

It is a truism that the faster one travels, the less one heeds or imbibes the land. On the district road, we travel close to the western front of the Etendeka Mountains, but at 60 or 70 kilometres per hour they become the backdrop of the world, wallpaper on our room of experience that is the road ahead. It is only when we leave it again and are slowed once more by the convoluted and uneven track that they loom once again as mountains, rugged and imposing, a dominant presence on the earth.

The track takes us into a surround of kopjes that are massifs of granite boulders. The place has recently had rain and around the rock the plains flow like a yellow tide. The short, thin grass is already brittle and dry, but it covers the ground so that, seen from a distance, the stone is gone and the savanna is a plain, as yellow as ripe corn. There are animals here, oryx and zebra, but to our dismay they run from the vehicle. The zebra are the most afraid, and many are already running at the sound of the engine when we first see them. At 700 metres, the oryx bolt and flee. Their anxiety at the approach of humans burns in my heart, because it is only the knowledge of death that makes them so afraid. These animals have all seen others shot. We set out on foot, but it is the same. Only when we use the boulders to screen our advance, or climb to photograph from between the rocks, do we get close.

The landscape is too arresting to pass through and we idle slowly back and forth between the kopjes. Their tops and sides are all manner of balanced and pointed stones and loom like chess pieces on a giant board. There is a bitter wind blowing inland from the coast, and as the light wanes a few oryx finally accept our presence and stand still, warming themselves in the last sun, their coats fluffed up like a dog whose fur has just been towelled. The moon rises full over a low saddle in the bouldered rock, and as it does two oryx walk across its rising face and stop and turn back to look at us. In our family, it is Beverly who believes in God and that such things are His blessing on us.

3 JULY

In a riverbed of camel thorn and leadwood trees, flocks of finches and pigeons drop from the trees to fly away, like seeds swept up in a gust of wind. In the hills, there are groups of Hartmann's zebra feeding on the slopes.

The zebra here are isolated enough to show little fear of humans and, although wary, watch us pass without moving off. It gladdens me when I must wait for the zebra to move out of the road rather than watch their distant dust as they flee. I have never believed that wilderness has its value in what can be reaped from it. Wilderness is the crucible of our beginning, and its worth is a spiritual measure that goes to the very root of being human. We sever our attachment to the earth at the peril of losing sight of what and who and where we are, at the peril of believing that we are God. But we are not God, and we were not able to predict that the already three-year-long drought in Namibia would extend through 2015. And, so, the policy that the wild animals of Kaokoland were worth money as meat or trophies has reaped a grim harvest. From what we have seen, there are not a great many left. It is the fundamental issue when seeing only material value in wilderness. Material value is perceived only by humankind and, as such, becomes corruptible by the foibles of men. When you have profited from the

406

death of wild game, year after year, the idea becomes entrenched and expected, and to stop it or turn it around is as difficult as slowing a runaway train.

To their credit, the Namibian conservancies held a meeting in May and suspended the policy of killing their game, but their meeting came a year after it should have, and only by the courage of one man who stood up against his friends and community and said, 'No more. It is time to stop.' Chris Bakkes put himself on the line when he published his article 'End of the Game' in *The Namibian*. The response was angry and called Bakkes to account, but Bakkes's words precipitated the meeting that suspended the killing of wild game.

5 JULY

Back in the Hoanib Valley, fog settles thick, limiting visibility to no more than 150 metres in any direction. All five lions appear, ghostlike in the gossamer veil of the fog. Sunrise is still an hour away. The lions are on a sandbar in the middle of the riverbed. They seem undecided, and although they spend a lot of time lying down and watching into the margins of fog, they get up and walk from time to time, moving 20 or 50 metres upstream each time. The grey fog pales to silver with first light and I hear vehicles approaching at speed from upriver. The lions hear them too, and after only a moment the first one gets up and sprints away from their approach. The others follow, running across the riverbed directly away from the approaching vehicles.

Until a year ago the Hoanib River Valley was a relatively quiet place, visited occasionally by adventurous overland visitors. The animals had it, for the most part, to themselves. The recent opening of a lodge in the lower reaches of the valley has undoubtedly brought greater disturbance, with aircraft overhead and vehicles using the valley. However, I have seen their guides at work and they are both cautious and considerate towards the game. The lodge's presence has, by default and through marketing, increased awareness of the Hoanib Valley as an exciting wilderness destination, and tourism traffic has increased. It is making the wilderness a busy place, and the reaction of the animals to the disturbance is distress. Action is required lest the Hoanib go the way of the Hoarusib, its neighbour to the north, where the last lions were poisoned to death by semi-nomadic pastoralists advancing down the river valley toward the sea, and where elephant numbers have dwindled in the last decades from herds in their twenties down to the last six. South of Puros, there are simply too many people for them to find peace.

I do not subscribe to the idea that wilderness should only be the province of those who can afford high cost. I am rather a strong advocate of the opposite, that wilderness should be available to ordinary people. This stems not only from opposing a sense of discrimination, but perhaps, more importantly, from the belief that a wilderness precious in a person's heart is a wilderness protected. When the common man loves a tract of wild land, then he will fight to keep it wild, and he can love it only by knowing it.

And so, the Hoanib River Valley's survival as a wilderness becomes a complex problem. People must be allowed to come here, so that it is known and loved and protected, but it must also be kept as a place where wild animals are comfortable and not stressed. In the future, this will require limiting the number of vehicles, as well as enforcing a code of conduct.

30 JULY

We travel north, then east, to a favourite area of ours: the Caprivi Strip. The Caprivi has to my mind always been a wilderness area waiting to be discovered. It remains so today.

A subtropical savanna of mature miombo woodland, the Caprivi was renamed Bwabwata National Park in 2007. The park is conceived as a multi-use facility, and is divided into areas for farming and conservation. I find the multi-use concept of the national park as difficult on my understanding as the new name is on my tongue. In the middle of the park, between the Kwando and Okavango rivers, we pass a sign welcoming us to the multi-use area of Bwabwata National Park. Almost immediately there are smallholdings, with trees chopped down to clear areas for cultivation, the wood for sale as raw poles or chopped into firewood beside the road. Rough fields of reaped sorghum replace the pristine forest, and thatching grass stands in neat stacks for sale. Women work the grass into bundles, tying them with the fronds of palms.

As we draw close to the places where boreholes provide a source of permanent water, settlement becomes denser and the forest sparser. Cattle, goats and dogs range the verge, searching for some succour in the depleted landscape. To call what we see a national park is a misnomer. Of all the activity that we witness in the multi-use area of Bwabwata National Park, only the cutting of thatching grass is sustainable: the grass replaces itself each year. But the trees cut for sale, as poles, will require at least twenty-five years to regenerate. The masses of hardwood, split into cords for sale as firewood, will take twice that long. What we see is human settlement of the pioneer type that strips the land of all its assets, with no apparent attempt at sustainability or replacement.

Without political will and courage, Bwabwata is doomed to be overrun by human settlement, to become like all of the land outside of designated protection, a place trodden to dust by the hooves of cattle and a shadow of its full majesty. Bwabwata deserves more than to pass into obscurity because there was no one prepared to defend the idea of a national park. Slash-and-burn farming has no place in a national park, and once human beings have settled in an area there is no going back. To imagine otherwise is naive. But more disturbing than the encroachment of settlement, couched in the guise of sustainable use, is the possession by professional hunters of some of the most singular reaches of African wilderness that I have encountered.

The Okavango River bisects the Caprivi Strip towards its western end. The eastern bank falls into the Bwabwata West section of the national park. The western bank is given over to human settlement, except for a small oblong of land adjacent to the Botswana border that is conserved as the Mahango section of the Bwabwata National Park.

The Mahango has long been a favourite of ours because, despite its relatively small size, it is rewarding to visit both for its variety of species and for its scenery. The floodplains of the Okavango teem with plains game, hippopotamus and herds of buffalo, while inland the coarse-grass woodland holds herds of rare sable and roan antelope. The fringe of giant riverine trees, including baobabs, supports herds of elephant and kudu.

Across the river, Bwabwata West rivals the best in Africa for its tall hills of miombo woodland that drop steeply to the often kilometre-wide floodplains of short green grass. These floodplains are fragmented by long, narrow islands, where stands of jackalberry trees provide a dark symmetry to the flats, their pools of water rimmed with wading birds. Pods of hippo lie on the edges, beside the deep water, in numbers that are rivalled only by the Luangwa in Zambia. Buffalo feed in and out of the reed beds in herds of hundreds, and lechwe and reedbuck are so prolific that, looking down from the heights of the hills, I am reminded of the pastures of New Zealand's sheep farms. It is an African Eden that competes head to head with the Chobe, where scores of multistorey hotels ferry a constant stream of tourists through the spectacle of an intact wild African river. In Bwabwata West, however, we encounter only two other vehicles, and in Mahango four. Why Namibia has not attempted to reap the potential that lies almost entirely untapped in so outstanding a portion of a national park is a puzzlement, the answer to which comes to me by chance.

We are completing the southern end of the river drive in Mahango when we come to a road leading off towards a large island. A pure white signboard with bold red writing proclaims, 'Private Property, No Entry'. The idea that there is private property within a national park is enough to make me want to drive the road to find out who it is that could enjoy such extraordinary privilege. I do not, but the sign bothers me. At the end of the river drive we turn around and happen upon a vehicle as it comes off the signposted road. I wave the driver to a stop. It is a Land Rover pick-up with the back converted to hunter seats. The driver is a big-boned Afrikaner in a camouflage shirt. On the dashboard, prominently displayed, is a SWAPO party flag. I ask him who the land belongs to.

'It's a hunting camp,' he grins at me, a full-teeth solid smile. 'It's where we slag them,' dragging his fingers across his throat.

'Really? Here inside a national park? What are you hunting?'

But he has picked up on my surprise. 'No, no, I am just joking. It's a camp for Ministry of Environment staff,' he lies.

When I press him to be specific about the camp, he cuts me off with a terse 'We gotta go now'.

The next morning, across the river in Bwabwata West, we drive off the narrow track to allow an oncoming vehicle to pass. It is a battered pick-up laden with passengers. In the railed-in back, seven men stand calf-deep in meat. They grin and wave in greeting, their foreheads and chests slick with perspiration. The pick-up leans heavily as it passes, the mass of bloody meat hacked into slabs. A kilometre further we come out onto the open floodplain to find the rest of their party sitting in the shade of a tree, a dismembered hippopotamus lying in bloody slabs at their feet. 'A hunter shot it,' they tell me.

Later in the afternoon, when we walk out to photograph a group of buffalo wading a lagoon at the edge of the trees, we discover an impala hanging in a tree. Its body has been cut in half, and its head, forequarters and ribs are strung by a rope around its horns. It hangs just above the height of my head, out of reach of the hyena: it is a bait for leopard, for which the hunter will wait in ambush to shoot. It is called trophy hunting. But shooting a leopard lured to bait, or a hippopotamus lumbering slowly out of the floodplain to feed, is not hunting; it is just killing.

I grew up around guns, with hunting dogs around my feet. I hunted duck and geese and game birds from the day I was big enough for the shotgun not to knock me over. But I learnt hunting in the company of men who counted the day, not what you carried home, as precious. I was taught to acknowledge the sadness that comes with killing, and learnt that the sadness only moves to guilt when you waste what you have killed. I last hunted a very long time ago, but I have not changed my mind. To hunt for the pot, be it out of need or luxury, is one of the most intimate ways of approaching the wilds. It is one of the rare occasions in our life when we become the animal again, and in that place the day becomes exceptional and memorable. But the essential premise is to hunt, which is quite different from shooting. It is then joined by the sadness of killing, which demands of the hunter that he respects the life he has taken and eats what he has killed. Trophy hunting, to kill something to display as an esteem of one's prowess, is far removed from hunting. It is steeped in ego. But to kill an animal for no reason other than to hang it in a tree is to move into a place too abominable to respect.

Bwabwata West and Mahango are without doubt among the jewels in Namibia's crown. To find them in the hands of hunters is astounding. It begs investigation. Imagine if the same was practised in Etosha? How has it come about that hunting, which provides considerable income in fees but almost nothing by way of employment, and only rudimentary infrastructure, has taken possession of a national asset that could be providing hundreds of jobs, as well as education, training, infrastructure and even greater fees? Could it be meat? Namibians love meat, and from our observation of the numbers of vultures descending from the sky, on separate carcasses in Bwabwata West, meat is plentiful. But even that plenty, by the time it is divided among a burgeoning community, is unlikely to sway common sense. Something is seriously amiss, and it is the community and the people of Namibia who are being wronged.

1 AUGUST

Returned to the Damaraland Desert, it is 5°c in the bed of the Huab River this morning. A pearl-spotted owlet is calling, a long, drawn-out series of whistles that marks the end of the night. My fingers are stiff with cold. We follow elephant tracks downstream.

The upper reaches of the Huab River Valley pass through country as lovely as Africa can be. The hills, hemming the riverbed, fall away to reveal tall mountains that stand alone to the north and south. The river valley becomes shallow and wide. Climbing out a low bank, we crest a saddle to look out over a vast plain of short dry grass. With the binoculars I can see ostrich, oryx and springbok. The hills of its surround are black stone, with sand dunes lying smooth in their lee. As soon as the animals become aware of the vehicle they run. It sets a cold stone in my stomach. Humankind must forever take; we seldom leave alone.

It has come to me in my time here that the desert river courses, running from the high mountains to the sea, are a few unique arteries that would flourish under determined conservation. The Ugab, Huab, Hoanib and Hoarusib – four dry rivers whose parallel course through the arid regions of the west brings a miraculous diversity of life, so surprising in such a desolate place. They are remarkable now; properly conserved, who knows what status they will enjoy in a hundred, two hundred, five hundred years' time?

Of the four, only the Hoanib is flourishing, and already that reputation is bringing too much pressure for it to bear. The Hoarusib, under advancing settlement and unsupervised tourism, is altering its natural systems, turning under the tread of humankind. The Huab, from where I stand, with a view to satisfy any king, is flowing to the same slow alteration that will eventually bring the demise of all that is wild about it.

The sun is a lightless red orb in the purple haze of the dust-laden sky when we discover the carcass of a young elephant that had been shot. What remains is being picked over by a few jackals and a pair of glossy starlings. We call ourselves civilised, but civilised is a state of mind, not technological advancement. Civilised is a place of courtesy and regard for other living things, of humility with courage, of standing upon that history where what is right has been defended at our own expense. What thought is it that brings men to kill what little life there is out here?

We camp on the riverbank a few hundred metres beyond the reach of the stench. The rising moon is hidden behind a tall kopje. A pair of brown hyena come out from behind the bush that shelters us from the wind. They stop 20 metres from Beverly and watch. In the light of the yellowing sky, the hyena, with their long, shaggy manes, have the aspect of animals of demonry, of wolves.

2 AUGUST

The hot east wind musters itself with the dawn. It hacks dead branches from the acacias and lifts the dust so thick from the unprotected earth that we can see less than a hundred metres ahead. Grit filters through every opening until our hair, our clothes and everything we touch is layered with sand. It crunches between my teeth. The cotton soil is so dry and powdery that our passage stirs dust in such clouds that we can see little at all. Finally, it forces us to abandon the riverbed as our route west, and we climb out into the surround of the stony mountains to the north. On a saddle where we stop to reinflate our tyres and rinse the grit from our mouths, the wind lifts the sand from the ridge of a dune in a 15-metre-high plume, like the mane of a giant horse charging through the mustard-coloured cloud of the valley through which ridges show as vague black silhouettes, like ships in a fog.

At the Springbokwasser entrance gate to the Skeleton Coast National Park the temperature is 33°c. On the journey into the coastal plain I use the windscreen wipers for the second time since our arrival in Namibia, to slap away the constant droplets of the encroaching fog. It is 13°c when we reach the coast and the afternoon is prematurely dark, the monotone land even more sombre beneath the grey luminescence of a fog bank.

3 AUGUST

The sun is determinedly thinning the fog, but even in the brighter light the land has a post-apocalyptic austerity, with nothing to relieve the eye. Each view of isolation is remarkable for its otherworldliness: low clusters of grey rock in a sea of ochre sand, a whole valley-like depression pink with garnet gravel, its rim an icing of black slate broken down into tiny stones. Nothing grows.

Beyond Torra Bay, quite suddenly, there is a reed bed beside the road, and in the small depression of the valley in which it lies, a swathe of green runs towards distant dunes to the east. There is a small herd of oryx grazing the grass and a single coot and a teal in a tiny pond beside the road. When we step down from the Land Rover I walk directly onto the fresh tracks of a lion. Staying well clear of the concealment offered by the reed beds, we climb a low ridge and glass the country all around. There are more patches of green to the north and a herd of springbok grazing on tussocked bushes in between, but we see nothing of the lion.

Nowhere have I seen the miracle of life that water brings more clearly evidenced than here. The land all about is the most desolate and untenanted that the earth can become. I sit silent, with only the wind to whisper the dirge of the absolute loneliness of an inanimate earth. And yet here, where water flowing underground is forced to the surface, a few hundred metres from the sea, there is enough life to sustain lion. It is a most peculiar island, cut off from the world, not by a bridge of water, but by a land too hostile for life. And yet life has come here. The oryx we see are clearly resident in this small oasis. It shows in their coats, which are shaggy against the cold, and, either because of isolation or restricted diet, far paler than other oryx.

In the next low valley to the north water trickles across the road. A few springbok are drinking and coots fussing in the last heatless sun. Standing before the Uniab Springs, in the midst of a vast and desolate earth, it occurs to me that it is only with water that life is begun.

4 AUGUST

Our room at Terrace Bay, currently the only place to stay in the Skeleton Coast National Park, is a tired prefabricated building from the 1960s. On our way out, we pass a separate enclave of bungalows, boldly designated as 'Semi VIP'. I am not sure who or how one qualifies? Perhaps the curious title is a legacy from the days when President Sam Nujoma used to come here to fish. On the pinboard at reception there is a photograph of him with another fisherman holding a big steenbras. In a broad-brimmed, brown felt hat, and with his white-fringed beard, he looks rather like a Quaker minister. Above his photograph is another, smaller picture, of a child sitting in the back of a blue pick-up. The faded legend beneath the image tells that in December 1995 Neelsie van Tonder and family caught forty-two steenbras in five hours. Nine of these were giants of 16 kilograms. From the photograph, in which a child is cornered by a mound of dead fish, it is clear that Neelsie van Tonder did not put a single fish back.

The morning is a rarity, without fog. With special permission, we drive north to Möwe Bay. The road to the north is blown over in sections with white sand. The tracks of large trucks run deep through the soft dunes. Of all the settlements along the Skeleton Coast, Möwe Bay has achieved a character lacking in the others. The buildings are as forlorn in their weathered, mouldy and rusting decay, but there is a cheerfulness here, as if it were a place chosen happily as home.

Bernard Awob, the sector warden, has a direct gaze and with his short, grizzled beard beneath the jaw of his round face reminds me of a bear. Two hours pass as we pore over maps of the Skeleton Coast National Park. Bernard knows the park intimately; he has been here for many years. On the few matters of park policy that we discuss, his answers are considered and insightful. He tells me that the heavy vehicles using the road are heading to a diamond-mining operation at the Kunene Mouth in the extreme north of the park.

'Inside the park?' I ask, raising my eyebrows.

'Yes, the whole area they are mining is Skeleton Coast Park. The government is quite clear. If there is an opportunity to create revenue then they will do it. Last month I took people wanting to establish a lodge at the Hoarusib Lagoon. Here,' he says, stabbing his finger at the coast in the middle of the restricted area. 'On Thursday, I am taking a group to Cape Fria.'

Tracing his finger along the central-eastern boundary, he tells me that he is finding more and more game inside the park. He feels that the animals are coming here to find sanctuary from the hunting in the community conservancies that border the park. He is talking of the area we visited over full moon, where we photographed the oryx and observed the game running away upon sight of the vehicle.

6 AUGUST

The fog lies dense in the dawn, every surface of metal or glass dripping water. The temperature is 10°c. We are camped on a plain of quartz gravel. Our world is a small halo in the ghostly surround of a luminescing earth, trapped beneath the close dome of a silver-white sky, where only a low ridge of black shale rock stops the sense of spinning, of floating free.

13 AUGUST

At 4:10 am the alarm beeps us awake. I look at the windscreen, my first indicator of the density of fog outside. If it is wet with water droplets, then the fog is to the ground, and we will not be able to fly. It is clear. I bundle myself into layers of clothing and clamber out into the darkness. The whole sky is solid cloud, but it sits a few hundred feet above the ground. By 5:45 am we are at the airstrip watching the sky grow light with our lanky pilot, Matthias Röttcher. Matthias owns the skydiving business in Swakopmund, and is one of the most experienced pilots in the area. The cloud remains solid, but its ceiling is still a few hundred feet above the ground and so we decide to go.

We are flying in a Cessna 206 with the side doors off, the body painted to resemble the coat of a cheetah. We climb briefly over the town and turn east, climbing into the whiteout of the cloud. Through the open door, with the rush of the aircraft, the air is icy cold. For a minute or two we sit in silent anticipation in the swaddled white world of the cloud. At 2 500 feet we break free into an open sky. The sun is just rising, the world below us banished beneath the clouds. Far to the east and north, the silhouettes of mountains rise through the unbroken blanket of cloud. It takes us fifteen minutes to reach the edge of the fog, and we turn south towards the Namib Desert.

The few kopjes and hills that rise out of the desert plain stand like boulders in a stream of pure white, and in their eddy the thinning fog gives a translucent glimpse of the land. At the Kuiseb Canyon the fog's edge is abrupt and contrasts starkly with the dark folded shale of the hills. Sand dunes run in undulations into the hills, and we turn twice over the graphic landscape, working furiously in the cold wind. As we fly southward, the twisting, sharp ridges of the Namib dunes rise in strong counterpoint to the pale wash of cloud, the earth red, like single smears of paint on a white canvas. The fog tendrils run like the foam of a wave across a flat shore. Seldom have I seen the desert so beautiful.

15 AUGUST

The time has come to turn our Land Rover toward home. On the ground the Kuiseb Canyon lies like a Sunday-morning duvet, a crumple of rounded shale hills. When we stop to walk out and look down on the eroded ground, the world is silent and the heat beats back at us from the stone. Where the land becomes plains again, at the foothills of the Gamsberg, a yellow grass soothes the view of the savanna. Isolated mountains rise from the plains, like an assortment of cakes in a confectioner's shop, their geological layers exposed: black-stone slopes capped with a broad band of white rock. Yet, the view from the heights of the hills is not the same, even though nothing has changed since our last visit. My mind, looking toward the pinpoint of the distance, sees not the trees thinning into the arid plains, nor the solitary hills cast now like ore in the sun, but rather the city beyond in the unseen distance where I must once more become of the world of men.

I have grown away from it, found a fulfilment that I cannot easily translate. Where I have arrived takes time, and that is a luxury that the busy world of men does not offer up. It is the urgency that I dread, that keeps me silent in my reverie. I will step into those shoes but I have had a barefoot freedom and mine have often been the only footprints on the shore. How can I talk to another about how it is to live outside of convention, in the company of wilderness? When I do try, it is as if I am speaking a dialect that is difficult on the tongue and tiring on the mind. It is easier to talk of other things, of the trite and the everyday, and I mouth the words and I feel lonely in my soul.

Strangely, those who do live that way, by their own parameters, in wild country, are often seemingly ignorant of their individuality and regard their life as neither special nor extraordinary. I have enjoyed their company for that quality alone, but also often find myself enriched by the thoughts that flow from such unassuming independence. It suits us most to be ourselves, but few are able to find that truth in all the jumble of expectation to conform and judgement that society demands. I drive into the dust-shrouded desert and wonder how different and rich the world would be if each of us held to our own ideas.

21 AUGUST

What is it, I wonder, that binds some of us so fiercely to wilderness? Perhaps to understand it one must look to where the flame is first kindled. All humans need country as part of their identity. It is fundamental to our individual persona, a place to which we belong. For most, it is the place of our birth, but not always. Whatever its root, it is the environment itself that defines us – be it urban, rural or wild. We dress according to its precedent but it moulds us more deeply than that. It cradles us in love. It is our most comfortable place. Land becomes our spiritual home, defining more than just country, making us individual. It is important to each of us, and in that regard, we nurture it and defend it. It is both in defence and nurturing that passion emerges.

I have examined my passion for wilderness and find it has two distinct sources. The first is the most difficult to define, but pre-empts the second. It has been given many names but none define it adequately because it is a feeling. It is a feeling that comes not quickly across the surface, like laughter, but rather something that reaches deeper in me, to the very crux of my being. It is like the coals of a fire. No matter what ash collects over them, when you scrape it away the coals are there, warm and glowing red. When I know it most closely it fills me with a physical sensation that feels warm and gives me a sense of contentment beyond anything else. It is that which makes me wish to share it with others, so that they too may know peace. But it is elusive and fleeting, something that comes, and when it does, it makes me want to stay the motion of my time, not to move, for fear of losing what is one of life's most precious gifts, for in it I feel whole.

It is in those moments that humility overtakes me and I can see myself clearly, as just one grain of sand on the whole earth. But in that humility, there is, too, so strong a sense of belonging that my inner self expands to embrace everything around me. It is in those rare moments that there comes the most elusive of human desires: meaning. I have found it nowhere else. I have seen it in others, in a boy sitting with his hands stretched out to the warmth of a fire, his sleeping bag rolled out on the ground beneath the stars, in a woman on a hilltop, walking away from my company to be alone with the expanse of the earth that falls away from her feet, all around, in an elder listening to the song of a bird she cannot see, tears caught in the folded skin of her cheeks for the unexpected sweetness to be so sharp a probe to her soul. In every instance that I have witnessed it, including in myself, it evokes a need to move away from company, to be alone. It must be so if it is to be real, for it is something intensely private, beyond words.

My passion for wilderness came to me as a youth at a time when wilderness was plentiful. I have in the space of my lifetime watched that plenty be whittled away until what is left is rare. It is that rarity, and the speed with which it has come about, that fuels my passion for wilderness, my desire to ensure that it endures intact.

There are two things that stand against the existence of the last great wilderness areas we have witnessed in making this book: the number of people in the world and the idea that material worth supersedes all other criteria. All the wild land that is left on earth has one thing in common: hostile weather. Extreme cold, extreme heat, extreme dry, extreme wet are the factors that keep most men at bay. But I say most men, not all. I remember a guide in Alaska telling me that Alaska's extreme climate would keep it wild. That is a view that many take, but it is a view that sees only the here and now. The truism is that, as the population grows, the percentage of people drawn to wild places, to frontiers, remains the same. It simply becomes more people.

Once change has come to land, history is very clear in its record that it never reverts. It is changed forever. Forever, like infinity, is a concept that is almost impossible to grasp. We think of forever in terms of our lifetime and we fail to see how ridiculous that is. To some degree, it is selfish, but mostly it is blindness. We are unable to see beyond the scope of the only real frame we have for time: our lives. But we must, for wilderness is rare, and our time to recognise that is now.

There is a greater selfishness, however, that threatens wild land, like a guillotine blade poised in the sky. It is the notion that wild land has no worth. It is monetary value that is seen then as the exclusive definition of worth, and this too is blindness. We are convinced of it through indoctrination, through a lifetime's striving to have more, and it blinds us to other values we know but fail to define.

Wild land in its pristine, or as natural as possible state, is rarer and therefore more precious than anything we would take from it. No matter what it is that is extracted from the earth, it is finite and will one day all be gone. And what will we be left with then? We do not see the earth in terms of a timescale that holds the future. If we did, then how could we possibly consider it credible to destroy what little wild land remains? Wilderness is now the rarest commodity on earth. Television has brought it into our living rooms as an adventure story, but it is a real place. What lives and is wild on our planet exceeds anything humankind could ever imagine. It is among the greatest privileges in life to know any facet of our wild earth. And there is nothing more valuable that we could give back than to keep it that way.

22 AUGUST

A clap of thunder wakes me. A patter of rain beats briefly on the roof, then stops. I wait. The first crease of dawn is a cold dark grey. The thunder crashes again, but no rain follows its loud threat. In the dune sea of the Namib, I lie in bed tense in anticipation of the gift it promises in this waterless land. Outside, the desert too is waiting, will wait, and, if the rain does not fall today, it will wait longer.

Acknowledgements

In the making of this book I am indebted to many and would like to express my sincere and lasting gratitude to:

My and Beverly's parents, who took us into the mountains and game reserves from an early age and imbued in us a love of the wilds, and particularly to my father who was so much a part of this book but never got to read the final text; to the Pickford and Ward-Bester families, and Terence Mulligan, who were with us every step of the way; to Mike and Nina Gregory, whose friendship and invaluable support were so much more than the sum of their parts; to all our friends, for their enduring belief and encouragement.

In South America:

To Doug and Kris Tompkins who were an inspiration to us in their unwavering commitment to wild land; at Parque Patagonia to Luigi Soles, Christían Galvez and the team who gave us a deeper insight into the remote lands of Patagonia; to Virna Almeida and Diego Actis of Tierra del Fuego and to Orlando Beltrán and Captain Luis Paredes Samoloval for taking us into the heart of Chilean fjordland; to Ciro Barrientos Rademacher and Luisa Espinoza Gallardo for inviting us into rural Patagonia and the life of the gauchos.

In the polar regions:

To the expedition companies: One Ocean, PolarQuest and Quark for their considerations towards making our journey possible and especially to Cheli

Larsen, Pam Le Noury, John Rodsted, Mette Eliseussen, Aaron Lawton, Boris Wise, Nate Small, Simon Boyes, Ewan Blyth, Sophie Ballagh, Mark Tatchell, Henrik Løvendahl and Sue Werner, for going the extra mile; to our friend Paul Goldstein of Exodus Travels in whose company the world expands.

In North America:
To Peter and Jill Rinearson who made us part of their family; to Lon and Nora Kelly for helping us piece together many conservation puzzles; to Bruce McLean, Jill Pangman and their son Caelan for such generosity in sharing their lifelong knowledge of the Yukon; to our guides Hugh Rose and Steve Springer and to Michael Wald of Arctic Wild.

In Europe:
To Bertram Rickmers, Jessica Klatten, Deike Rickmers and Jasper Rickmers, and to Fried and Ebba Busse, without whom this project would not have begun.

In Asia:
To Wangden Tsering of SnowLion Tours for his meticulous research and especially to Tenzin and Gongpo who shared so much; to Bhuchung, our guide in the remote south.

In Australia:
To the Younglesons for sharing their home in the south, and to Anne and John Koeyers, Pat and Peter Lacy, Don MacLeod, Donny Imberlong and Matthew Wayner for revealing the northern country; to Tourism Western Australia for supporting our aerial photography and to Michael Collett, Eddie Feleppa, Ronni Feleppa and Mark Raissis of Odyssey Expeditions, for the journey of exploration along the Kimberley coast.

In Africa:
To Jockel and Monica Grüttemeyer, and Paul Liechti of Namibia Tracks and Trails for opening the scope of Namibia beyond all expectations; to Dr Margaret Jacobsohn and Garth Owen-Smith for long fireside insights into the ways of the desert and its peoples; to Namibia's Ministry of Environment and Tourism: Colgar Sikopo, Uatirohange Tjiuoro, Rehabeam Erckie, Shayne Kötting, Isaskar

Uahoou, Kapofi Immanuel and Bernard Awob; to Dr Kolette Grobler and Sakkie Jason of the Ministry of Fisheries and Marine Resources; to Rian Jones and Tony Delport, custodians of the bird islands; to Danie 'Jakkals' van Ellewee and Uri Adventures; to Wilderness Safaris, Louis Nortje, Clement Lawrence and Emsie Vervey; to Ruth and Johan Klein for time at Onanis, and to Matthias Röttcher for our remarkable hours over the desert.

To our sponsors Land Rover, Nikon, Michelin and Rickmers Line for their commitment to the idea that it is vital to conserve wild land.

To my principal editors Alfred LeMaitre and my wife, Beverly, who took an axe to my overgrown tree and fashioned it to stand in the garden of writing; to Sean Fraser and Wesley Thompson for the final edits.

To my publisher Louise Grantham and her close team Russell Clarke and Nicola van Rooyen for saying yes and seeing it through; to Russell Stark for the cover design; Kevin Shenton for the interior design and setting and my godson Justin J Fox for the maps.

And above all, to Beverly – my muse, my critic, my sounding board, my tonic, my refuge and my friend.

Peter Pickford